PHOTOVOLTAIC SYSTEMS

PHOTOVOLTAIC SYSTEMS
Analysis and Design

A.K. Mukerjee
Chief Scientific Officer (Retd.)
Centre for Energy Studies
Indian Institute of Technology Delhi

Nivedita Thakur
Associate Fellow
The Energy and Resources Institute (TERI)
New Delhi

PHI Learning Private Limited
Delhi-110092
2026

In fond memory of ***Shri Asoke K. Ghosh*** *(October 1942 – February 2024), Founder Chairman and Managing Director of PHI Learning, whose vision endlessly inspires.*

The Legacy Continues. . . .

Published by Pushpita Ghosh, PHI Learning Private Limited, Rimjhim House, 111, Patparganj Industrial Estate, Delhi-110092 and Printed by Gopsons Printers Pvt. Ltd., A-14, Sector-60, Noida, Gautambudh Nagar, Noida, Uttar Pradesh-201301.

₹795.00

PHOTOVOLTAIC SYSTEMS—Analysis and Desig
A.K. Mukerjee and Nivedita Thakur

ISBN-978-81-203-4417-4 (Print Book)
ISBN-978-93-5443-532-4 (e-Book)

The export rights of the book are vested solely with the publisher.

Contents

Preface

Renewable energy sources have lately acquired a prominent place in energy policy decisions of countries all over the world, prompted by the spiralling prices of petroleum products. It is widely perceived that the conventional energy sources, by and large, are likely to dwindle very fast in the near future. With this problem in mind, universities and engineering colleges have progressively introduced courses in their curricula to deal with the challenges of fast depleting conventional resources and to supplant these with their renewable counterparts. The Centre of Energy Studies at IIT Delhi was established in the 1970's to encourage the study and research in energy conservation, maximizing energy extraction from conventional fuels and, of course, the study of renewable sources of energy, photovoltaic (PV) technology being a major constituent. Photovoltaic systems form a significant part of direct energy conversion, that is, generation of electricity directly from sunlight, which is considered to be inexhaustible. Photovoltaic devices are the core components of these systems. Since these two are inseparable, their study, in some detail, is necessary to understand the science and engineering of the conversion process, its efficiency and economics.

While teaching students at the Master of Technology level, we felt that there was a shortage of books and materials that discussed the design of PV systems in detail. Therefore, we made an effort to systematically compile relevant works from the existing literature and blend it with a mixture of research and design work done in-house to present a course that can not only help the students but also new researches in the field. The effort culminated in the present book which, most probably, would be suitable at the B.Tech. level too. A large number of excellent treatises are available to describe the physics and technology of the solar cells but not many works explain the optimal use of these products for the ultimate function they were made for, that is, producing electricity. This latter function has been attempted in this book with a little explanation of the devices.

Chapter 1 provides an introduction to PV systems. Chapter 2 describes the nature and measurement of solar radiation. Chapter 3 introduces the solar cell parameters and their determination using several established techniques. A general PV system is described in

Chapter 4. The effect of tilt on the solar modules and improvement in solar radiation collection by tracking, are also explained. A very important aspect of PV systems is the storage device which more or less dictates their deployment. A discussion on the storage battery is taken up in Chapter 5 to determine the type suitable for a particular application. Chapter 6 gives a brief description of the types of inverters likely to be used in PV systems with the design of one based on sine pulse width modulation topology.

Maximum power point trackers are used to extract nearly all the power that a PV array can supply. Chapter 7 illustrates the principles and operation of some of these electronic devices, while Chapter 8 completes the discussion by introducing more recent advances in this topic. Chapter 9 is devoted to the design of several PV systems which enable engineers to calculate the size of a PV array, in terms of its power, and the capacity of the battery needed to store the generated power during sunless conditions. The requirements and standards pertaining to uploading of the power generated by solar arrays, to the power grid for onwards transmission to distant locations, are considered in Chapter 10. Without the design and construction of inductors and transformers, it is very difficult to build a PV system. In order to do so, Chapter 11 has been added.

A.K. Mukerjee
Nivedita Thakur

Acknowledgements

Several persons have helped me to first work on the subject of photovoltaic technology and then write this book. First of all, I wish to record my gratitude to Prof. K.L. Chopra, of the Thin Film Laboratory at IIT Delhi, who provided every opportunity to work there and later helped me to establish the Photovoltaic Laboratory, where a major part of the work presented in this book was accomplished. I must thank Prof. G.N. Tewari for urging me to write a book on PV systems. I am thankful to Prof. Viresh Dutta for extending every facility in his laboratory to me during my work on the book. I am grateful to Dr. S.N. Garg and Dr. Vamsi Krishna for going through the manuscript and for offering their valuable suggestions for improvements. I also want to record my appreciation to Dr. Deepak Kumar, Mr. Vikas Kumar, Mr. Nitesh Kumar, Mr. Amit Tewari and other members of the staff and students of the Photovoltaic Laboratory, for their help. I also wish to thank Mr. Harish Pal for keying in the manuscript and for drawing a large number of diagrams.

Finally, I thank the management of Bhagwati Institute of Technology and Science, Ghaziabad, for extending all the facilities to complete this book.

A.K. Mukerjee

CHAPTER 1

Introduction

1.1 History of Solar Cells

The development of solar cells started almost 140 years ago, when Becquerel in 1839 observed that some voltage developed when light fell on an electrode in an electrolyte [1]. Forty years later Adams and Day discovered a similar effect in selenium after photoconductivity was found in the metal by Smith [2]. Later, work on this material and cuprous oxide established the firm foundation of photovoltaic effect which was commercially exploited in exposure meters for cameras used in photography. By 1914 efficiencies in the range of 1% was achieved with the realisation that energy barrier height was involved in both selenium and copper/copper oxide cells. Later other materials like silicon were experimented with finally yielding a conversion efficiency of 6% in the mono-crystalline form of the semiconductor by Chapin and coworkers [3].

This heralded the modern era of solar cells when researchers started looking seriously for improvement in the efficiency. By 1958 the efficiency of silicon solar cells had reached a value of 14%. Reynolds reported in 1954 another combination of cadmium sulphide and copper sulphide (CdS/Cu_2S), in the form of a hetero-junction that yielded a conversion efficiency of 6% [4].

1.2 Overview of Solar Photovoltaic (SPV) Technologies

The Solar Photovoltaic (SPV) cells described above are connected in series and parallel combinations to form SPV modules. These, by optimal combination, form a solar array. Solar cells are thus, the basic unit of SPV arrays and are a type of diode generating electricity when illuminated. The output of the solar cells, in terms of current and voltage and thus, in-turn power, varies depending upon operating conditions such as solar radiation, temperature, wind speed, etc. Thus, different solar cells behave differently with location and season. There are several materials which go into making various types of solar cells; silicon being the primary and most widely used material. Crystalline silicon solar cells make for what are known as the first generation solar cell technology. Such solar cells are about 20–24% efficient and enjoy about 85–90% of the market share. Generally, one crystalline silicon solar cell generates about 0.5 V, thus, typically 36 cells are connected in series to form a solar module to charge standard 12 V batteries.

The second generation solar cell technology comprises mainly of the amorphous silicon solar cells. These fall in the category of thin-film solar cell technology that also include materials such as copper indium diselenide (CIS), cadmium telluride (CdTe) and copper indium gallium diselenide (CIGS). Such solar modules have lesser efficiency (in the range of 5–8%) compared to crystalline silicon solar cells but have lower manufacturing costs.

Concentrator solar PV technology (CPV) involves optical concentrating systems such as mirrors and lenses focusing the solar radiation on special cells and use active cooling methods to dissipate the generated heat. Solar trackers are recommended to be used with concentrator SPV modules. This technology is yet to be fully commercialized.

1.3 Types of SPV Systems

Solar photovoltaic energy primarily involves two fields of research, viz. Technology (solar cell materials) and the Balance of Systems (that includes the supporting components in SPV power plants). The first category was discussed in the previous section. Solar power plants can be broadly categorized as:

- Stand-alone SPV systems
- Grid interactive SPV systems
- SPV Hybrid systems

Standalone or off-grid systems include common applications such as Solar lanterns, solar home and street-lighting systems, solar water pumping, stand-alone power plants for powering appliances in remote areas away from access to grid electricity. Such systems are designed either with or without battery back-up options.

Grid interactive systems are those that are tied-up to the grid and aptly compliment the grid electricity in places with intermittent power supply. The electricity demand of the load is primarily met by the SPV array. Depending upon the relative output power, excess power from the solar array is fed to the grid. These are generally used for residential systems, community centres such as schools, nursing homes, etc. Building integrated SPV can be standalone or grid-interactive type.

An SPV based hybrid system involves two or more power sources including solar PV. Other sources such as wind, diesel generator, grid, etc. are intelligently combined with SPV to provide reliable power to the critical loads. Such systems optimally use the available resources and help suitably supplement SPV power whenever required.

1.4 SPV Energy Status in India

Companies and organizations such as Bharat Heavy Electricals Ltd. (BHEL) [5], Central Electronics Limited (CEL) [6] and Rajasthan Electricals & Instruments Limited (REIL) and many other manufacturers started playing major roles to initiate and promote solar energy in India. The Ministry of New and Renewable Energy (MNRE) [7] has taken the lead in commencing various projects and schemes to promote and scale-up solar photovoltaic installations and utilization.

Aid from other financial institutions like World Bank and Asian Development Bank has also come for the development of renewable energy sources. So far solar water heating devices are the most popular clean energy systems for domestic and industrial use. Figure 1.1 shows the photograph of a typical solar water heating plant installed at the roof top of a building. Figures 1.2 and 1.3 show the photographs of the associated controller and the water tank of a typical water heating plant. However, this chapter focuses mainly on the background of solar photovoltaic systems and their development.

Figure 1.1 A typical solar water heating plant.

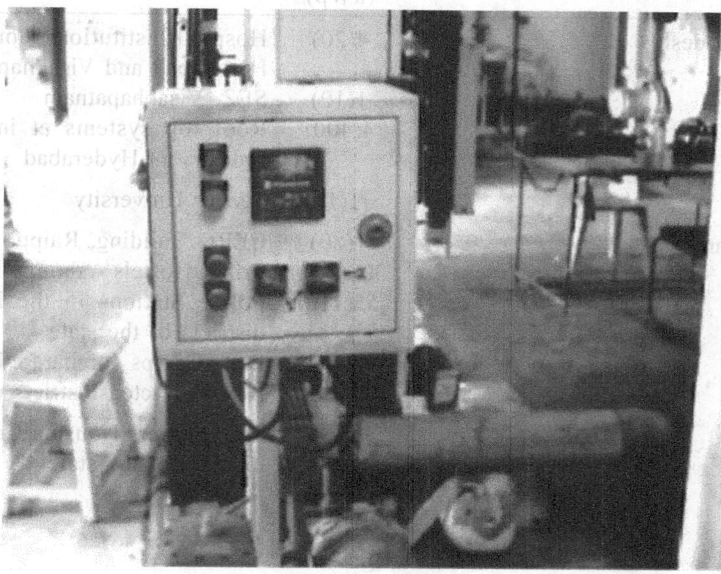

Figure 1.2 Controller associated with a typical solar water heating plant.

Figure 1.3 Water tank of a typical solar water heating plant.

1.4.1 SPV Achievements in India

Table 1.1 gives an overview of the state-wise installed off-grid solar systems by MNRE. This includes home lighting systems (HLS), solar street lighting systems (SLS), solar lanterns (SL) and off-grid power plants.

Table 1.1 State-wise list of off-grid projects sanctioned by MNRE up to 31st December 2009

S.No.	*State*	*Systems Sanctioned*				*Beneficiaries*
		HLS (No.)	*SLS (No.)*	*SL (No.)*	*Power Plants No.* (kWp)	
1	Andhra Pradesh				4(20)	Hospital/Institutions/Housing colonies Hyderabad and Visakhapatnam
					1(10)	SEZ, Visakhapatnam
					4(400)	Roof top systems at institutions and Industry in Hyderabad
2	Assam				1(1)	Tezpur University
3	Chhattisgarh				1(80)	CERC Building, Raipur
					300(361)	Tribal Hostels/Ashrams in the state
					55(110)	Police Stations in the Naxal affected districts of the state
					47(141)	Base Camps for paramilitary forces in Naxal affected districts of the state
4	Gujarat	3058				Salt Workers Housing Colonies in eight districts of the state

(*Contd.*)

Table 1.1 *Contd.*

S.No.	*State*	*Systems Sanctioned* *HLS (No.)*	*SLS (No.)*	*SL (No.)*	*Power Plants No. (kWp)*	*Beneficiaries*
5	Haryana				1(41,65)	Akshaya Urja Bhawan, Panchkula
		987				Police station in the state
		1290				Villages with 50% scheduled caste population in Jind, Bhiwani, Faridabad and Gurgaon districts 11 villages with 50% SC Population in Hissar district.
					25(1)	Roof Top Solar Systems for BPO at Gurgaon
					2(200)	Roof Top Solar Systems for Industry in Gurgaon
6	Karnataka				25(50)	Bus Station in the State
					100(30)	
7	Madhya Pradesh				113(113)	Range Officers of Forest Department
8	Maharashtra				1(2.1)	Tribal Research Institute, Bhopal
			100			Municipal Council, Paltan (Satara)
9	Manipur	365	438			Police Stations in State
10	Orissa		80			Industries/educational institutions in Bhubneshwar
11	Puducherry		100			Selected Villages in the U.T.
12	Rajasthan		90			Textile Park, Ajmer
			30		1(10)	Engineering College, Ajmer
		10000				Cluster of villages in the state
13	Sikkim	750	30	800	1(1)	Unelectrified villages, Housing Colonies/ Tourist spots in the state
14	Tamil Nadu				1(10.5)	Engineering college, Kanchipuram
15	Tripura				1(1)	Rajya Sainik Board Office, Agartala
16	Uttar Pradesh				20000	Unelectrified villages in the state
			1304			88 SC and minority populated villages in Aligarh, Barabanki, Hardoi, Siddath Nagar and Gonda districts.
					1(100)	SEZ, Noida
			383			SIRD, Lucknow and its regional and district centres in the state
					5(7.5)	Educational Institutions in Gonda district
17	Uttarakhand		895			129 villages in Pauri/Tehri/Chamboli/ Utarkashi/Nainital & Pithoragarh Districts
					1(100)	University of Petroleum and Energy Studies, Dehradun
18	West Bengal			14000		Cyclone AILA affected in Sunderbans
		19783				SC/ST population in Sunderbans area of North and South 24 Paraganas, Bankura, Purulia and West Midnapur districts.

The MNRE website lists the achievements (in terms of installations of renewable energy systems) for the year (2009) with cumulative figures. These are regularly updated. Table 1.2, referred from the ministry website [7], gives an overview.

Table 1.2 MNRE achievements in renewable energy based systems installation

No.	*Sources/Systems*	*Achievements during 2009-10 (up to 31.03.2010)*	*Cumulative Achievements (up to 31.03.2010)*
I.	**Power From Renewables**		
A.	**Grid-interactive renewable power**		
1.	Biomass Power (Agro residues)	153.30 MW	865.60 MW
2.	Wind Power	1565.00 MW	11807.00 MW
3.	Small Hydro Power (up to 25 MW)	305.27 MW	2735.42 MW
4.	Cogeneration-bagasse	295.30 MW	1334.03 MW
5.	Waste to Energy	4.72 MW	64.96 MW
6.	Solar Power	8.15 MW	10.28 MW
	Sub Total (A)	**2331.74 MW**	**16817.29 MW**
B.	**Off-Grid/Distributed Renewable Power (including Captive/CHP Plants)**		
7.	Biomass Power/Cogen.(non-bagasse)	50.80 MW	232.17 MW
8.	Biomass Gasifier	13.28 MW_{eq}	122.14 MW_{eq}
9.	Waste-to-Energy	15.88 MW_{eq}	46.72 MW_{eq}
10.	Solar PV Power Plants	0.16 MWp	2.46 MWp
11.	Aero-Generators/Hybrid Systems	0.22 MW	1.07 MW
	Sub Total (B)	**80.34 MW_{eq}**	**404.56 MW_{eq}**
	Total (A + B)	**2412.08 MW**	**17221.85 MW**
II.	**Remote Village Electrification**	1013 villages and hamlets	5348 villages/ 1408 hamlets
III.	**Decentralized Energy Systems**		
12.	Family Type Biogas Plants	1.06 lakh	42.40 lakh
13.	SPV Home Lighting System	72,886 nos.	5,83,429 nos.
14.	Solar Lantern	82,999 nos.	7,92,285 nos.
15.	SPV Street Lighting System	8,680 nos.	88,297 nos.
16.	SPV Pumps	106 nos.	7,334 nos.
17.	Solar Water Heating – Collector Area	0.62 Mln. sq.m.	3.53 Mln. sq.m.

MW_{eq} = Megawatt equivalent; MW = Megawatt; kW = kilowatt; kWp = kilowatt peak; sq. m. = square metre.

[a]1. Although the potential is based on surplus agro-residues, in practice there are several barriers in collection and transportation of such agro-residues to the generation site and biomass power generation units prefer to use fuel-wood for techno-economic reasons. A potential of 45,000 MW_e from around 20 mha of wastelands assumed to be yielding 10 MT/ha/annum of woody biomass having 4000 kcal/kg with system efficiency of 30% and 75% PLF has not been taken into account. In order to realize this potential a major inter-Ministerial initiative involving, among others, Environment & Forests, Agriculture, Rural Development, and Panchayati Raj would be

required. Further, a Biomass Atlas is under preparation which is expected to more accurately assess state-wise renewable energy potential from agro-residues.

2. Potential based on areas having wind power density (wpd) greater than 200 W/m^2 assuming land availability in potential areas @ 1 per cent and requirement of wind farms @ 12 ha/MW, not all of which may be technically feasible for grid-interactive wind power. In line with international practice for setting up grid-interactive wind power systems on sites having wpd greater than 300 W/m^2, potential would be 5000 MW. Further, preliminary surveys do not at this juncture suggest a sizeable grid-interactive off-shore wind power potential.
3. Technically feasible and economically viable hydro potential is generally accepted at 40% of the total estimated potential. Accordingly, the technically feasible and economically viable small hydropower potential could be around 6000 MW.
4. With new sugar mills and modernization of existing ones, technically feasible potential is assessed at 5000 MW$_e$, not all of which may be economically viable. Furthermore, several sugar companies/ cooperatives are unable to develop bankable projects on account of their financial and liquidity positions.
5. With expansion of urban population post census 2001, current technically feasible municipal waste-to-energy potential is assessed at 1700 MW$_e$, not all of which may be economically viable. However, subsidy disbursement under the municipal waste to energy programme had been kept in abeyance on the orders of the Supreme Court in the case of a PIL, in May 2005. This site has now been vacated for setting up 5 pilot projects.
6. Not all of this renewable energy potential may be suitable for grid-interactive power for technical and/or economic reasons. Further, estimate excludes potential for solar power which is dependent on future developments that might make solar technology cost-competitive for grid-interactive power generation applications.

The ministry supports various research and development (R&D) projects on utilization of solar photovoltaic energy in universities, recognized colleges, autonomous societies, industries, etc.

1. Development of polycrystalline silicon.
2. Development of device fabrication processes and improvements in crystalline silicon solar cell/module technology.
3. Development of thin film solar cell technology (based on amorphous silicon films, cadmium telluride films and copper indium diselenide thin films, organic, dye sensitized and carbon nano tubes).
4. It also supports development of photovoltaic systems and components used in manufacture of such systems. The Ministry has identified thrust areas for R&D support in solar photovoltaic technology during the 11th plan.

The following from the MNRE website, provides an overview of the thrust areas in photovoltaics in which MNRE is encouraging R&D activities:

MNRE has been supporting R&D and technology development in solar photovoltaic technology for more than three decades. During the 11th Plan, it is envisaged that the cost of solar photovoltaic modules can be brought down to about ₹ 120 per W$_p$. In order to achieve this goal the key areas of R&D and technology development have been identified. Research, design and development efforts during the 11th plan are proposed to be focused on development of the following:

Polycrystalline silicon material

1. To undertake R&D to make poly silicon material using alternative methods to achieve direct electricity consumption of 125 kWhr/kg or lower. The process will be scaled up to pilot plant production, based on process reactor, which can demonstrate a capacity up to 100 tons per annum. The quality of poly silicon material produced at the pilot plant will be suitable to make high efficiency (>15%) silicon solar cells and the combined trace impurities of heavy metals, carbon and boron will be limited to parts per billion (ppb) level.
2. To design, develop and demonstrate at pilot plant scale poly silicon deposition reactors, based on traditional poly silicon deposition methods, to reduce the energy consumption to less than 125 kWhr/kg of polycrystalline silicon material produced, with the combined trace impurities of heavy metals, carbon and boron limited to ppb level.

Crystalline silicon solar cells and modules

1. To reduce the direct consumption of silicon wafer (3 gm per Wp) in commercial production by reducing the wafer thickness and increasing the efficiency of crystalline silicon solar cell to average 18% and more.
2. To develop and produce multicrystalline silicon ingots/wafers and produce solar cells with conversion efficiency of 17% and more in commercial production.
3. To undertake R&D on alternative device structures to make crystalline silicon solar cells to demonstrate very high efficiency (22–24% on small size laboratory devices).
4. Improving the effective PV module life to 25 years and more, with total degradation within 10% of the initial rating under standard test conditions (STC).
5. Design and development of low cost, low weight, non-glass type PV modules with effective module life of 10 years or more, with total degradation within 10% of the initial rating under STC.
6. Study and evaluate new materials for use in PV modules.
7. To develop low resistance metal contact deposition materials and processes.

Thin film solar cell modules

Thin film solar cell modules have potential to reduce the cost of solar modules due to consumption of less material and energy in the fabrication processes. Amorphous silicon thin film solar cells were the first to be developed. In the recent years pilot plants and a few commercial plants based on other thin film solar cell modules like CdTe, CIGS and silicon, have been set up. It is proposed to take up research and development work and pilot plant demonstration of thin film based modules in the country during the 11th five year plan, with the following objectives:

1. To undertake R&D on different processes and device structures to make laboratory scale small area (2 cm × 2 cm) devices of efficiency more than 10% using CdTe, CIGS and silicon thin films.
2. Development of poly crystalline thin film integrated modules (1 sq ft or more) at pilot plant scale using different materials (CdTe, CIGS, silicon films) to achieve efficiency of more than 8% and life of integrated module above fifteen years.

New materials for solar cells

In the recent years worldwide R&D efforts are being made to study new materials, which are easy to deposit and consume significantly less energy. Thin film modules based on organic materials, dye sensitized and doped with nano-materials has potential to produce solar cells. However, these device structures are in the early stages of development. It is proposed to establish R&D centres in the country to study these concepts, and pursue research in these emerging PV devices.

1. To study and characterize new materials to determine their suitability for the fabrication of solar cells.
2. Design and development of new thin film device structures based on dye sensitized (liquid and solid state) organic, carbon nano tubes, quantum dots, etc., materials. The laboratory scale efficiency of 5–10% to be achieved.

Concentrating solar cells and modules

Apart from the improvements in the performance of flat plate PV modules, it is possible to reduce the material consumption and lower the cost through the use of concentrating PV systems. In order to gain experience on performance of the concentrating system and develop systems suitable for use in India, the following activities are proposed during the 11th plan:

1. Design and development of concentrator solar cells (concentration ratio of 200 suns and more) and modules (efficiency ~25–30%) and testing of concentrating PV system in Indian conditions.
2. Development of two axis tracking system suitable for high concentration PV system.
3. Design and development of heat sink for mounting of solar cells under high concentration.
4. Design and development of optical systems to achieve concentration ratio of 200 suns and more, with minimum optical aberration.
5. Development of silicon and gallium arsenide (GaAs) based solar cells suitable for use under high concentration (200 suns or more).

Storage system

At present stand-alone systems use lead acid batteries. However, with low power consuming LED based systems, use of NiMH batteries is also coming up. One of the major constraints in battery storage system is the limited life of storage batteries. There is an urgent need to enhance the battery cycle life to get at least 10 years of operating life. Further, it is also necessary to develop non-lead acid batteries. In addition, alternative methods of storage, especially to store large quantity of power for a few hours would substantially improve the viability of grid interactive PV systems. Therefore, the following tasks have been identified for the 11th Plan:

1. Development of long life (5000 cycles or more) storage batteries suitable for use in PV systems/applications.

2. Development and testing of new storage systems up to MW scale. It should be possible to store electricity for about 8–10 hours, with storage losses limited to about 10%.

Balance of system and PV systems

1. Design and development of high efficiency (more than 50%) motor-pump set having output power of 2 HP, 3 HP and 5 HP to lift water from a depth of about 30–60 metres.
2. Design and development of small capacity inverter cum charge controller, with efficiency of 90% or more, suitable for use in solar lighting systems including LED based lighting systems.
3. Design and development of LED based PV systems for indoor and outdoor lighting applications
4. Design, development and field-testing of inverters and grid synchronizing system components (peak efficiency more than 96% and part load @ 30% efficiency above 88%) used in residential grid interactive roof top PV systems.
5. Field-testing and performance evaluation of grid interactive rooftop residential PV systems.
6. Design, development and testing of inverters and grid synchronizing system components (peak efficiency more than 96% and part load efficiency above 88% at 30% load) for grid connected PV systems of 500 kWp capacity or more.
7. Field-testing and performance evaluation of large grid interactive PV power plants.

Testing and characterization facilities

1. Upgrading the testing and characterization facilities for PV materials, devices, components, modules and systems.
2. Setting up of testing facilities for concentrating PV systems.
3. Study and evaluation of new materials, device structures and module designs etc.

While this section has been dedicated to the Indian scenario of SPV power, the following section gives a brief overview of the international situation.

1.5 Solar Photovoltaic Energy—The international scenario

The SPV market is growing like never before in the world. According to Renewable Energy (REN21) Global Status Report [8], SPV is the fastest growing power generation technology in the world. As presented in the report, around 21 GW of grid-tied capacity has been achieved worldwide already. According to the International Energy Agency (IEA) [9], for more than a decade now, the SPV market has been growing with an average annual growth rate of approximately 40%. Around 16–22% of SPV modules are made by the 'Thin Film' technology. Figure 1.4 [9], shows the rapid growth of the cumulative installed capacity of SPV till year 2008.

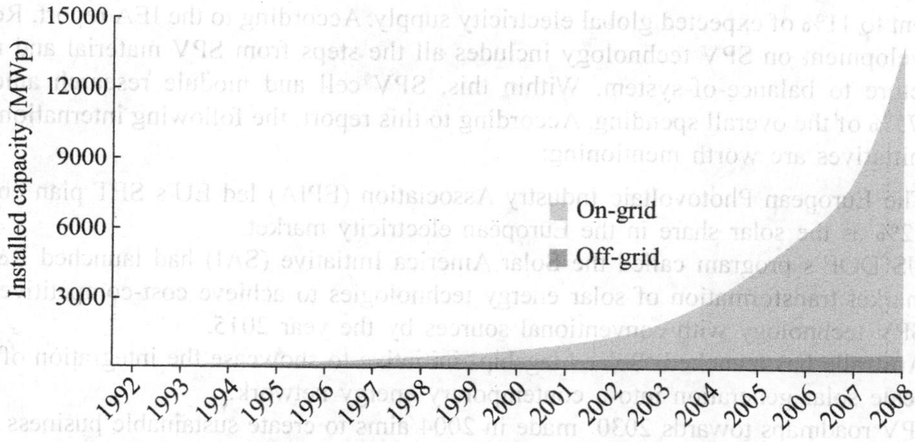

Figure 1.4 Year-wise global SPV installed capacity.

Source: Technology Roadmap—Solar Photovoltaic Energy Figure 2, page 9. © OECD/IEA, 2010.

The total systems cost includes SPV modules and the balance-of-system (such as inverters, mounting structure, batteries and other related electronic components). According to IEA [9] the turn-key prices, without batteries, were around USD 4000 per kW for utility scale in 2008. As for the countries contributing to the SPV market, Germany is the global leader in SPV installations followed by Spain. Figure 1.5 [9] shows the country-wise percentages of SPV capacity till 2009.

With the global SPV market doubling from 2007 to 2008 and system prices falling by 40% from 2008 to 2009, SPV is set to achieve grid parity by the year 2020 and there might be 3000 GW of installed SPV capacity worldwide by the year 2050 that will generate energy

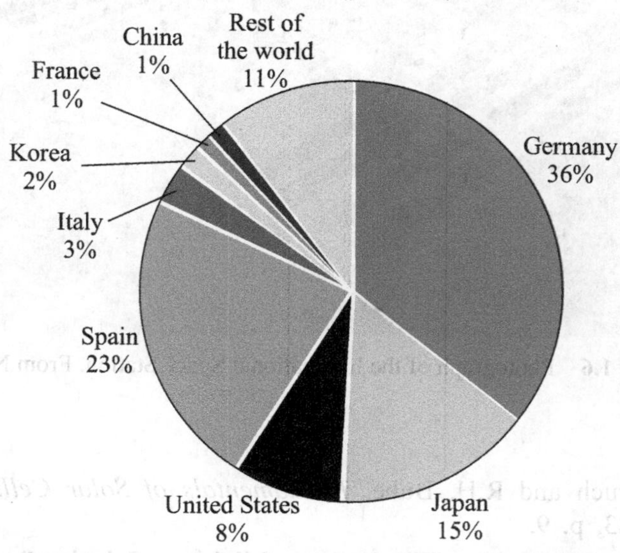

Figure 1.5 Region-wise status of the global SPV market.

Source: Technology Roadmap—Solar Photovoltaic Energy Figure 3, page 10. © OECD/IEA, 2010.

equivalent to 11% of expected global electricity supply. According to the IEA report, Research and Development on SPV technology includes all the steps from SPV material and module manufacture to balance-of-system. Within this, SPV cell and module research amount to around 75% of the overall spending. According to this report, the following international SPV based initiatives are worth mentioning:

1. The European Photovoltaic Industry Association (EPIA) led EU's SET plan envisions 12% as the solar share in the European electricity market.
2. US DOE's program called the Solar America Initiative (SAI) had launched R&D and market transformation of solar energy technologies to achieve cost-competitiveness of SPV technology with conventional sources by the year 2015.
3. Australia has launched 'Solar Flagship' initiative to showcase the integration of utility-scale solar generation into a contemporary energy network.
4. 'PV roadmaps towards 2030' made in 2004 aims to create sustainable business in each SPV related field, starting from raw material production to module manufacturing to balance-of-system components.

In addition to the terrestrial use of solar cells, the space sector is almost entirely dependent on photovoltaic technology. For example, the communications satellites derive electrical energy from PV panels. Also the International Space Station is equipped with an array of sixteen main PV panels to supply its electricity needs, as shown in Figure 1.6.

Figure 1.6 Photograph of the International Space Station. From NASA.

References

[1] A.L. Fahrenbruch and R.H. Bube, *Fundamentals of Solar Cells*, Academic Press, New York, 1983, p. 9.

[2] W.G. Adams and R.E. Day, "The action of light on Selenium", Proc Roy. Soc., 25 (1877), p. 113.

[3] O.M. Chapin, C.S. Fuller and G.L. Pearson, "A new silicon p-n junction photocell for converting solar radiation into electrical power", *Journal of Applied Physics,* **25**, 1954, p. 676.

[4] D.C. Reynolds, G. Leies, L.L. Antes and R.E. Marburger, "Photovoltaic effect in cadmium sulfide", *Phys. Rev.*, **96**, 1954, p. 533.

[5] Bharat Heavy Electricals Limited (BHEL), www.bhel.com/bhel/home.php

[6] Central Electronics Limited, www.celindia.co.in/

[7] Ministry of New and Renewable Energy (MNRE), www.mnre.gov.in/

[8] www.ren21.net/globalstatusreport

[9] Technology Roadmap–Solar Photovoltaic Energy, 2010, IEA www.iea.org/,© OECD/IEA, 2010.

CHAPTER 2

The Sun

The conversion of sunlight to electricity requires a good knowledge of the strength and constituents of the incident radiation. For example, a monocrystalline solar cell will respond favourably to a certain band of wavelengths which might not be suitable for a CdS/Cu_2S device. This chapter presents a discussion on the variation of the solar spectrum as it passes through the earth's atmosphere, its strength, measurements of these parameters and the conditions under which such measurements are carried out. The large angle of inclination of the sunrays on the surface of the earth obviously means that the sunlight has to travel more through the atmosphere, resulting in more attenuation as compared to the vertical rays. This has been quantified and referred to as *air mass* (AM). Different components of radiation such as direct, diffuse and global have been given defferent definitions since their values are different for places close to the equator than those at far north.

2.1 Solar Spectrum

The total energy need of the earth for one year is almost the same as the energy received from the sun in one hour. This means that in one hour, the energy that comes from the sun is 5000 times more than the total production of energy on earth from all sources. The huge energy production in sun is the result of thermonuclear reaction which converts hydrogen into helium. The difference between the masses m of the two gases is converted to energy according to Einstein's formula $E = mc^2$, c being the velocity of light. As a result of this reaction intense heat is produced that keeps the surface temperature of the sun at almost 6000 K while that at the interior is much higher. The sun can be thought of as a black body, similar to what Planck postulated, radiating energy in all directions. Planck's formula for blackbody radiation is

$$W_\lambda = \frac{2\pi h C^2 \lambda^{-5}}{e^{\frac{hC}{\lambda kT}} - 1} \quad \text{(W/m}^2\text{/wavelength in metres)} \tag{2.1}$$

where
Planck's constant (h) = 6.63×10^{-34} J s
Boltzmann's constant $k = 1.38 \times 10^{-23}$ J/K
wavelength λ is in metres and velocity of light (c) = 3×10^8 m/s.

This is the energy density on the surface of the sun. However, by the time the sunlight reaches the atmosphere of the earth, its value reduces to 1367 W/m^2 called the *solar constant.* It is defined as the energy received on a unit surface perpendicular to the sun's direction, in free space, at the earth's mean distance from the sun. However, because of the variation of distance between the earth and the sun, over a period of one year, this constant changes by ±3.35%. The maximum of 1400 W/m^2 occurs on January 3 while a minimum of 1310 W/m^2 is received on July 6. Although the sun resembles a black body at 6050 K, it remains a complex radiator. The variation of temperature across the sun's surface, Fraunhofer absorption lines and the solar atmosphere, modifies the black body spectrum.

Figure 2.1 gives the plots of Planck's black body radiation formula, given in Equation (2.1), for several different temperatures along with the smoothened out part of the solar spectrum outside the earth's atmosphere. At higher temperatures, the peaks of the curves shift towards blue and at lower ones these move towards yellow. Even though most of the solar energy in the outer space lies between 0.25 and 3.0 μm, the colour of the normal sunlight is determined by the portion visible to the human eye.

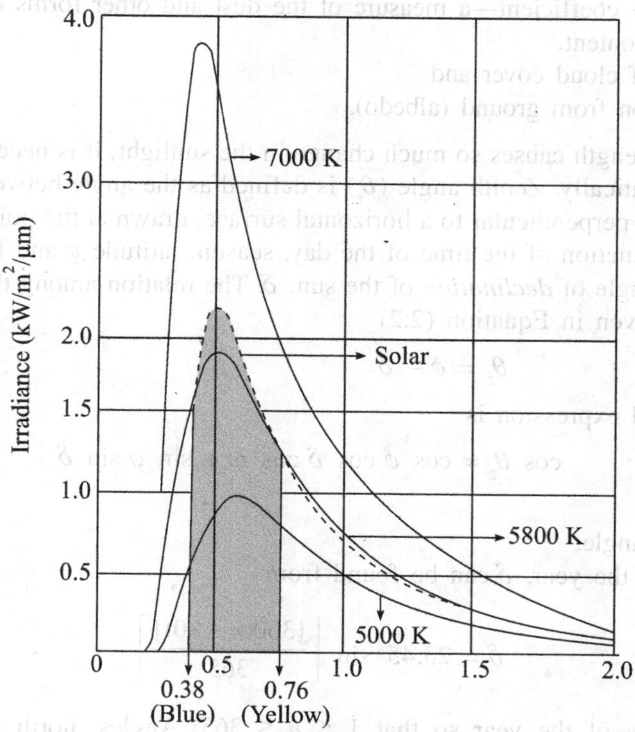

Figure 2.1 Black body spectra for different temperatures. The shaded area represents the sun's visible spectrum.

All light sources are not perfect black bodies, nor is the sun. However, practical black bodies can be simulated by heating filaments to high temperatures. To be a perfect black body, a radiating source must also be a perfect absorber of radiation which is not the case with ordinary light sources. The discharge lamps certainly do not fall into this category as these emit light characteristic of the gas under discharge. For example, a sodium vapour lamp will emit light only at wavelengths of 0.5895924 μm and 0.5889950 μm called *D*1 and *D*2 lines, respectively. Only the mercury vapour lamp has enough different wavelengths to simulate white light. For this reason, only heated filament sources are preferred to simulate solar radiation, of course with suitable filters. Data of the sun's spectrum is thus necessary, not only for generation of the solar spectra, for testing purposes inside laboratories, but also for finding new materials that will favourably respond to the received radiation to yield maximum power.

2.2 Impact of Earth's Atmosphere on Sunlight

Both the spectral distribution and the intensity of the solar radiation, incident on the surface of the earth, depend on various factors. The main among them are (a) the length of the path traversed by it and (b) the composition of the atmosphere which largely depends on

(i) the water content
(ii) the turbidity coefficient—a measure of the dust and other forms of scattering
(iii) the ozone content,
(iv) the depth of cloud cover and
(v) the reflection from ground (albedo).

Since the path length causes so much change in the sunlight, it is necessary to define and quantify it mathematically. Zenith angle (θ_z) is defined as the angle between the direction of the sunlight and the perpendicular to a horizontal surface, drawn at the point of measurement. Zenith angle is a function of the time of the day, season, latitude ϕ and longitude. It is also dependent on the angle of *declination* of the sun, δ. The relation among the three, valid only at solar noon, is given in Equation (2.2)

$$\theta_z = \phi - \delta \tag{2.2}$$

A more general expression is

$$\cos \theta_z = \cos \phi \cos \delta \cos \omega + \sin \phi \sin \delta \tag{2.2a}$$

where

ω is the hour angle.

On any day of the year, δ can be found from

$$\delta = 23.45^\circ \sin \left[\frac{\{360(n-80)\}}{365}\right] \tag{2.3}$$

where n is the day of the year so that $1 \le n \le 365$. Angles, north of the equator, are considered positive while those in the south are taken as negative. The tropic of cancer is situated 23.45° north of the equator. Since the position of the sun decides the value of δ, its variation over the year should be traced. On the first day of the spring, $n = 80$, taking January 1 as $n = 1$, and that of autumn, $n = 260$, the sun is directly over the equator. On these two days, called *equinox*, the day and night are equally long. This occurs on March 21 and September 21. In both the cases, from Equation (2.3), the declination angle $\delta = 0$. The *summer solstice* and the *winter solstice* occur respectively on June 21, when the day is longest, and December 21, when the day is shortest, in the Northern hemisphere. All these positions of the sun are shown in Figure 2.2.

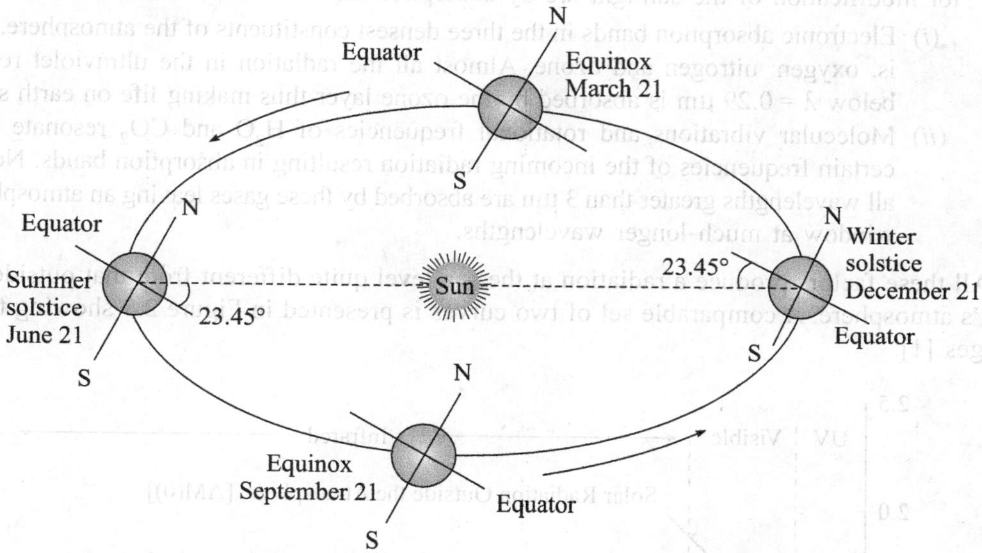

Figure 2.2 Earth's orbit around the sun showing solstice and equinox positions.

The earth revolves around the sun, once every year, in an elliptical orbit with the sun at one of its foci. The distance between the earth and the sun, in metres, on day n of the year is given by the expression:

$$S = 1.5 \times 10^{11}\left[1 + 0.017\sin\left\{\frac{360(n-93)}{365}\right\}\right] \tag{2.4}$$

However, the orbit is quite close to a circle so that the distance between the two bodies can be expressed in a value that is an average of the minor and major axes of the ellipse. For this reason, the position of the sun in Figure 2.2 is shown near the centre of the ellipse instead of one of its foci. In addition to revolution around the sun, the earth also rotates about its polar axis which is inclined to the plane containing both the bodies at an angle of 23.45°. This tilt causes longer days in summer and shorter ones in winter in the northern hemisphere. The maximum tilt angle of 23.45° occurs on June 21, when the day is longest and the sun appears to be directly above latitude of the same angle called the *tropic of cancer*. Similarly, on December 21, the day is shortest, and the sun is positioned above the same latitude below the equator, called *tropic of capricorn*.

The following processes alter the spectrum of the sunlight as it passes through the atmosphere:

1. The blue coloured appearance of the sky is due to Rayleigh scattering that takes away some energy from the direct sunlight.
2. Scattering by particulate matter and aerosols occurs mostly at lower wavelengths.
3. The change in refractive index of the atmosphere due to temperature variations and turbulence also affects the incoming radiation. The other two mechanisms responsible for modification of the sunlight are by absorption bands. These are attributed to:
   (*i*) Electronic absorption bands in the three densest constituents of the atmosphere, that is, oxygen, nitrogen and ozone. Almost all the radiation in the ultraviolet region below $\lambda = 0.29$ µm is absorbed by the ozone layer thus making life on earth safer.
   (*ii*) Molecular vibrations and rotational frequencies of H_2O and CO_2 resonate with certain frequencies of the incoming radiation resulting in absorption bands. Nearly all wavelengths greater than 3 µm are absorbed by these gases leaving an atmospheric window at much longer wavelengths.

All these factors produce a radiation at the sea level quite different from that outside the earth's atmosphere. A comparable set of two curves is presented in Figure 2.3 showing these changes [1].

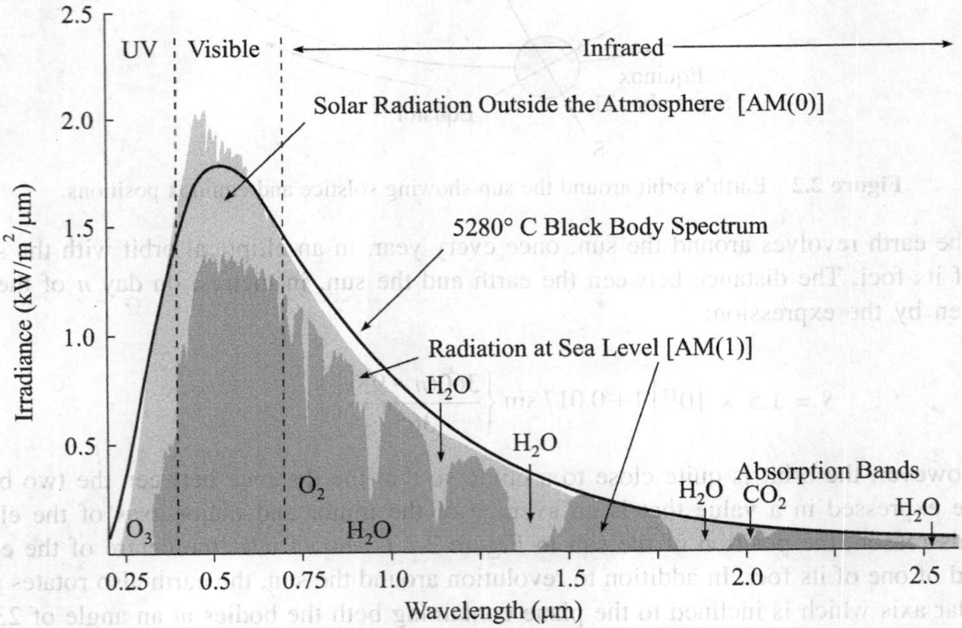

Figure 2.3 Spectral distribution of solar radiation.

The amount of solar power striking a given area is called *solar irradiance*. A measure of the intensity of the sunshine, it is given in units of watts (or kilowatts) per square metre. *Insolation* (*incident solar radiation*), also called *irradiation*, is the amount of solar energy received on a unit area over time and is measured in kWhr/m².

This value is equivalent to peak sun hours. Insolation can be given in various units like the British thermal units (Btu) per square foot (Btu/ft²), Langleys (L) or megajoules per square metre (MJ/m²). One unit can be converted into another by the following conversion factors:

$$\text{kWhr/m}^2 = \text{Langley}/86.04 = 317.2\ \text{Btu/ft}^2 = 3.6\ \text{MJ/m}^2 \tag{2.5}$$

Outside the earth's atmosphere a nearly constant 1.367 kW/m² of solar radiant power, called the *solar constant*, impinges on an area of 1 m². This is the value obtained by integrating the area under the graph in Figure 2.3. The two curves represent the radiation patterns one just outside the atmosphere and the other on the surface of the earth. Clearly, the earth's atmosphere is a strong absorber of certain wavelengths which also reduces the power of the radiation reaching the surface. The silicon PV modules convert to electricity the solar power from 0.3 to 0.6 μm. These wavelengths constitute the highest energy region of the solar spectrum. On a sunny day, the total irradiance striking the earth can be about 1,000 W/m².

The solar radiation that passes through directly to the earth's surface is called *direct solar radiation*. The radiation that has been scattered out of the direct beam is called *diffuse solar radiation*. The direct component of sunlight and the diffuse component of skylight falling together on a horizontal surface make up *global solar radiation*. The sum total of the direct, the diffuse and that reflected from the ground called *albedo*, falling on a tilted surface, is called the *total radiation*. It must be noticed that the albedo cannot fall on a horizontal receiving surface in absence of any nearby inclined reflecting structure higher than the receiver.

The design of any photovoltaic or photo thermal system strongly depends on the amount of energy falling on the collector. The higher the energy received, the more is the system's conversion output. The energy must be maximised in the winters since the demand peaks then, at least for hot water generating plants near the tropics, but is negligible in summers. The collection of energy in the months of winter is greater if the collector is tilted to an angle of 15° *more* than the latitude of the site.

This is in sharp contrast to the energy acquired when the tilt is less than that of the latitude by 15°. A plot of the energy received on a collector at these two different angles, as compared to the energy collected by the same device when tilted at an angle equal to the latitude of the place, is shown in Figure 2.4. At latitude +15°, more energy is received from November to March when the season is coldest and hot water is needed most. During the summer months it is of no consequence as no hot water is required in Northern India and other countries near the equator. However, this is not the case with photovoltaic systems which need maximum power available throughout the year. To do so, the PV array must be placed at an angle equal to the latitude of the place. In fact 0.9 × (latitude angle) yields a better result. If the tilt is increased by more than +15°, the insolation is approximated by the dashed curve which shows unacceptably lower levels of collection, than the reference, from November to January while peaking near April.

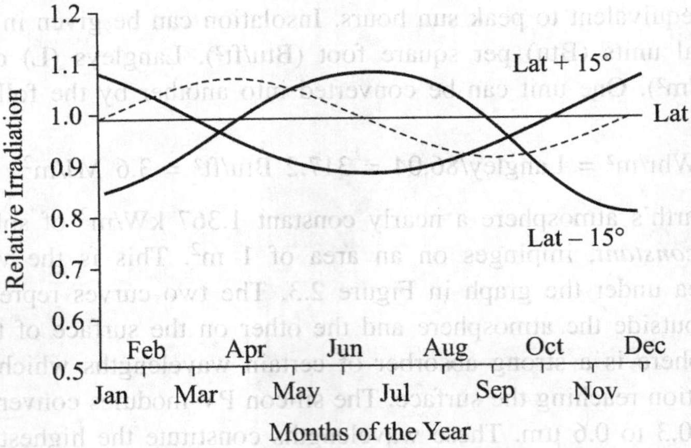

Figure 2.4 Relative variation of irradiation over the year.

Table 2.1 gives the radiation received in New Delhi throughout the year for three different tilt angles for fixed arrays as well as for east–west single axis tracked arrays. In this case the arrays are tilted at the desired angle in the north–south direction and move from east to west along with the sun. Data for two axes tracked arrays are also included. The tracking procedure is described in some length in Chapter 4. Similar data for other Indian cities are given in reference [2].

Table 2.1 Average Daily Peak Sun Hours in kWhr/m^2 for New Delhi, India, Latitude: 28° 58′ N, Longitude: 77° 20′, Elevation: 210 m [1]. This is also called (EHFS × 1 kW/m^2). EHFS stands for 'Effective Hours of Full Sunshine'

	Array tilted at Latitude		*Array tilted at Latitude* –15 deg.		*Array tilted at Latitude* +15 deg.		
Month	*Fixed Array*	*One-Axis Tracking*	*Fixed Array*	*One-Axis Tracking*	*Fixed Array*	*One-Axis Tracking*	*Two-Axis Tracking*
Jan	5.83	7.38	5.04	6.38	6.28	7.87	7.92
Feb	7.04	8.97	6.37	8.09	7.31	9.23	9.24
Mar	7.31	9.02	7.05	8.6	7.18	8.83	9.05
Apr	6.94	9.17	7.12	9.23	6.42	8.5	9.3
May	6.87	9.36	7.38	9.83	6.08	8.25	9.86
June	6.19	8.53	6.76	9.15	5.38	7.32	9.23
July	4.20	5.94	4.5	6.31	3.75	5.17	6.34
Aug	5.30	7.27	5.53	7.44	4.83	6.6	7.46
Sep	5.70	7.44	5.66	7.23	5.46	7.13	7.45
Oct	6.57	7.99	6.09	7.34	6.69	8.09	8.13
Nov	6.43	8.56	5.62	7.49	6.88	9.05	9.08
Dec	5.73	7.11	4.87	6.06	6.26	7.68	7.77
Yearly Avg.	6.18	8.06	6	7.76	6.04	7.81	8.4

2.2.1 Air Mass

The concept of the zenith angle and the solar noon is fundamental to the further understanding of the modification of the solar radiation in its passage through the atmosphere. The solar noon at any place in the northern hemisphere is measured by measuring the length of the shadow of a vertical rod, on a horizontal surface, throughout the day. The length of the shadow when it points towards the *geographic north* is the shortest. The time when this happens is the *solar noon.* It is necessary to take the *geographic north* and *not* the *magnetic north,* found by a compass, as the two may differ from place to place.

The zenith angle θ_z is useful for finding the path length of the solar radiation through the atmosphere normally expressed in terms of a quantity called *air mass.* The air mass at any time is defined as the ratio of the path length the sun rays travel in the atmosphere and the path length when the sun is at the zenith; this is given by sec θ_z. When $\theta_z = 0$, sec $\theta_z = 1$, the value of the air mass is AM (1). AM (0) refers to the air mass, that is the spectra and the intensity, in space. The average solar spectrum on earth is close to AM (2) but for calibrating solar cells an irradiation of AM (1.5) is accepted as standard. Figure 2.5 shows the quantities involved.

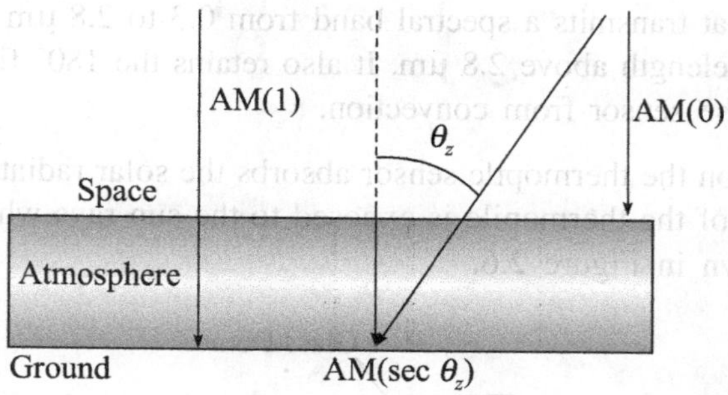

Figure 2.5 The zenith angle and its effect on the relative air mass.

2.3 Measurement of Solar Radiation

Most radiation measurement instruments use a thermopile as the sensing element. A thermopile consists of a large number of thermocouples connected in series to increase the output voltage per degree centigrade thus, increasing the sensitivity of the instrument. The absorption of thermal radiation by one of the thermocouple junctions, called the *active* or *hot junction,* increases its temperature. The differential temperature between the hot junction and a reference junction kept at a fixed temperature produces an electromotive force directly proportional to the difference in temperature created. This is called *thermoelectric effect* which has three constituents:

1. Seebeck,
2. Peltier and
3. Thomson effects.

These three are related and reversible. The thermocouple is well known and has been used extensively over the last 100 years for measurement of temperature and process control. The principle governing the operation of thermocouples is the Seebeck effect.

2.3.1 Pyranometer

A pyranometer is used to measure the global solar radiation. As stated earlier, the solar radiation spectrum extends approximately from 0.3 to 2.8 μm. Pyranometers usually cover that spectrum with a spectral sensitivity that is as "flat" as possible, that is, the same radiation power at any wavelength, falling on the sensor produces the same electrical power output.

For a flux density or irradiance measurement it is required by definition that the response of the sensor to beam or direct radiation varies proportionally to the cosine of the angle of incidence.

This means that there is maximum response when the solar radiation falls on the sensor perpendicularly (normal to the surface, sun at zenith, 0° angle of incidence), zero response when the sun is at the horizon (90° angle of incidence, 90° zenith angle), and 0.5 at 60° angle of incidence. It follows from the definition that a pyranometer should have a directional response to the incoming radiation that is close to the ideal cosine characteristic.

The main components of a pyranometer are:

1. A thermopile sensor with a black coating. This sensor absorbs all solar radiation in the range of 0.3 to 50 μm and has a near-perfect cosine response.
2. A glass dome that transmits a spectral band from 0.3 to 2.8 μm while filtering out any radiation of wavelength above 2.8 μm. It also retains the 180° field of view. The dome further shields the sensor from convection.

The black coating on the thermopile sensor absorbs the solar radiation which is converted into heat. The hot end of the thermopile is exposed to the sun rays while its cold end is kept in the shadow as shown in Figure 2.6.

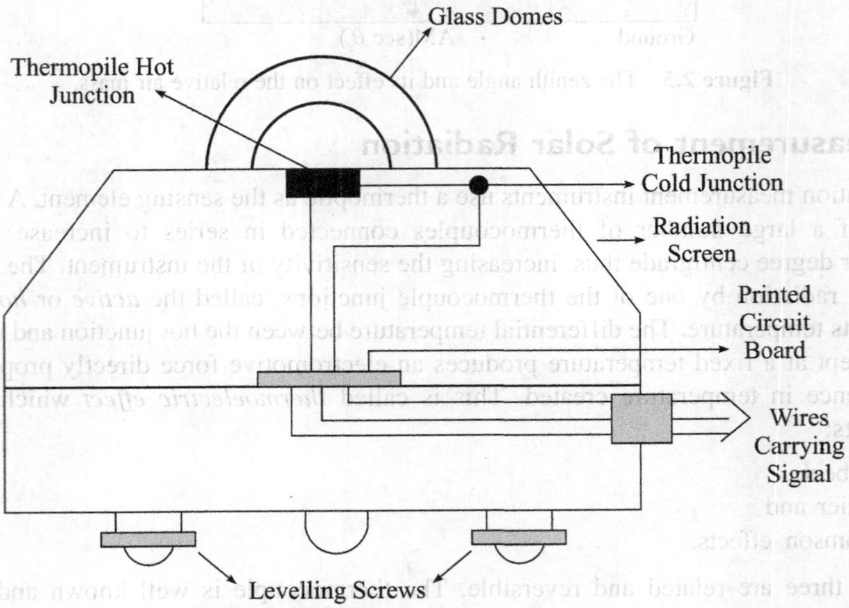

Figure 2.6 Side view of a pyranometer.

Pyranometers are standardised according to the ISO 9060 standard that is also adopted by the World Meteorological Organisation (WMO). This standard discriminates three classes. The best is paradoxically called *secondary standard*, the second best *first class* and the last one *second class*. Calibration is done relative to World Radiometric Reference (WRR). This reference is maintained by PMOD (Physikalisch-Metorologisches Observatorium Davos) in Davos, Switzerland.

Since most pyranometers rely on thermopiles to measure the temperature difference between a radiation-receiving surface and a shielded thermal mass, the temperature-sensing element typically consists of 50 or more copper-constantan thermocouples. This is done because the output of a single thermocouple is approximately 22 μV/°C, and a large number of junctions are needed in series to produce a usable output voltage for temperature measurement with good resolution. Further the rectangular geometry of these thermopiles tends to create a non-symmetrical angular response along the axis of the instrument. For this reason some manufacturers [3] use precision thin-film platinum resistance thermometers as the legs of an active electronic bridge circuit in a configuration similar to *bolometers*. This circuit produces a signal that is proportional to the temperature difference between the radiation-receiving surface and the shielded thermal reference. The unbalance signal is then amplified, in a thermally stable amplifier, to give an output that has low impedance and is independent of changes in ambient temperature. The inherent linearity of the platinum resistance thermometer scheme means that no thermistors are required to compensate for changes in ambient temperature.

The modern pyranometer is manufactured by using wire wound plated thermopiles. Over the years, there have been efforts to increase the linearity of response, durability, the adherence to the Lambert cosine response law and independence from ambient temperature effects. A precision pyranometer operates without deviations at various inclinations and is, therefore, often used to test solar panels; there it is mounted in the plane of the collector.

A commercially available pyranometer is shown in Figure 2.7. From its construction it can be seen that the window of the instrument allows light from all directions, at an angle of almost 180°, which includes the direct as well as the diffuse radiation. To measure only the diffuse part of the radiation a semicircular strip of aluminium is used to block the direct radiation by casting a shadow on the thermopile sensor when the pyranometer is mounted on a shadow band stand.

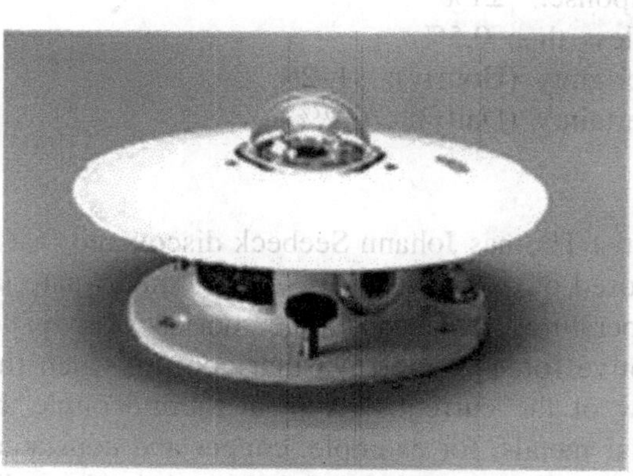

Figure 2.7 A commercially made pyranometer (*Courtesy*: Eppley laboratory).

The photograph of the instrument on a shadow band stand is shown in Figure 2.8 [4].

Figure 2.8 Pyranometer mounted on a shadow band stand. (*Courtesy*: Eppley Laboratory).

The specifications of the pyranometer are given below as an indication of the quality expected from accurate instruments. This instrument is classified as an ISO Secondary Standard or WMO High Class [4].

1. Response Time (95%): 10–15 s
2. Zero Offset to 200 Wm^{-2} net radiant loss to sky: 4–6 Wm^{-2}
3. Zero Offset to 5°C/hr change in ambient temperature: 1–2 Wm^{-2}
4. Resolution: less than 1 Wm^{-2}
5. Non-Stability: less than 0.5%/yr (typical)
6. Non-Linearity: ± 0.5% from 0–2800 Wm^{-2}
7. Directional Response: less than 10 Wm^{-2}
8. Spectral Selectivity: 1%
9. Temperature Response: ±1%
10. Tilt Response: less than 0.5%
11. Achievable Uncertainty (Hourly): 1–2%
12. Achievable Uncertainty (Daily): 1–2%

Thermocouple

In 1821, a German scientist Thomas Johann Seebeck discovered that an electric current flows in a closed circuit composed of two conductors, of dissimilar materials, when their junctions are kept at different temperatures. A thermocouple consists of two such conductors connected in series. The electromotive force or emf (V) that appears when the circuit is open results from an interrupted flow of the current. A typical thermocouple arrangement is shown in Figure 2.9. Two dissimilar metals, for example, copper and constantan are shown as Metal 1 and Metal 2. The two are joined on the left hand side where the radiation heats it. The other end of Metal 1 is connected to a terminal made of Metal 2 to form the cold junction. On the other side the wire made of Metal 2 is connected to a terminal made of the same material so that no junction is formed. If these two are connected, a current will flow in the circuit. However, if the two are connected to the input of a high impedance voltmeter the emf generated between the two junction may be read. Normally, the cold junction is kept in a container of melting ice whose temperature is at zero degrees centigrade.

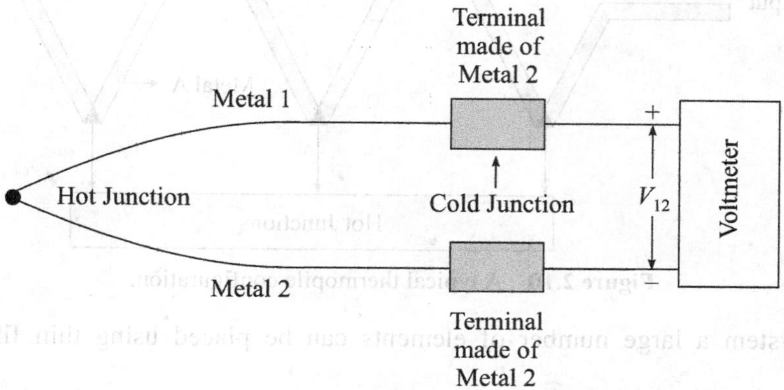

Figure 2.9 A thermocouple.

The open circuit electromotive force (emf) so created, V_{12}, is called *Seebeck voltage.* The generated emf V_{12} (V) is directly proportional to the differential temperature ΔT (K) between the two junctions, that is,

$$V_{12} = S_{12}\ \Delta T \tag{2.6}$$

where S_{12} is called the *Seebeck coefficient.*

Thermopile

When two or more thermocouples are connected in series, the configuration is called a *thermopile.* The schematic of such a device is shown in Figure 2.10. All the junctions of one side of the individual thermocouple are kept exposed to the radiation while the other junctions with the next thermocouple, that is, the cold junction, are placed away from the radiation. The voltage that is generated is taken from the mid-section of the last strip of metal.

From the arrangement of the thermopile it can be seen that the hot junctions will be spread over a longer distance compared to a single junction of a thermocouple. Thus, the heat that is produced may not be uniformly applied to all the junctions although it is necessary to keep all the junctions at nearly the same temperature.

An alternate geometry containing eight thermocouples is shown in Figure 2.11. In this the thick black ring, cut at intervals to make series connections, is made of one material on which wires of the same and another material are connected. The two dissimilar metals of the outer ring form the cold junctions while the junctions at the centre constitute the hot junctions marked H. Since these are huddled together in a small circle they can be easily kept at the same temperature.

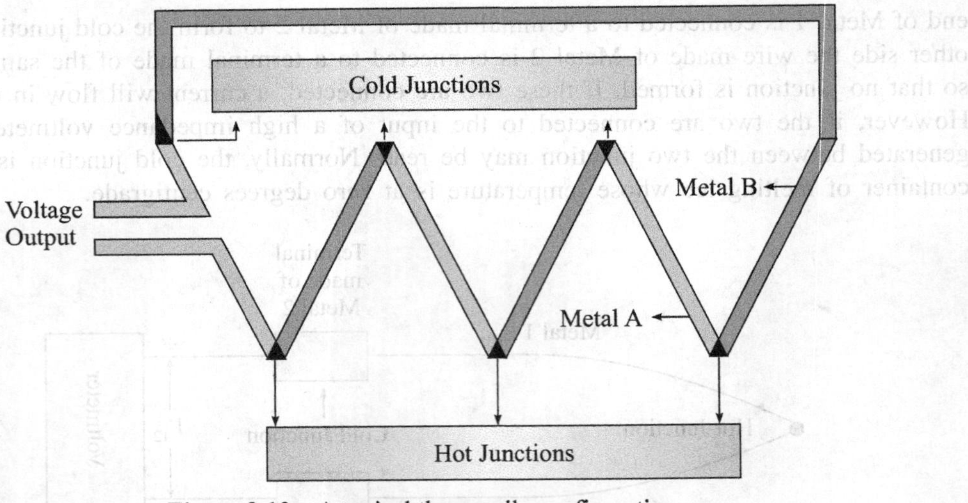

Figure 2.10 A typical thermopile configuration.

In this system a large number of elements can be placed using thin film deposition techniques.

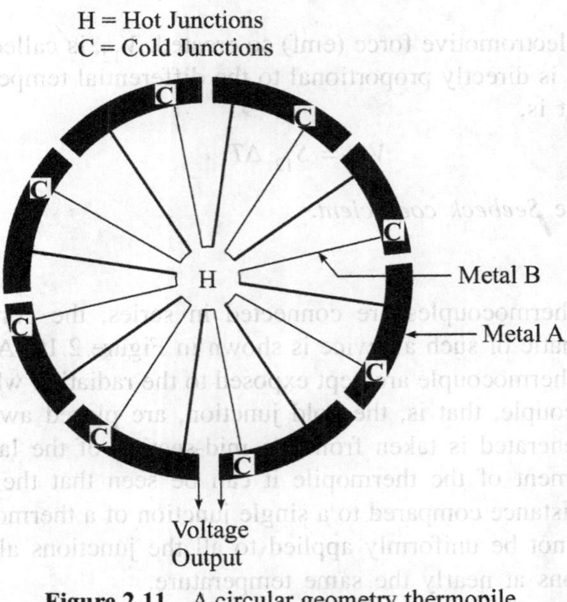

Figure 2.11 A circular geometry thermopile.

2.3.2 Pyrheliometer

Direct radiation at normal incidence is best measured by a pyrheliometer. Figure 2.12 shows the schematic of a pyrheliometer. A long tube, blackened from inside to stop reflection, is used to capture the direct solar radiation.

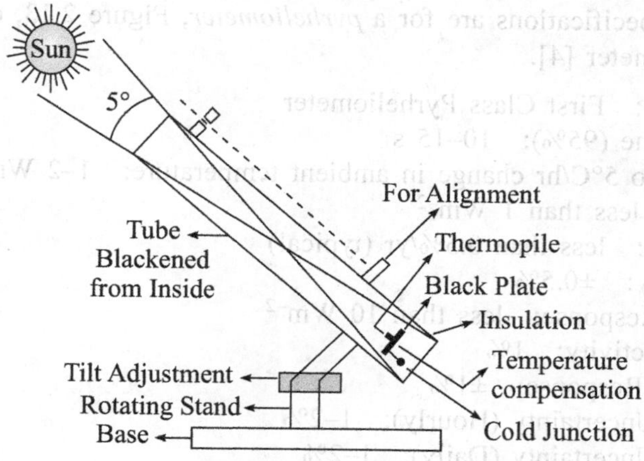

Figure 2.12 Schematic of a pyrheliometer.

The maximum angle subtended by the incoming radiation is limited to 5° so that scattered part of the radiation does not fall on the thermopile, or any other sensor, placed near the bottom of the tube. The sensor is mounted on a black plate that absorbs all the incident rays so that its temperature rises. Any light coming in at an angle of more than 5° will be absorbed by the blackened interior of the tube.

The cold junction of the thermopile is kept away from the sunlight. The difference in temperature creates an emf that is measured by a suitable external voltmeter. Temperature compensation is made by a thermistor based integrated circuit. As the sun moves the instrument has to follow it otherwise the sensor will be shaded. This can be manually done by adjusting horizontal and vertical movements from the joints marked "Tilt Adjustment" and "Rotating Stand" or by mounting the instrument on a dual axis sun tracker. To determine the correct initial direction, to let the sun rays fall on the sensor, a fore sight and a target are placed on the tube marked "For Alignment". A small hole on the fore sight allows a thin beam of sunlight to fall on the target. This spot is brought close to the centre of the target by moving the tube. Such an external focusing arrangement is needed because it is not possible see the sensor while exposing it to direct sunlight.

The pyrheliometer is normally mounted on an automatic solar tracker for continuous readings. A manufactured instrument is shown in Figure 2.13.

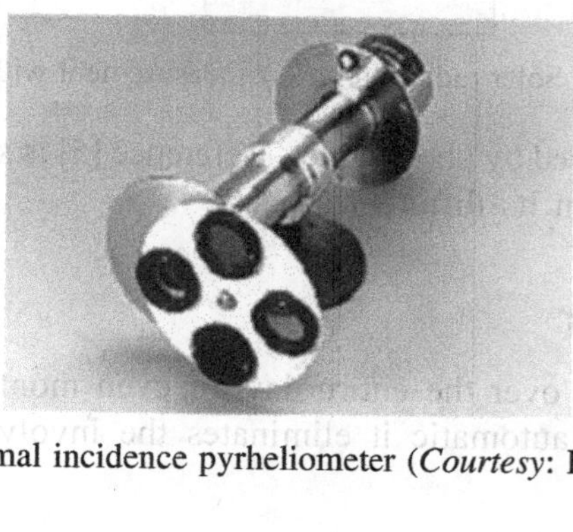

Figure 2.13 Normal incidence pyrheliometer (*Courtesy*: Eppley laboratory).

The following specifications are for a *pyrheliometer*, Figure 2.13, classified as an ISO First Class Pyrheliometer [4].

1. Classification: First Class Pyrheliometer
2. Response Time (95%): 10–15 s
3. Zero Offset to 5°C/hr change in ambient temperature: 1–2 Wm^{-2}
4. Resolution: less than 1 Wm^{-2}
5. Non-Stability: less than 0.5%/yr (typical)
6. Non-Linearity: ±0.5%
7. Directional Response: less than 10 Wm^{-2}
8. Spectral Selectivity: 1%
9. Temperature Response: ±1%
10. Achievable Uncertainty (Hourly): 1–2%
11. Achievable Uncertainty (Daily): 1–2%
12. Suitable Application: Working Standard or Network Measurements

2.3.3 Solar Cell Based Technique

An approximate method for measuring the global radiation is based on the measurement of the short circuit current of a mono crystalline silicon solar cell calibrated against a secondary standard such as a pyranometer. Very sophisticated measuring techniques like an operational amplifier, in a current to voltage mode, may not be used here. Instead a galvanometer with a proper shunt, providing low input impedance, is directly connected across the solar cell so that the current that is measured is quite close to the short circuit value. Since the short circuit current of a solar cell is proportional to the irradiance, the deflection of the meter provides an approximate value of the light intensity. This instrument is light, portable and does not require a cell to power it which makes it attractive for general field use. The schematic of such an apparatus is shown in Figure 2.14.

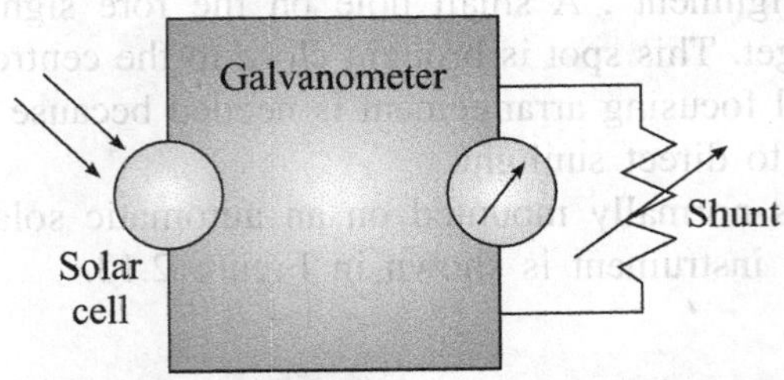

Figure 2.14 Solar radiation power measurement with a solar cell.

Solar cells are often replaced by photodiodes. Reference [5] is suggested for a comprehensive coverage of solar radiation in its different aspects.

2.4 The Data Logger

For recording radiation data over the entire day, or even months, an instrument called the *data logger* is used. Being automatic it eliminates the involvement of human beings for routine jobs.

A data logger or data recorder is an electronic instrument that records data over time at a location either with a connected instrument or sensor. Normally, these are built around microcontrollers or computers depending on the complexity and versatility of the system. They generally are small, battery powered, and portable, with internal memory for data storage. To view and analyze the collected data, some data loggers come with software to activate computers to which they may be attached while others have a local interface device like keypad and LCD so that they can be used as a stand-alone devices. The software enables capabilities for programmable data acquisition, processing and display.

Data loggers record temperature, relative humidity, pressure, voltage, water level, solar radiation etc. among other parameters. Effectively they have replaced the older strip chart recorders which did not have data processing capacity anyway. Further these instruments record data in units of the parameters instead of plotting them on a strip of chart paper to be decoded later. Decoding the data often lead to inaccuracies resulting from personal errors. With the introduction of silicon technology, it has become possible to miniaturize these instruments so that they can be battery operated, yet remain functionally same and record information for later retrieval by a computer. All communication with the data logger is controlled by software through an interface cable; normally a USB one. Having done the initial work the data logger is taken to the site of observation to be connected to different recording instruments such as a clock, thermometer, pyranometer and pyrheliometer for recording radiation data at different times of the day.

After the data have been recorded at the site of the experiment, the logger is brought to the computer, where they are downloaded using the data logger software. Most data logger software packages allow the user to view data in a number of different formats like graphical and tabular. The tabular format provides the user with the raw data which can either be exported to a spreadsheet application such as Microsoft Excel, for further operations or printed for records.

Figure 2.15 shows the typical configuration of a data logger. The raw DC input from an external source is stabilized to a value of 5 V to power the whole circuit. The data from the

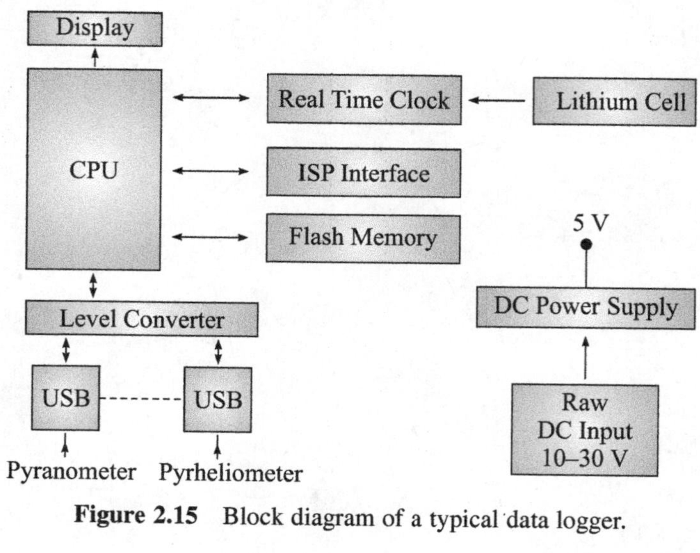

Figure 2.15 Block diagram of a typical data logger.

pyranometer and pyrheliometer are connected to the input ports of the microcontroller with a real time clock providing the time at which the data are recorded. When the USB ports are connected to the computer, the stored data are downloaded using the software.

References

[1] *Stand-Alone Photovoltaic Systems: A handbook of Recommended Design Practices,* Photovoltaic Design Assistance Centre, Sandia National Laboratories, Albuquerque, New Mexico, U.S.A., 1995, p. 16.
http://photovoltaics.sandia.gov/docs/PDF/Stand%20Alone.pdf

[2] Chetan Singh Solanki, *Solar Photovoltaics: Fundamentals, Technologies and Applications,* PHI Learning, New Delhi, 2009, pp. 442–458.

[3] *TSP–700 Total Solar Pyranometer*, Yankee Environmental Systems, Inc., Airport Industrial Park, 101 Industrial Blvd., Turners Falls, MA 01376, USA.
www.yesinc.com

[4] The Eppley Laboratory Inc., 12 Sheffield Avenue, PO Box 419, Newport, Rhode Island, USA 02840.
www.eppleylab.com

[5] John A. Duffie and William A. Beckman, *Solar Engineering of Thermal Processes,* 2nd ed., John Wiley & Sons, Inc., New York, 1991.

CHAPTER 3

Solar Cell Characteristics

3.1 Semiconductor Physics

3.1.1 Intrinsic Semiconductors

The fourth column of the periodic table contains elements such as germanium and silicon, which are the two most extensively used materials for transistor fabrication. Silicon has prevailed over germanium as the preferred material for reasons considered later. They contain four electrons in the outermost shell or orbit which do not satisfy the energy needs of those levels. In their pure elemental state these atoms tend to form a crystal, when brought closer under certain conditions, which acquire a pattern known as *face centred diamond cubic lattice* in the three dimensions, a structure identical to that of the diamond form of carbon.

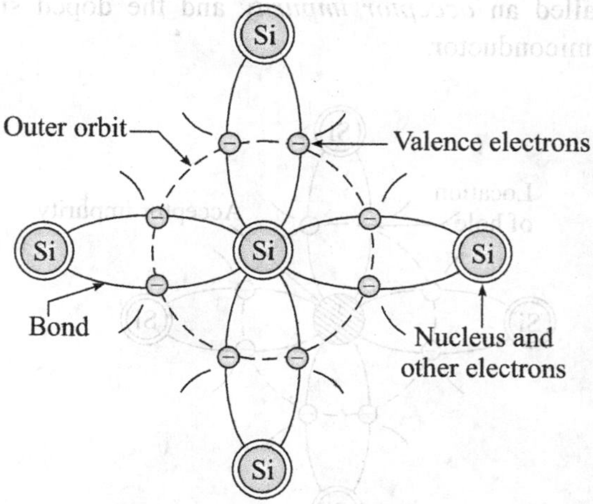

Figure 3.1 A section of the perfect silicon crystal.

Since these atoms cannot exist in isolation, one atom shares four electrons with four neighbouring atoms to form a stable structure as shown in Figure 3.1. When eight electrons occupy the outer shell, the energy required for stability is met. This sharing of electrons is called *covalent bonding* which ensures that there is no free electron at 0 K. However at higher temperatures adequate energy may be infused to break the bonds to release some electrons. Thus, it is clear that at low temperatures both germanium and silicon in their pure form, have insulating properties, and are called *intrinsic semiconductors*. The resistivity of germanium is 50 Ω cm with 10^{13} charge carriers per cubic centimetre while those of silicon are 240,000 and 10^{10} respectively.

3.1.2 Extrinsic Semiconductors

The pure crystal is not used to fabricate a diode or a transistor but imperfections are introduced in the crystal structure by adding controlled amounts of impurities to regulate its conduction and nature. This process is known as *doping*. The added impurities generally belong to the third and the fifth columns of the periodic table shown in Table 3.1

Table 3.1 A portion of the periodic table

Group 3	*Group 4*	*Group 5*
Boron$^{5}_{10.8}$	Carbon$^{6}_{12}$	Nitrogen$^{7}_{14}$
Aluminium$^{13}_{27}$	Silicon$^{14}_{28.1}$	Phosphorus$^{15}_{31}$
Gallium$^{31}_{69.7}$	Germanium$^{32}_{72.6}$	Arsenic$^{33}_{75}$
Indium$^{49}_{114.8}$	Tin$^{50}_{118.7}$	Antimony$^{51}_{121.8}$

Suppose an impurity like indium with three electrons in its outer orbit is added to a crystal of silicon, then the resulting picture in Figure 3.2, shows that there is a lack of an electron to provide proper covalent bonding with one of the silicon atoms so that it creates a location for a *hole* which is ready to capture or accept any electron passing by. This impurity is therefore called an *acceptor impurity* and the doped silicon is also called an *acceptor* or a *p-type* semiconductor.

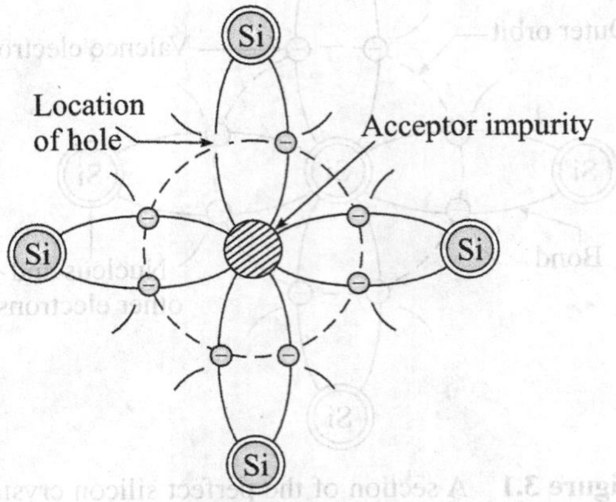

Figure 3.2 Acceptor impurity in crystal.

A similar placement of an atom of an element like antimony from the fifth column of the periodic table inside a crystal of silicon creates an excess electron even after satisfying the energy requirements of all the surrounding silicon atoms. This combination of elements, as shown in Figure 3.3, can now donate this extra electron for improving the conduction and is therefore called a *donor* or simply an *n-type* semiconductor.

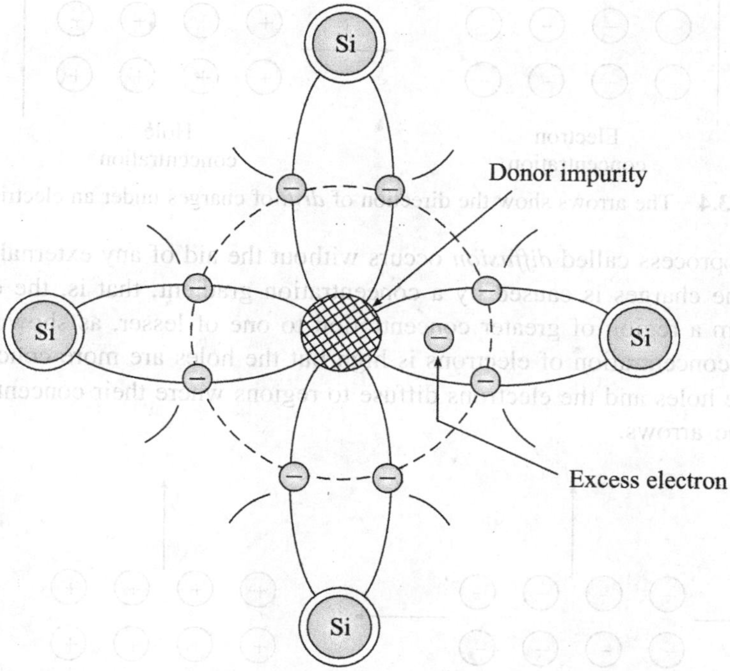

Figure 3.3 Donor impurity in silicon crystal.

There are two types of movements of the charge carriers in the semiconductors called *drift* and *diffusion*. Both the electrons and holes move randomly in their respective *n*- and *p-type* materials. When an electron passes by a hole, which is positively charged, they attract each other and merge together until some form of energy breaks their bonding. The electron then moves away, leaving behind the hole, and continues to do so till it is captured by another hole. This movement causes a picture to emerge where both the charge carriers, electrons and holes, seem to travel of course with different mobility. When an electric field is applied across the two types of doped semiconductors, the electrons are attracted by the positive polarity while the holes move towards the negative side, opposite to that of the electrons. Such movement of charges under an electric field is called *drift*. Figure 3.4 shows the movement of the electrons and holes when an electric field is applied across a space containing them. The positively charged holes move in the direction of the negative plate while the electrons go towards the positive plate. The velocity and the energy of the charges depend on the magnitude of the electric field.

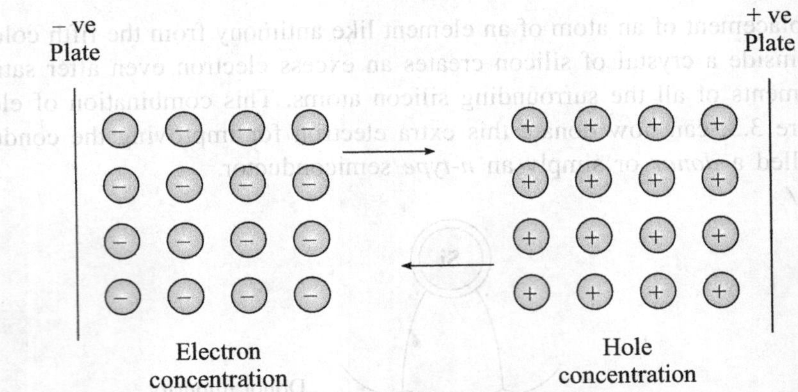

Figure 3.4 The arrows show the direction of *drift* of charges under an electric field.

The second process called *diffusion* occurs without the aid of any external field. Here the movement of the charges is caused by a concentration gradient, that is, the charges tend to move away from a region of greater concentration to one of lesser, as shown in Figure 3.5. On the left the concentration of electrons is high but the holes are more concentrated on the right side so the holes and the electrons diffuse to regions where their concentrations are less as shown by the arrows.

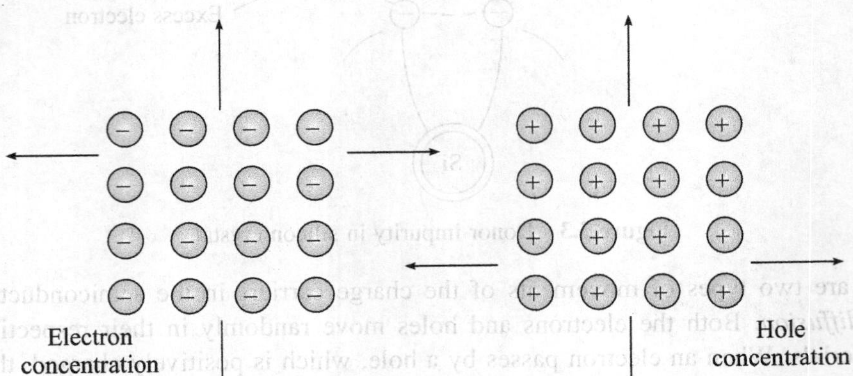

Figure 3.5 *Diffusion* from high to low concentration regions.

The movement of an electron through the lattice to fill a hole results in hole movement. This becomes clear from a study of Figure 3.6. Suppose the location of the hole, to start with, is at point *Y* near the acceptor impurity. The electron at point *A* is let loose by breaking the covalent bond by heat and than it moves towards *Y*, where it is trapped by the positive charge of the hole. So now the hole apparently travels from *Y* to *A*. This process is repeated by the electron from *B* and then to *C* and so on until the position of the hole ends in *G*. Thus, the movement of the electron ends up in moving the hole from *Y* to *G* so that the hole seems to move.

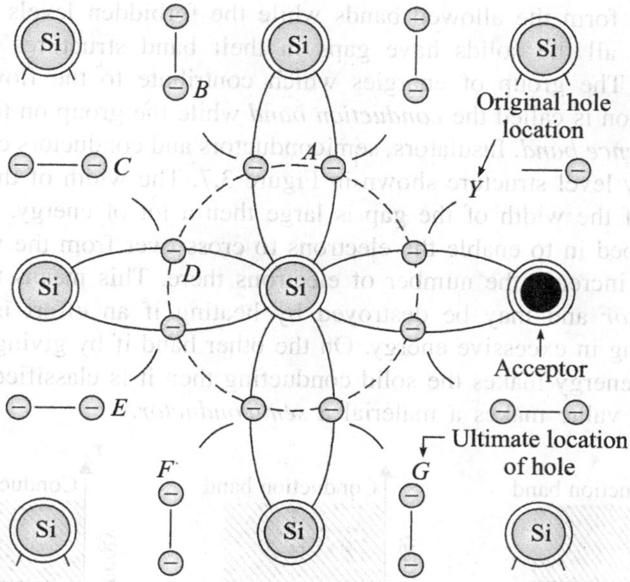

Figure 3.6 Electrons move in the lattice to fill holes so that holes appear to move.

It is clear that the hole moves in the opposite direction to that of the electron, and significantly, it takes place within the valence shell of the atom. When such a movement takes place in the presence of an electric field, it becomes an example of *drift*. It is worth remembering that the average hole velocity is approximately one-half that of electron velocity. To summarize the understanding of a semiconductor it must be remembered that the silicon, germanium or the impurity atom is stationary within the crystal while the conduction process is carried out by the free electrons and holes resulting in the electric currents. In an *n-type* semiconductor the electrons are *majority* carriers while the holes are *minority* carriers. Similarly the holes are majority carriers in a *p-type* semiconductor, electrons being the minority carriers.

3.1.3 Energy Level Diagrams

Atomic theory states that every electron that resides in an orbit around the nucleus continues to do so until it is given sufficient energy to break away from the nucleus. If an electron, situated in an orbit close to the nucleus, is given energy from an external source it jumps to an orbit farther away from the nucleus. Thus, an electron can be moved to a higher energy orbit by giving it energy equivalent to the difference of energy between its initial and final orbits. This energy can be in the form of heat or light or any other such as collision. However, there are certain energy levels that are forbidden for occupation to any electron, in which case the particle needs enough energy injection to jump over these to a higher energy level.

When a large number of atoms are squeezed together to form a solid, the energy of individual atoms loses its importance because of others in the vicinity and a collective picture emerges. The energy levels of individual electrons are so close to each other that when atoms, in which they reside, form a solid, a band of energy levels emerges. Those on either side of the forbidden levels form the allowed bands while the forbidden levels themselves form a *forbidden gap*. Not all the solids have gaps in their band structures but insulators and semiconductors do. The group of energies which contribute to the flow of electricity by movements of electron is called the *conduction band* while the group on the other side of the gap is called the *valence band*. Insulators, semiconductors and conductors can be differentiated based on this energy level structure shown in Figure 3.7. The width of the gap is the key to this classification. If the width of the gap is large then a lot of energy, in one form or the other, must be pumped in to enable the electrons to cross over from the valence band to the conduction band to increase the number of electrons there. This means that such a solid is normally an *insulator* and may be destroyed by heating if an effort is made to make it conducting by putting in excessive energy. On the other hand if by giving a negligibly small amount of external energy makes the solid conducting then it is classified as a *conductor*. A gap of intermediate value makes a material a *semiconductor*.

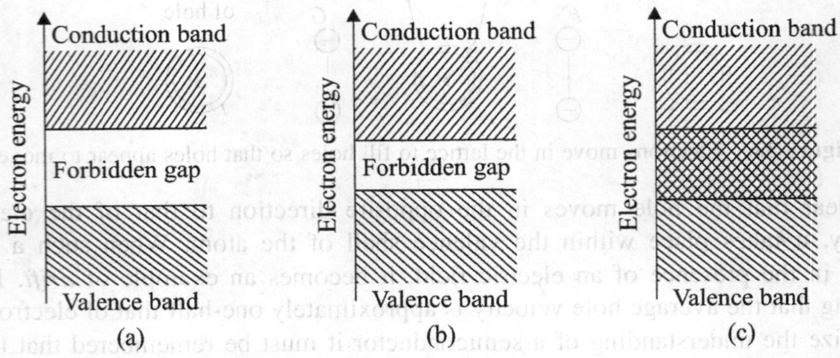

Figure 3.7 Classification of materials based on the forbidden gap: (a) An insulator, (b) A semiconductor and (c) A conductor.

The band gap of the semiconductor germanium is 0.72 eV while for silicon generally accepted value is 1.1 eV. The electron volt (eV), is a unit of energy equivalent to 1.6×10^{-19} (W s). These gaps are small enough for these materials to become conducting at room temperature because an appreciable number of covalent bonds between atoms are broken, by the available heat, to release sufficient number of electrons for conduction.

In fact, germanium is so sensitive to heat that transistors made from this element had a problem of thermal runaway that led to its discontinuation for device fabrication. If we examine the other two elements of the fourth column of the periodic table we find that their use is not suitable for making electronic components. Carbon, in the diamond form, for example, has an energy gap of 7 eV and therefore behaves as an insulator up to very high temperatures. Gray tin in its crystalline form has a low band gap of only 0.1 eV and work only at very low temperatures. These two materials cannot be used for transistor or diode fabrication.

The broken inter-atomic bonds not only generate additional electrons for the conduction band but they also create holes there. These holes create traps for electrons that effectively reduce mobility. The addition of pentavalent impurities to a crystal of silicon or germanium creates an additional energy level 0.01 eV below the conduction band, as shown in Figure 3.8, so that they contribute electrons there without generating holes. This is made possible as the thermal energy at room temperature is about 0.02 eV. Another advantage of doping is that even at low temperatures the semiconductors continue to conduct unlike intrinsic ones whose resistivity increases as the temperature gets lower because the electrons fall to the valence band leaving only a small number in the conduction band.

A similarity can be seen when a trivalent impurity is introduced in a semiconductor crystal. The energy level of such an impurity settles just above the valence band and contributes holes to it. Energy level diagrams are plots of electron energies and so the energies of holes are highest near the valence band and reduce as they travel towards the conduction band where they are lowest.

To conclude, an *n-type* semiconductor has an excess of electrons in the conduction band that are primarily responsible for contribution to electrical conductivity. The holes are present in larger numbers in the *p-type* material and electrical conduction in them is a valence band hole drift phenomenon.

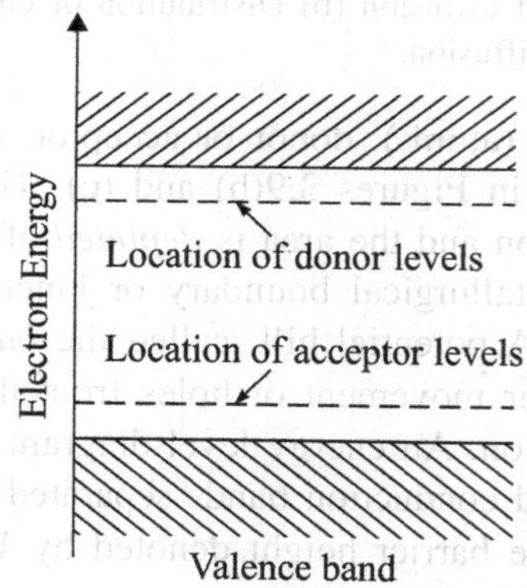

Figure 3.8 The location of energy levels of donor and acceptor impurities in an extrinsic semiconductor.

3.2 *p–n* Junctions and Formation of Solar Cells

When a *p-type* and an *n-type* semiconductor material are bonded together, a *p-n* junction is formed. Methods of junction formation are discussed in books on devices. Immediately after junction formation, electrons from *n-type* cross over to the *p-type* region by the process of *diffusion* mentioned earlier. Similarly, holes from the hole-rich *p-type diffuse* to the *n-type* material. Recombination is possible among electrons and holes on either side of the junction between the two types of semiconductors. This results in ionized acceptors in the *p*-region and ionized donors in the *n*-region. Since both the *n-* and *p-type* materials were initially neutral this charge transfer results in a positive *n*-region and a negative *p*-region. The diffusion of charge carriers continue to take place until enough charge builds up to produce electrostatic forces to repel any further charges from crossing over from either side. The negatively charged *p*-region repels electrons from crossing over there while a positively charged *n*-region stops any hole from entering. In equilibrium the *p-n* junction is depicted in Figure 3.9(a).

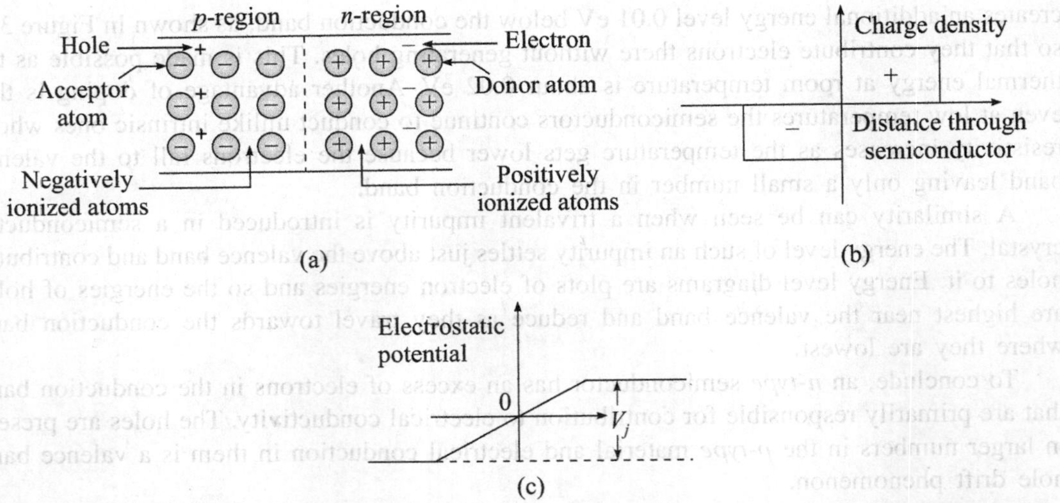

Figure 3.9 (a) After the completion of diffusion (b) Distribution of charges and (c) The electrostatic potential developed after diffusion.

Because of the ionized carriers (atoms), donor or acceptor, charges build up leading to an electrostatic potential as shown in Figures 3.9(b) and (c). This potential sweeps out all charges in the vicinity of the junction and the area is *depleted* of loosely bound carriers. For this reason the region near the metallurgical boundary or junction is called the *depletion, space charge* or *transition region*. A potential hill, called the *barrier height*, of large value is produced here to deter the further movement of holes from the *p*-region to *n*-region and for electrons to go from *n* to *p*-region. An energy level diagram for the *p-n* junction can be envisaged involving the valence and conduction bands separated by the band or energy gap, subjected to the potential hill or the barrier height denoted by V_j. This picture can be seen in Figure 3.10.

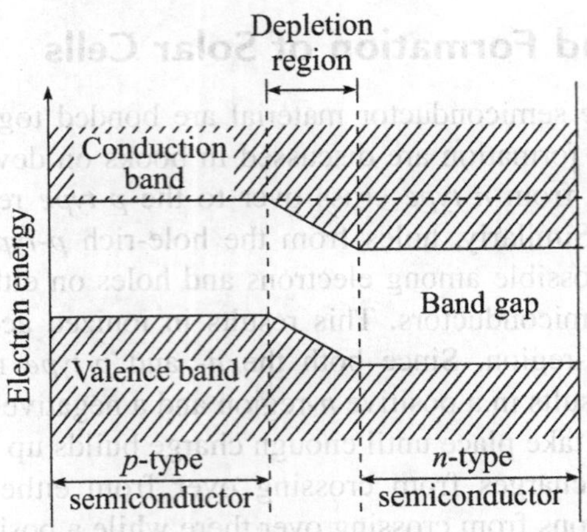

Figure 3.10 The depletion region and the barrier height qV_j.

Since the *p*-region has lost some of its holes but has gained some electrons it is no longer electrically neutral. Its energy will be higher than that of the *n*-region by an amount qV_j (φ), where q is the charge on an electron and V_j is the potential through which the electron has travelled. In order for another electron to move to the *p*-region it must acquire energy equal to or more than this. Similarly, a hole must have energy more than this potential hill confronting it in order to move to the *n*-region.

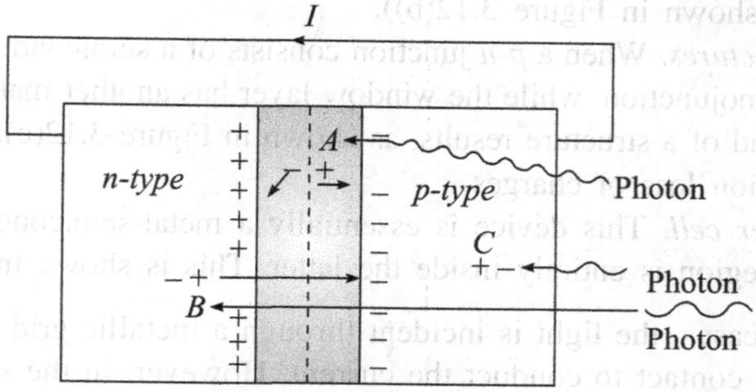

Figure 3.11 The junction potential separates charges generated by light.

Referring to Figure 3.11, the full implication of this potential generated by the process of diffusion can be appreciated. When exposed to light, the photons enter the *p-n* junction from the right. The energy of the photons is sufficient to break the covalent bonds between atoms to release some electrons to create electron-hole pairs. Very weak photons whose energies are insufficient to break the bonds will be absorbed while those strong enough will achieve their goal. Usually, the number of electron-hole pairs generated is proportional to the number of photons entering the device. Consider the pair created at point *A*. This pair is subjected to a strong electrostatic potential inside the *depletion* region and is promptly separated with the hole going to the negatively charged *p*-region with the electron *drifting* to the positively charged *n*-region. The pair at the point *B* is close to the space charge region but is not affected by the electrostatic field there unless the positively charged hole diffuses into it and is immediately pulled to the *n*-region. However, the electron-hole pair produced at *C* might not be so fortunate because they may not reach the depletion region as they can recombine with opposite charges on their way. The net effect of the process of charge separation is that the *p*-region has excess positive charge while the *n*-region becomes negatively charged. If a wire is used to connect the *n*- and *p*-regions externally then the two types of charges will flow to restore the initial neutral condition or in other words an electric current (I) will flow through the wire. This is the mechanism of photo generated current which is nothing but a solar cell. The depletion region is fundamental to the solar cell action because of its ability to separate positive and negative charges. It should be noticed that the charge build-up on either side of the junction resembles a parallel plate capacitor. This is a very important part of the solar cell property and will be described later in detail. Also because the *p-n* junction is a diode the solar cell must exhibit the properties of a diode that will be used to determine its characteristics.

3.2.1 Types of Solar Cells

The four most common types of solar cell junctions are shown in Figure 3.12. They are:

1. *Homojunctions*. Here the material is same for both *n* and *p* doped types that form the junction, for example, *n-* and *p-type* silicon as shown in Figure 3.12(a).
2. *Heterojunctions.* The *n* and *p-type* materials that form the junction are different in this case. Cadmium sulphide and copper sulphide (CdS/Cu_2S) are the two materials that can be cited here (shown in Figure 3.12(b)).
3. *Heteroface structures.* When a *p-n* junction consists of a semiconductor of band gap E_{g1}, similar to a homojunction, while the window layer has another material of a *larger* band gap E_{g2} this kind of a structure results, as shown in Figure 3.12(c). This helps to reduce the recombination loss of charges.
4. *Schottky barrier cell.* This device is essentially a metal-semiconductor junction where the depletion region is entirely inside the latter. This is shown in Figure 3.12(d).

In the first three cases, the light is incident through a metallic grid on the surface which provides an electrical contact to conduct the current. However, in the second case there can be illumination through the back wall also. In the fourth case, light enters through the semi transparent metal on the front wall.

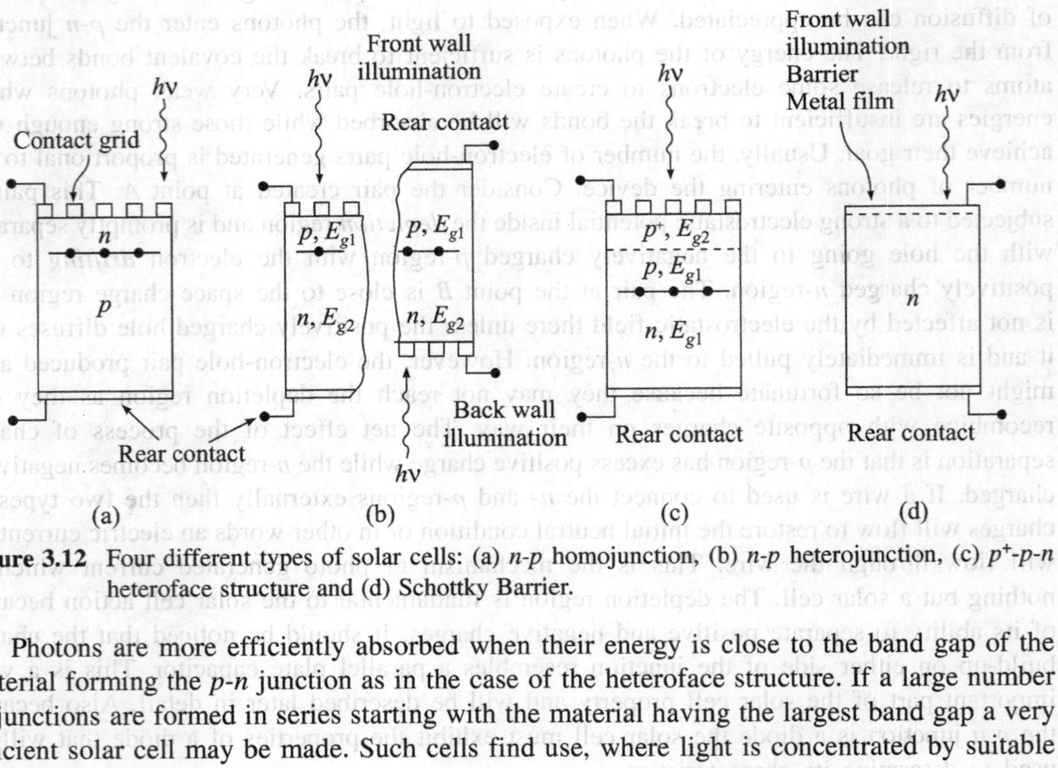

Figure 3.12 Four different types of solar cells: (a) *n-p* homojunction, (b) *n-p* heterojunction, (c) p^+-*p-n* heteroface structure and (d) Schottky Barrier.

Photons are more efficiently absorbed when their energy is close to the band gap of the material forming the *p-n* junction as in the case of the heteroface structure. If a large number of junctions are formed in series starting with the material having the largest band gap a very efficient solar cell may be made. Such cells find use, where light is concentrated by suitable lenses.

3.3 Solar Cell Characteristics and Their Measurements [1]

3.3.1 Current versus Voltage (*I*–*V*) Characteristics

The *p-n* junction discussed so far forms a junction diode shown symbolically in Figure 3.13.

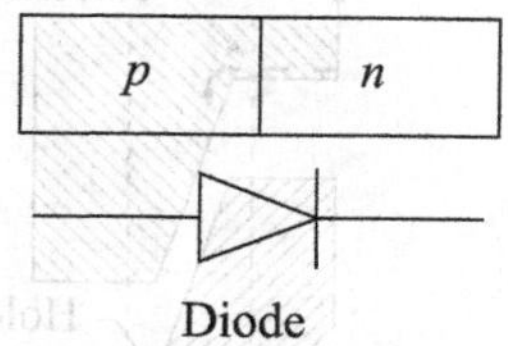

Figure 3.13 The *p-n* junction, which acts as a diode, and the symbol of a diode.

This device has the property of passing current only in one direction. Because of this property alone it is used extensively in electronic circuits for detecting radio frequency signals for communications as well as high voltage and high current rectifiers for converting the power line frequency of 50 or 60 Hz to direct current (DC) for driving DC loads. There are other useful properties to be discussed later. The mechanism of signal rectification can be understood from Figure 3.14.

Figure 3.14(a) shows the transition region of a normal *p-n* junction with its barrier height not influenced by any external electric field. This barrier height dissuades majority carriers from either side from crossing over to the other side. However, it aids the minority carriers to cross over. For example, the holes in the *n*-region, produced by thermally broken bonds, find an accelerating force in the potential hill and therefore slide down to the *p-type*. Figure 3.14(b) shows the *p-n* junction subjected to a *reverse bias*. A reverse bias is a condition, where an electric field from a battery is applied across the junction such that the *p-type* is connected to the negative terminal while the *n-type* is attached to its positive one. The negative polarity on the *p-type* adds a further field to the already existing potential hill which, in effect, creates a higher potential for any electron trying to enter this region. A similar argument holds good for the holes too. In this condition, the junction blocks the flow of electrons, and therefore the conventional current whose direction is opposite to the flow of electrons. Figure 3.14(c) shows a *forward* biased junction where the positive terminal of the battery attracts the electrons from the *n*-region by effectively lowering of the barrier height thereby creating a situation conducive for current flow. In this state a *forward* current flows through the junction. It must be remembered that the major source of minority carriers is thermally broken covalent bonds.

Although the reverse bias stops current flow, it cannot be increased beyond a limit. The minority carrier generation is limited by the heat supplied and hence the current due to it *saturates* after a while on increasing the bias. This is called the *reverse saturation current*. On further increasing the reverse bias, the minority carriers acquire large velocities and therefore greater energies. This enhancement of their energies can knock off electrons from atoms in their path. This secondary emission of electrons can create further additional electrons by collisions.

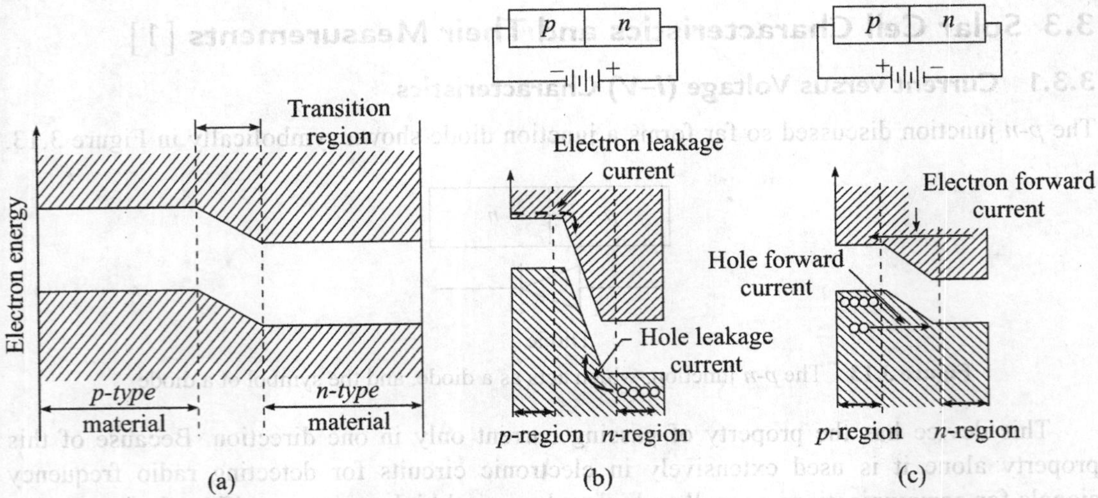

Figure 3.14 The energy band diagrams of a *p-n* junction: (a) Without any electric bias, (b) Under reverse electric bias and (c) Under forward electric bias.

This process continues until there is an *avalanche* breakdown when the current increases suddenly without any increase in the voltage across the *p-n* junction. This phenomenon is utilized in electronics for stabilizing voltages by devices normally known as *zener diodes* although the zener breakdown is different and appears to be a 'field emission' process where the electrons are pulled out of atoms by a strong electric field in the junction region.

The forward volt-ampere characteristic of a junction diode, shown in Figure 3.15, is known as Shockley's diode equation which is written as:

$$I = I_R \left(e^{\frac{qV}{nkT}} - 1 \right) \tag{3.1}$$

where

I = junction current in amperes
q = charge on an electron, 1.602×10^{-19} C
I_R = saturated value of the reverse current in amperes
V = potential difference in volts
T = absolute temperature in Kelvin
k = Boltzmann's constant, 1.38×10^{-23} J/K
n = diode factor which varies from one to two depending on junction fabrication

For small values of the forward bias voltage V the current I is nearly zero due to (–1) present in the equation. As the voltage increases the forward current I goes up exponentially and attains a steady slope depending on the bulk resistance of the *p-* and *n-type* materials. It is thus advisable to use some current limiting resistance to protect the junction from damage. Similarly the reverse breakdown current should also be limited by appropriate methods.

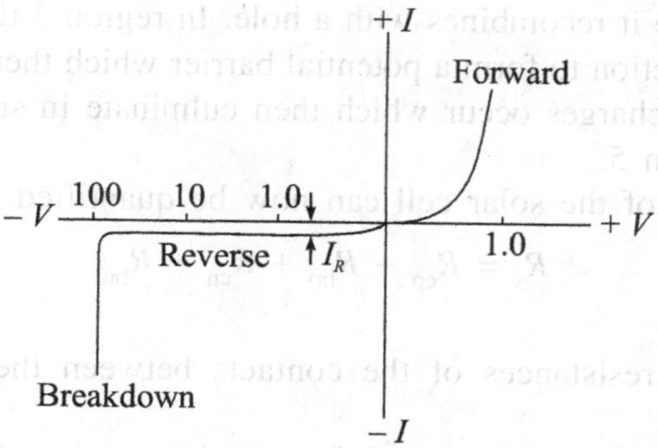

Figure 3.15 A typical current–voltage plot of a *p-n* junction.

3.3.2 Calculation of Solar Cell Parameters and Efficiency [2, 3]

Different electronic processes occurring in a *p-n* homojunction solar cell are shown in Figure 3.16. Starting from the left the region marked 1 contains one of the contacts between the *p-n* junction and the wire used to take out the current.

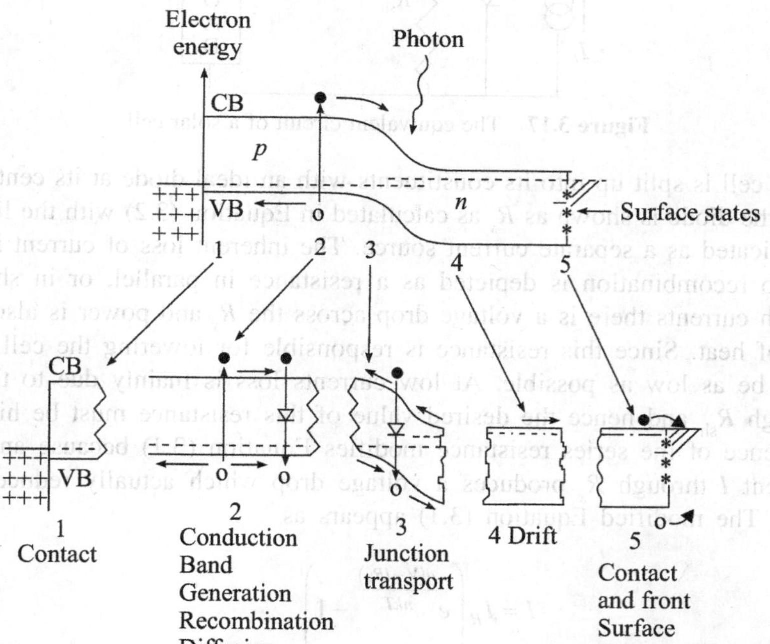

Figure 3.16 The basic functions occurring in different places of a *p-n* homojunction solar cell.

Region 2 indicates the transition of an electron from the valence band to the conduction band, after an electron-hole pair generation, on receiving energy from external sources. The number of electron-hole pairs depends on the intensity of illumination and the optical absorption coefficient α (λ). The electron diffuses through the material but loses energy and goes back to the valence band where it recombines with a hole. In region 3 the electrons and holes are transported across the junction to form a potential barrier which then stops further movement. In region 4 only drift of charges occur which then culminate in surface recombination near the front contact in region 5.

The series resistance of the solar cell can now be quantified from Figure 3.16 as:

$$R_s = R_{cp} + R_{bp} + R_{cn} + R_{bn} \tag{3.2}$$

where

R_{cp} and R_{cn} are the resistances of the contacts between the wires and the p and n materials, and

R_{bp} and R_{bn} are the bulk resistances of the p and n semiconductors.

From the above discussion the equivalent circuit of a solar cell can be drawn below in Figure 3.17.

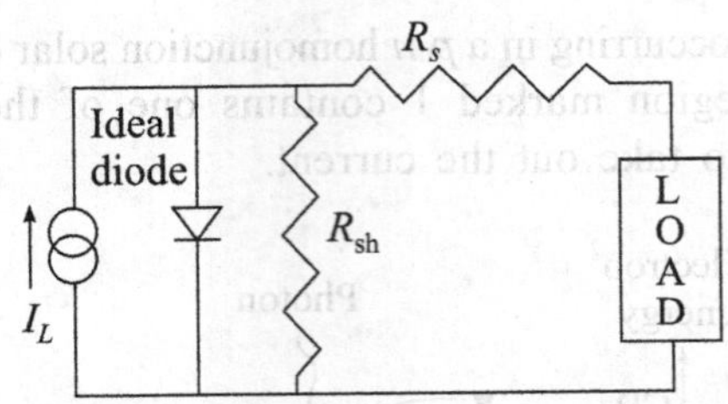

Figure 3.17 The equivalent circuit of a solar cell.

The solar cell is split up into its constituents with an ideal diode at its centre. The series resistance of the diode is shown as R_s as calculated in Equation (3.2) with the light generated current I_L indicated as a separate current source. The inherent loss of current in a solar cell due mostly to recombination is depicted as a resistance in parallel, or in shunt, with the diode. At high currents there is a voltage drop across the R_s and power is also lost as $I_L^2R_s$ in the form of heat. Since this resistance is responsible for lowering the cell efficiency its value should be as low as possible. At low currents loss is mainly due to the leakage of current through R_{sh} and hence the desired value of this resistance must be high.

The presence of the series resistance modifies Equation (3.1) because any flow of the forward current I through R_s produces a voltage drop which actually reduces the applied forward bias. The modified Equation (3.1) appears as

$$I = I_R\left(e^{\frac{q(V-IR_s)}{nkT}} - 1\right) \tag{3.3}$$

When the solar cell is exposed to light the light generated current I_L goes in the opposite direction to the forward current, as explained above, such that Equation (3.3) appears as

$$I = I_R\left(e^{\frac{q(V-IR_s)}{nkT}} - 1\right) - I_L \tag{3.4}$$

Figure 3.18 shows a plot of the two Equations (3.3) and (3.4). The maximum voltage of the cell when no current is drawn from it is called the *open circuit voltage* V_{oc} while the maximum current that can be drawn from the cell when the voltage across is zero is called the *short circuit current* I_{SC}. It is quite clear from the diagram that the power delivered by the cell is zero at the points where the light generated curve intersects the current (I) or the voltage (V) axis, that is, at the points I_{SC} and V_{oc} because either the current or the voltage there will be zero. However, between these two extremes both the current and the voltage have finite values so that their product, the power, is also finite. There is a point where this power passes through a maximum which is called the maximum power point and is denoted

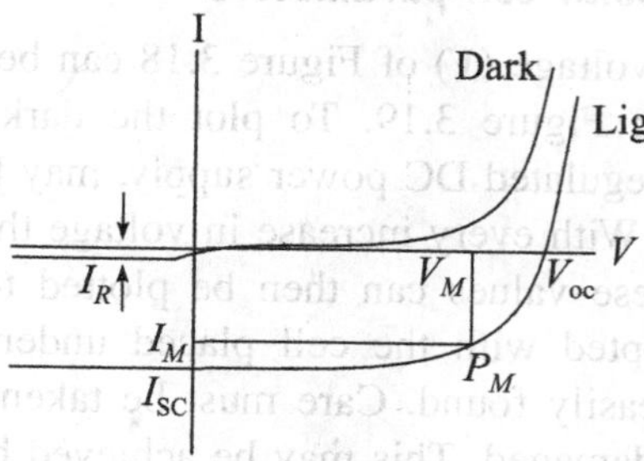

Figure 3.18 The current I versus voltage V curve when a solar cell is placed in dark and illuminated conditions.

by P_M. The corresponding current and voltage values are called the maximum current I_M and the maximum voltage V_M points. A quantity called the *fill factor* (*ff*) is usually introduced to relate P_M to I_{SC} and V_{oc} by the expression

$$ff = I_M \times V_M / I_{SC} \times V_{oc} = P_M / I_{SC} \times V_{oc} \tag{3.5}$$

The efficiency of a solar cell can therefore be written as

$$\eta_s = ff \times I_{SC} \times V_{oc} / P_S = P_M / P_S \tag{3.6}$$

where

P_S is the power of the illumination falling on the cell. Some representative values of these quantities for a solar cell are:

$V_{oc} = 0.5$ to 0.8 V, $I_{SC} = 10$–40 mA cm^{-2} of cell area, $ff = 0.6$–0.8 and $\eta_s = 6$–28%.

All these quantities are measured under the test conditions laid down by the National Aeronautics and Space Administration (NASA) under the direction of the Department of Energy (DOE), USA. A standard solar cell made of the same material as the cell under test is used to calibrate a solar simulator used to illuminate the cell. The material of both the cells has to be the same so that their spectral response does not differ. The standard cell is calibrated to give a known short circuit current for a known intensity of illumination. The other conditions called STC are:

1. The cell temperature during measurement is 28 ± 2°C
2. An irradiance of 100 mW cm^{-2}

3. While measuring the short circuit current with a resistor there should not be a voltage drop of more than 20 mV across it.
4. The open circuit voltage of an unloaded cell should be measured with a voltmeter of internal resistance of at least 10^4 Ω/V.
5. The test cell area is the entire area of the solar cell including the contacts and the upper grid.

The irradiation used is usually AM 1.5.
The IEC 60904 is also used as another standard.

The I–V plotter to measure solar cell parameters

The plots of current (*I*) versus voltage (*V*) of Figure 3.18 can be obtained in many ways, the simplest of which is shown in Figure 3.19. To plot the dark current and voltage plot a variable DC source, such as a regulated DC power supply, may be used with a voltmeter and an ammeter in place as shown. With every increase in voltage the corresponding current may be recorded and tabulated. These values can then be plotted to get the *dark I-V* curve. A similar procedure may be adopted with the cell placed under a light source. From these curves the parameters can be easily found. Care must be taken to limit the current through the diode otherwise it may be damaged. This may be achieved by setting a suitable overload current limit in the power supply.

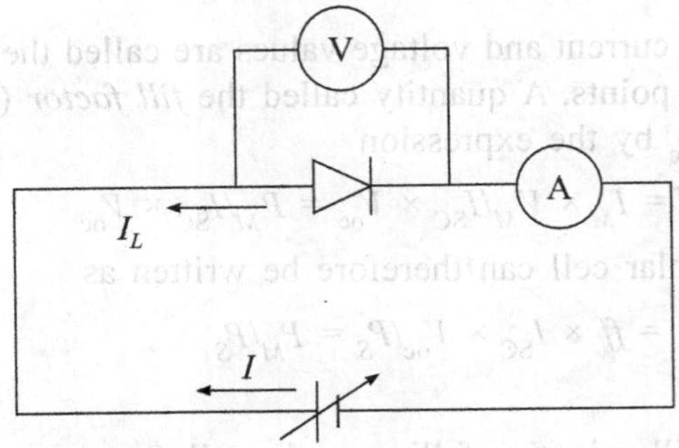

Figure 3.19 Current vs Voltage measurement arrangement of a solar cell.

Several modifications to this circuit can be made to display the curves automatically. The first one is to connect an *X-Y* recorder with its *Y*-input in place of the ammeter and the *X*-input replacing the voltmeter. The power supply may then be manually controlled or it may be replaced by a ramp signal capable of giving the required current. In all the cases the instruments must conform to the NASA-DOE or IEC 60904 specifications. A digital oscilloscope can also perform the same function provided at least one of the channels, preferably the voltage measurement one, is used in a differential mode to avoid a common earth point.

Alternately the current and the voltage signals can be recorded in a computer through a card containing analogue to digital converters. This will not only provide the on screen display but also enable further calculations to determine the diode parameters and efficiency. Since these options are fairly expensive some cheaper methods are discussed as follows.

Short circuit current measurement

It is necessary to have a measuring device for this to have a very small input resistance so that the voltage across it never exceeds 20 mV. A simple circuit for short circuit current measurement is a current to voltage converter [4, 5 and 6] as shown in Figure 3.20. The solar cell placed in this circuit is connected between the earth and the virtual earth of the operational amplifier (opamp). It is thus in a short circuit mode as the input resistance of the opamp is R/A, where A is the open loop gain of the opamp. Since A is very large, of the order of 10^5 or more, and the actual value of R is about 1 kilo-ohm, the input resistance is 0.01 Ω or less.

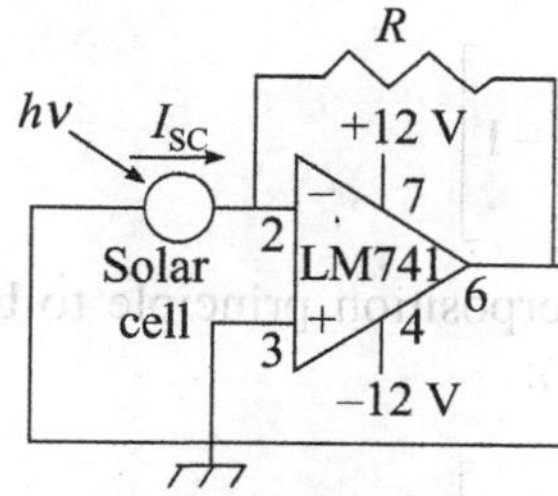

Figure 3.20 A current to voltage converter used for measuring the short circuit current of a solar cell.

This current flows through R consequently the output voltage $V_o = I_{SC} \times R$ volts. By measuring V_o and R, being known, I_{SC} can be found.

Open circuit voltage measurement

The primary condition that must be fulfilled during the measurement of this parameter is that the solar cell must not be loaded by the voltmeter, which NASA specifies to have a 10^4 Ω/V internal resistance. A voltage follower made from an opamp usually has a higher input resistance particularly when the opamp has a MOSFET input like the CA 3140 opamp. The input resistance of this device is 1.5×10^{12} Ω which far exceeds the NASA specification. The circuit diagram of a voltage follower to measure the V_{oc} of a solar cell is shown in Figure 3.21. The output voltage follows the solar cell open circuit voltage V_{oc} available at its non-inverting terminal marked 3. The 22 mega-ohms resistance is placed in parallel to the cell to conduct away any charge accumulating at the terminal. With the required input resistance of 10^4 Ω/V the maximum V_{oc} that can be measured is 22×10^6 Ω/10^4 Ω = 22×10^2 V = 2.2 kV. This is much higher than the voltage of about 10 V, which the opamp can take.

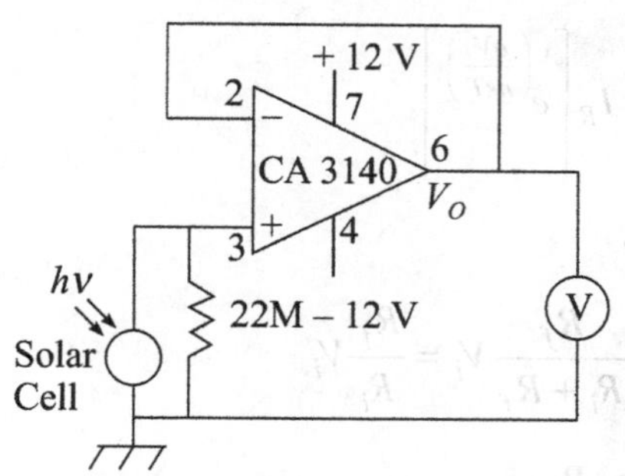

Figure 3.21 A voltage follower to measure the V_{oc} of a solar cell.

To succeed in producing an efficient solar cell, it is necessary to couple fabrication techniques with basic understanding of the device. It is, therefore, of fundamental interest to be able to correlate the cell performance with the junction behaviour and identify avenues of further improvement in conversion efficiency. The junction can be conveniently evaluated in terms of current (*I*)–voltage (*V*) characteristics and the temperature dependence thereof to obtain essential functional parameters such as reverse saturation current (I_R), the diode factor (n), the barrier height (ϕ), the open–circuit voltage (V_{oc}), the short-circuit current (I_{SC}), the fill factor (ff), the series resistance (R_s), the shunt resistance (R_{sh}) and the conversion efficiency (η). Re-writing Equation (3.3) without the R_s component

$$I = I_R\left[e^{\left(\frac{qV}{nkT}\right)} - 1\right] \tag{3.7}$$

Under illumination, assuming the superposition principle to be valid, the *I–V* characteristics can be written as from Equation (3.4):

$$I = I_R\left[e^{\left(\frac{qV}{nkT}\right)} - 1\right] - I_L \tag{3.8}$$

where I_L is the light-generated current.

By taking the logarithm of Equation (3.7), for $V >> kT/q$, we obtain

$$\log_e I = \log_e I_R + qV/nkT \tag{3.9}$$

Plotting $\log_e I$ versus V, we obtain a straight line with slope given by q/nkT, from where n can be calculated. Alternatively, from Equation (3.8), if we plot $\log_e (I + I_L)$ versus V, we again obtain a straight line with slope equal to q/nkT. In both cases, $\log_e I_R$ can be obtained by the intercept on the ordinate. Since silicon single crystal solar cell is also a diode the above analysis applies to it too.

An experiment on a diode will verify these postulates. A BY127 silicon rectifier was taken for this purpose [7]. Like the proportionality between $\log_e I$ and V as shown in Equation (3.9) it is also easy to prove that the forward voltage of a diode is proportional to the logarithm of the input voltage in a certain circuit configuration. Since the exponential term is much greater than one for appreciable values of V, Equation (3.7) becomes

$$I = I_R\left[e^{\left(\frac{qV}{nkT}\right)}\right] \tag{3.10}$$

Considering Figure 3.22,

$$V = \frac{R_f}{R_i + R_f}V_i = \frac{R_f}{R_i}V_i \tag{3.11}$$

When $\qquad R_i >> R_f$

where

R_f is the total diode forward resistance
R_i is the input resistance put externally to get the proper logarithmic conditions
V_i is the externally impressed voltage

Also,

$$V = R_f I \tag{3.12}$$

From Equations (3.11) and (3.12), we get

$$R_f \times I = R_f \times V_i/R_i \text{ or } I = V_i/R_i \text{ or } V_i = I \times R_i$$

Substituting I from Equation (3.10)

$$V_i = R_i I_R \, e^{\left(\frac{qV}{nkT}\right)} \tag{3.13}$$

or,

$$\log_e V_i - \log_e R_i I_R = \frac{qV}{nkT} \tag{3.14}$$

Hence

$$V = (nkT/q)\log_e V_i + C \tag{3.15}$$

$C = [(-nkT/q) \log_e R_i I_R]$ is a constant.

Equation (3.15) shows a straight line when V is plotted against $\log_e V_i$. It has a slope of (nkT/q) with an intercept C. Substituting the values for n, k and q, the slope is 1/20 for $T = 300$ K. If the base of the logarithm was changed to 10, the slope of the line would just be altered by a factor of 2.303. This means that if V is plotted against log V_f, it will give a straight line but with a different slope.

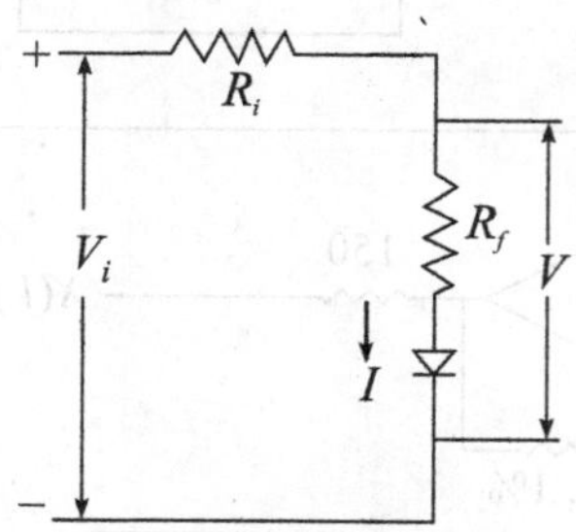

Figure 3.22 Conditions under which the diode displays a logarithmic nature.

Several diodes were examined for logarithmic characteristics. The best results were obtained with the BY127 diodes made by Bharat Electronics Ltd.

The characteristics of three diodes are shown in Figure 3.23 along with the measuring circuit.

All the three diodes show good logarithmic response for inputs of 3–300 V. Below 3 V, the curves deviate. It may also be observed that two out of the three diodes give identical curves above 3 V.

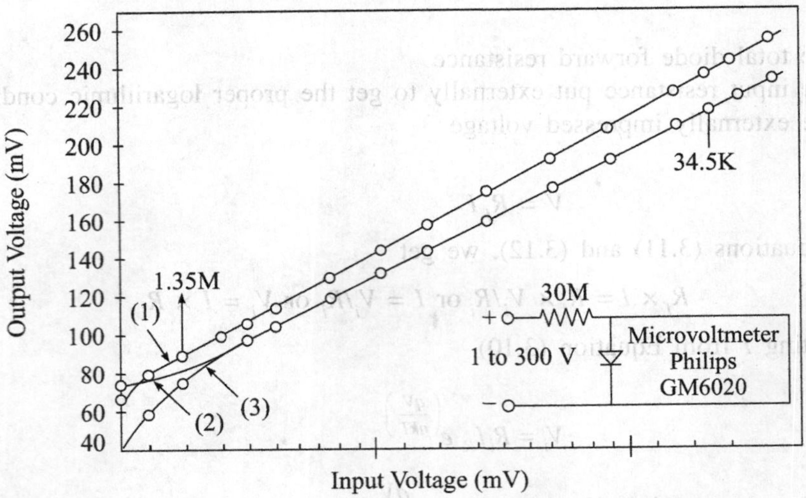

Figure 3.23 The logarithmic curves of BY127 diodes.

Since a logarithm of the current through a cell or the voltage across it can furnish valuable information about the cell parameters a logarithmic amplifier was designed and several cells were studied. The entire circuit is described in reference [8]. The block diagram of the circuit is given in Figure 3.24.

The basic principle of the *I–V* plotter is illustrated in Figure 3.24. A low frequency sinusoidal signal is applied to the solar cell through a resistor *R* (10, 33 or 100 Ω). The voltage developed across the resistor is proportional to the current passing through the solar cell

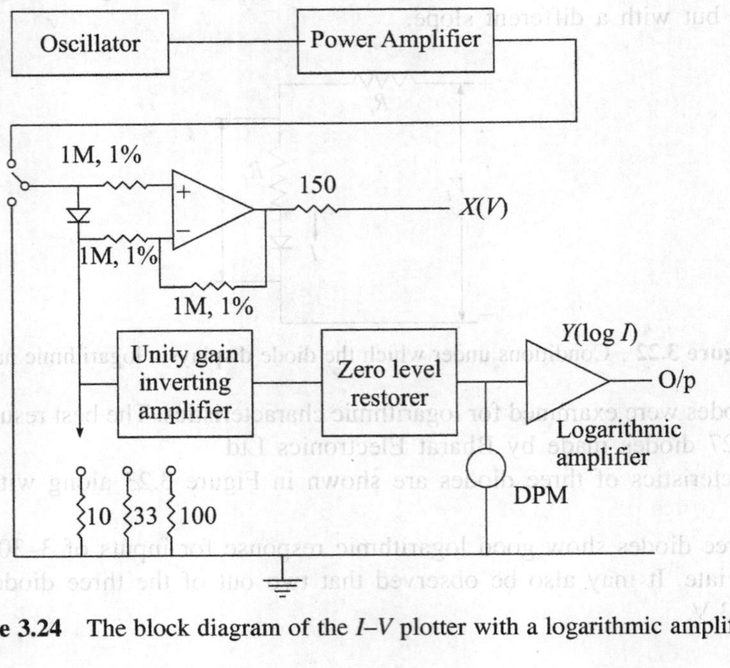

Figure 3.24 The block diagram of the *I–V* plotter with a logarithmic amplifier.

and is fed to the Y-input of the recorder/oscilloscope. The voltage across the sample is fed to the X-input of the recorder. To obtain the log I–V characteristics, the voltage output across the resistor is amplified by a logarithmic amplifier and is fed to the Y-input and the voltage across the solar cell is fed to the X-input of the X–Y recorder. In the $\log_e (I + I_L) - V$ characteristics of a cell under illumination, the voltage across the resistor R is offset by a voltage equal to $I_L \times R$, using an offset network to obtain a voltage proportional to $(I + I_L)$ which is then amplified by the logarithmic amplifier to provide a log $(I + I_L)$ output. It is important to note that the oscillator output is amplified by a power amplifier so that it can supply the short circuit current of the cell under test.

The logarithmic amplifier is designed using the principle given above and shown in Equation (3.9). The circuit diagram of an amplifier is given in Figure 3.25. The logarithmic amplifier is a DC amplifier, where stability is of primary importance so the drift characteristic with temperature must be very good. The LM108 is a very thermally stable operational amplifier used in military applications. However, its commercial version LM308 can easily replace it in the present case. When a voltage V is applied to the input of the circuit of Figure 3.25, it drives a current through the 100K input resistor as the other end of the resistor is connected to the virtual ground of the opamp. This current then passes through the diode $D1$, if V is positive or $D2$ if V is negative, so that the output voltage is proportional to this current in accordance with Equation (3.9). Thus, the output voltage is proportional to the logarithm of the input voltage [4]. Since the value of $R2$ determines the current through $D1$ and $D2$ which in turn is responsible for keeping the diode in the exponential region of the diode's I–V curve, its value is critical in determining the input voltage swing.

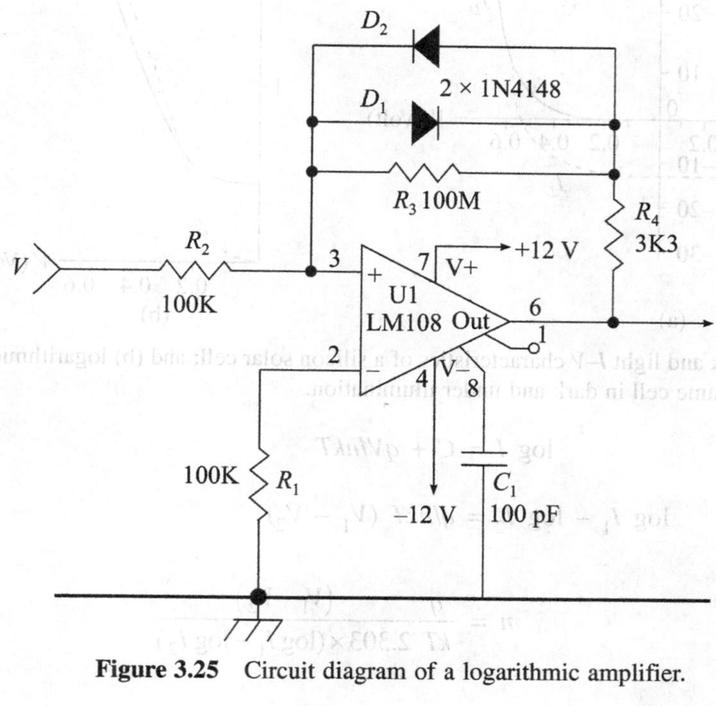

Figure 3.25 Circuit diagram of a logarithmic amplifier.

Using the instruments shown in Figure 3.24, in conjunction with the logarithmic amplifier shown in Figure 3.25, *I–V* and logarithmic characteristics in dark and under illumination have been determined for several solar cells using the logarithmic *I–V* plotter. Figure 3.26(a) presents the *I–V* characteristics of a silicon solar cell in dark and under illumination. The $\log_e I$ versus V in dark and $\log_e (I + I_L)$ versus V (under illumination) characteristics of the same cell, as obtained from the plotter, are shown in Figure 3.26(b). The relation between the natural logarithm and logarithm to the base 10 is:

$$\log_e N = 2.303 \log N.$$

If the curves are plotted as V versus $\log I$, they will still be straight lines but with a different slope.

Example 3.1: Calculate the diode factor (n) from Figure 3.26.

To calculate the diode factor (n), the linear portion of the $\log_e I$ versus V or $\log_e (I + I_L)$ versus V characteristics is chosen. The voltage change, $(V_1 - V_2)$, corresponding to a decade change in dark current, that is $\log I_1 - \log I_2 = 2.303\ (2 - 1)$, at the output of the logarithmic amplifier, is then determined. It is 140 mV. From Equation (3.9) the value of n is calculated as follows:

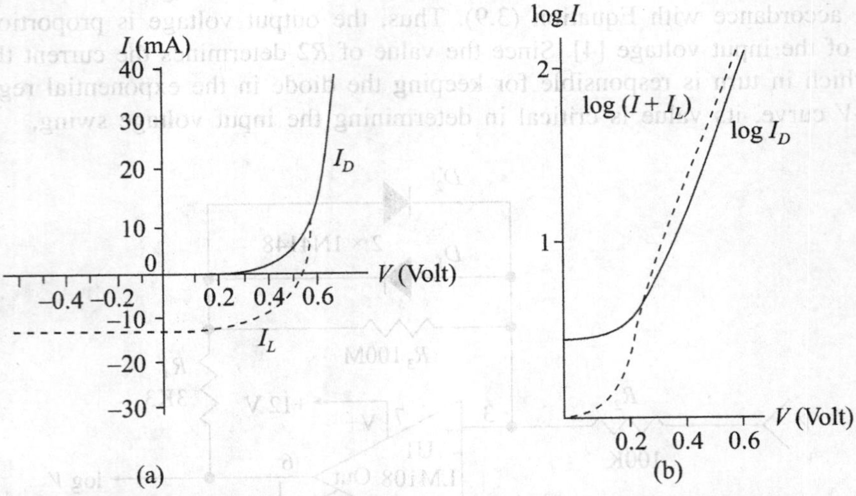

Figure 3.26 (a) Dark and light *I–V* characteristics of a silicon solar cell; and (b) logarithmic characteristic of the same cell in dark and under illumination.

$$\log I = C + qV/nkT$$

or,

$$\log I_1 - \log I_2 = q/nkT\ (V_1 - V_2)$$

or,

$$n = \frac{q}{kT} \frac{(V_1 - V_2)}{2.303 \times (\log I_1 - \log I_2)} \tag{3.16}$$

At room temperature $T = 300$ K, therefore, putting the values of q and k from Section 3.3.1, $q/kT = 39$. The factor 2.303 comes from the conversion of natural logarithm to that of base 10. Substituting in Equation (3.16)

$$n = \frac{39\times(0.59-0.41)}{2.303\times(2-1)} = 39\ (0.18)/2.303 = 3.05 \approx 3.1$$

Table 3.2 gives the values of the diode factor of the solar cell n_1 and n_2 which refer to two diode factors obtained in different voltage regions thus, showing that the solar cell exhibits *double diode* behaviour. The computed and the corresponding experimentally obtained values match closely.

Table 3.2 Experimentally obtained and computed values of the diode factor of several devices

Device	*Experimental values*		*Computed values*	
	From log I–V plot (in dark)	*From log $(I + I_L)$–V plot (under illumination)*	*From I–V characteristics (in dark)*	*From I–V characteristics (in light)*
Silicon solar cell	$n_1 = 3.1$ $n_2 = 2.1$	$n_1 = 3.0$ $n_2 = 2.1$	$n_1 = 2.9$ $n_2 = 2.1$	$n_1 = 3.1$ $n_2 = 2.1$
DR25	1.2	–	1.2	–
SR100	1.9	–	1.9	–
BY127	2.0	–	2.0	–
IN4148	2.0	–	2.0	–

The characteristics of several other junction devices have also been plotted. The computed and experimentally determined diodes factors, given in Table 3.2, match very well. These devices show single diode behaviour.

There are more expensive and versatile instruments available to plot these characteristics and determine the parameters not only for a single solar cell but also an entire module. The Oriel *I–V* test station [9] and the PV measurement [10] are such examples. In these instruments the cell or module is exposed to light from a solar simulator at the desired intensity and the measurements are automatically carried out, and the final results displayed on screen along with a hard copy. The only problem is that the module temperature increases during the measurements leading to a drift in the parameters. To overcome the heating problem Spire Corporation has introduced a system, SPI–SUN SIMULATOR 350i, which flashes a light source to illuminate the module for about 2 milliseconds during which all the measurements are carried out [11].

I–V plotter for PV modules and arrays

Another method exists to measure the *I–V* plots of PV devices ranging from a single cell to an array but only under illumination. *I–V* curve of such an arrangement is shown in Figure 3.27.

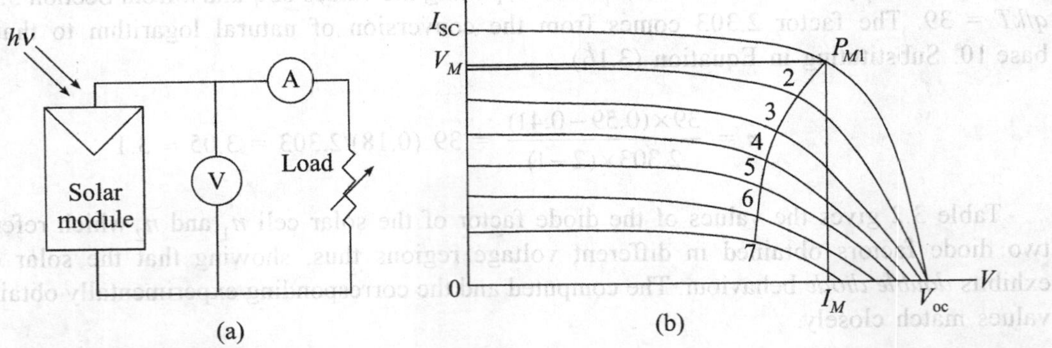

Figure 3.27 (a) Measurement circuit and (b) a typical set of *I–V* curves of a module.

Light is incident upon a solar module so that it generates current as shown in Figure 3.27(a). A variable resistance is connected as a load across the module through an ammeter to measure the current while a voltmeter monitors its output voltage. For different values of the resistance the current and the voltage are measured and tabulated. A plot of the current and the voltage, for different incident radiation powers, are shown in Figure 3.27(b). I_{SC}, V_{oc}, I_M and V_M are the short circuit current, open circuit voltage, the maximum current and the maximum power voltage while P_{M1} to P_{M7} are maximum power points. It is to be noted that while the I_{SC} is proportional to the insolation V_{oc} does not vary appreciably with it. In fact, for higher levels of radiation V_M lies between 0.7 and 0.8 of V_{oc} and I_M is almost equal to $0.8I_{SC}$. Since the current is zero at V_{oc} and the voltage is zero at I_{SC} the power levels at these points are zero since power is a product of current and voltage. It is clear then that the power will increase to a maximum in between these two extremes as is explained in detail in Figure 3.28.

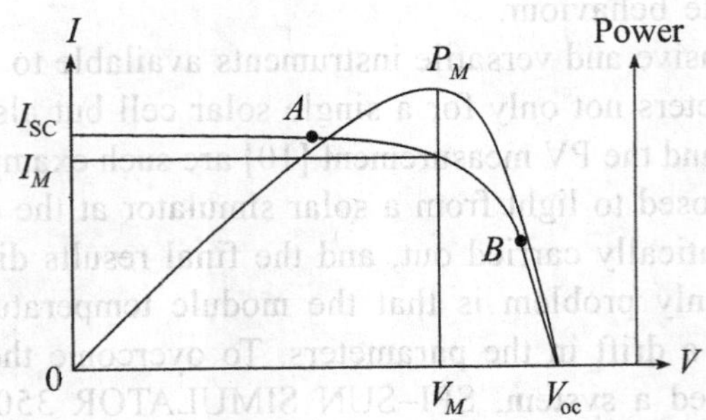

Figure 3.28 A power curve is superimposed on an *I–V* curve of a solar module.

In this diagram the current and the power are plotted against the voltage across the module. The power peaks at the maximum power point where all modules and arrays are supposed to be operated so that all the power produced by them is extracted since the cost of PV modules is very high. It is to be noticed that in the region I_{SC} to point *A* of the curve the current is almost constant while the voltage remains constant between the point *B* and V_{oc}.

Several methods to simulate the variable load exist in practice.

1. A large capacitor in a discharged condition is connected in place of the variable resistor of Figure 3.27(a) with a switch in between it and the module. When the switch is closed, current rushes in to charge the capacitor. As the capacitor charges up the current reduces, with a consequent increase in the module output voltage. Both these parameters are recorded as discussed earlier in Section 3.3.2. The charging rate of the capacitor depends on its capacitance the series resistance of the module.
2. Another method often used is an electronic load as shown in Figure 3.29.

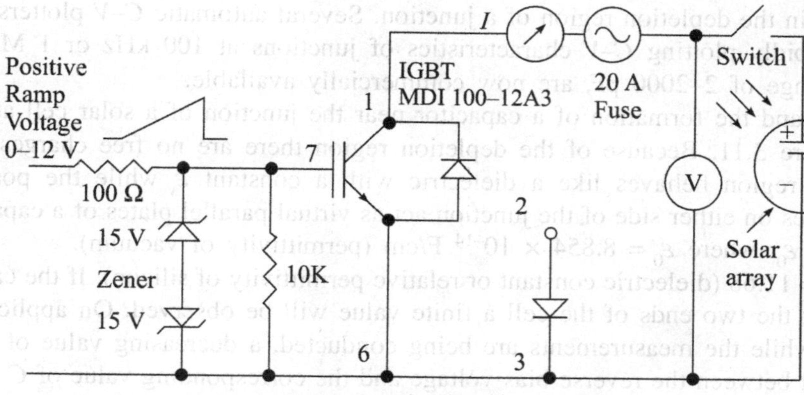

Figure 3.29 IGBT based electronic load.

The IGBT, MDI 100–12A3, can take up to 1200 V and a current of 135 A but is restricted to a current of 20 A, to keep low the voltage across it, from an array of V_M equal to 100 V. This can therefore be operated at a maximum power of 2 kW. Since this is the load itself it must dissipate that power. The IGBT must be mounted on a heat sink that is capable of dissipating this energy otherwise the device temperature might exceed the rated value. To confine the temperature to 50°C sufficient forced air cooling with fans must be provided. A slow ramp voltage, 0 to 12 V, from a function generator is given to the input that is protected by two back to back 15 V zener diodes. Alternately a variable voltage of the same range from a power supply can be used. When the switch is closed the array is connected to the collector of the IGBT which turns ON when its gate voltage ramp reaches about 4 V. The current and the voltage of this circuit are then recorded as usual to obtain the desired curve.

Conclusion

From the *I–V* plot of a solar cell or a module several parameters can be obtained which enable scientists to improve the efficiency of these devices. They are:

1. The short circuit current I_{SC}
2. The open circuit voltage V_{oc}
3. The maximum power point P_M
4. The maximum power voltage V_M
5. The maximum power current I_M
6. The fill factor *ff*

7. The device efficiency η, and
8. The diode factor from the log I–V curve.

3.3.3 The Capacitance versus Voltage (C–V) Plot

Capacitance versus voltage (C–V) and conductance versus voltage (G–V) measurements are now routinely employed to characterize semiconductor junctions and deduce the nature of the junction (linear or graded) and estimate junction parameters such as diffusion voltage (V_d), depletion width (W_d), and doping concentration and profile. Transient capacitance measurements, known as *deep-level transient spectroscopy* (DLTS), are used to gain information about deep levels present in the depletion region of a junction. Several automatic C–V plotters, with the capacity of rapidly plotting C–V characteristics of junctions at 100 kHz or 1 MHz in the capacitance range of 2–2000 pF, are now commercially available.

To understand the formation of a capacitor near the junction of a solar cell attention is turned to Figure 3.11. Because of the depletion region there are no free charges available there and the region behaves like a dielectric with a constant ε_s while the positive and negative charges on either side of the junction act as virtual parallel plates of a capacitor. For silicon $\varepsilon_s = \varepsilon_{si}\varepsilon_0$, where $\varepsilon_0 = 8.854 \times 10^{-14}$ F/cm (permittivity of vacuum).

While $\varepsilon_{si} = 11.68$ (dielectric constant or relative permittivity of silicon). If the capacitance is measured at the two ends of the cell a finite value will be observed. On application of a reverse bias, while the measurements are being conducted, a decreasing value of C will be noticed. A plot between the reverse bias voltage and the corresponding value of C will result in a curve shown in Figure 3.30(a). The junction capacitance is given by

$$C = \varepsilon_s A/W_d \tag{3.17}$$

where W_d is the width of the depletion region and A is the area of the junction.

Consider Figure 3.31. The variation of the charge density on either side of the junction is not equal with the distance but more for acceptors over a shorter distance x_p.

To maintain the charge neutrality the donors N_D are spread over a longer distance x_n. Hence

$$N_D x_n = -N_A x_p \tag{3.18}$$

This assumes that all the donors and acceptors are ionized within the transition region. The edges of the depletion layers at x_n and x_p are abrupt for which the following analysis is valid. These are called *abrupt junctions*. The electric field at the junction is obtained by the integration of Poisson's equation written below.

$$d\mathcal{E}/dx = \rho/\varepsilon_s \tag{3.19}$$

where ρ is the charge density and ε_s is the permittivity of the semiconductor.

Under the proper boundary conditions at x_n the junction built in voltage or the diffusion voltage V_d is given by

$$V_d = qW_d^2\,[N_A N_D/(N_A + N_D)]/2\varepsilon_s \tag{3.20}$$

Combining Equations (3.17) and (3.20), we get {[2], p. 109}

$$1/C^2 = (W_d/\varepsilon_s A)^2 = [2(N_A + N_D)/qA^2\,\varepsilon_s N_A N_D]\,(V_d - V) \tag{3.21}$$

This can be written as

$$1/C^2 = [(2/qA^2\ \varepsilon_s)\ (N_A + N_D)/N_A N_D]\ (V_d - V) \tag{3.21a}$$

A plot of Equation (3.21a) is shown in Figure 3.30(b). When $V = V_d$, the diffusion voltage, $1/C^2 = 0$. This is the experimental determination of the *diffusion voltage* from the C–V plot of a solar cell. Usually the distribution of the acceptors and donors are not as abrupt as considered above but are graded in some way. An analysis of a linearly *graded junction* gives a linear curve between $1/C^3$ and V, instead of the variation shown in Equation (3.21), like the one shown in Figure 3.30(c).

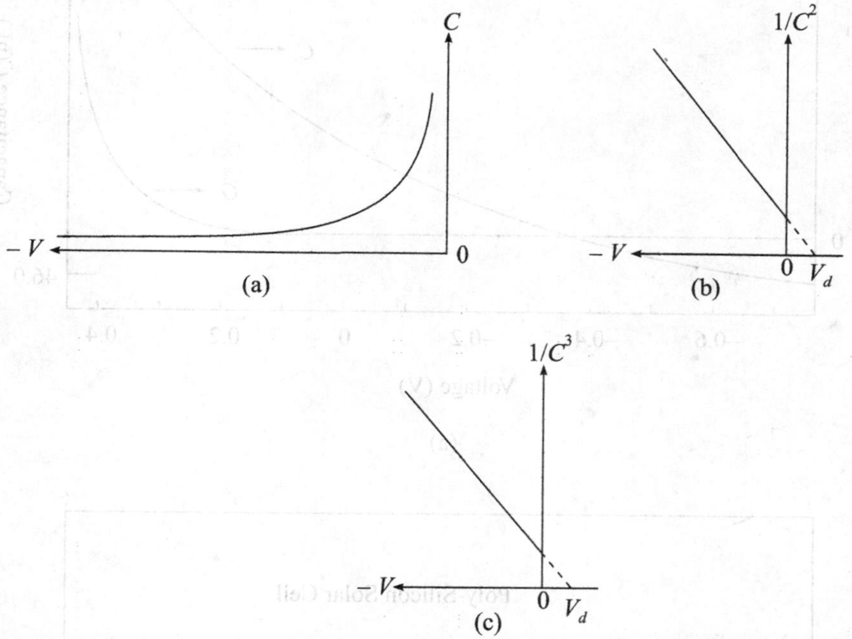

Figure 3.30 (a) The C–V curve of a solar cell, (b) $1/C^2$–V curve for an abrupt junction and (c) $1/C^3$–V curve for a graded junction solar cell.

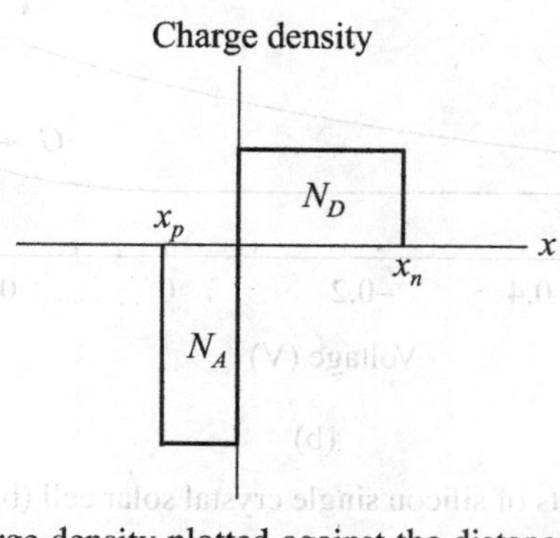

Figure 3.31 Charge density plotted against the distance from the junction.

Figure 3.32 (a) and (b) show the experimentally measured C–V and G–V characteristics of single crystal silicon and polycrystalline silicon solar cells. The C^{-2} versus V plots of the single crystal and polycrystalline silicon solar cells are shown in Figure 3.33 (a) and (b) from where the diffusion voltage for the single crystal one can be seen to be 0.7 V. Besides this cell definitely has an abrupt junction.

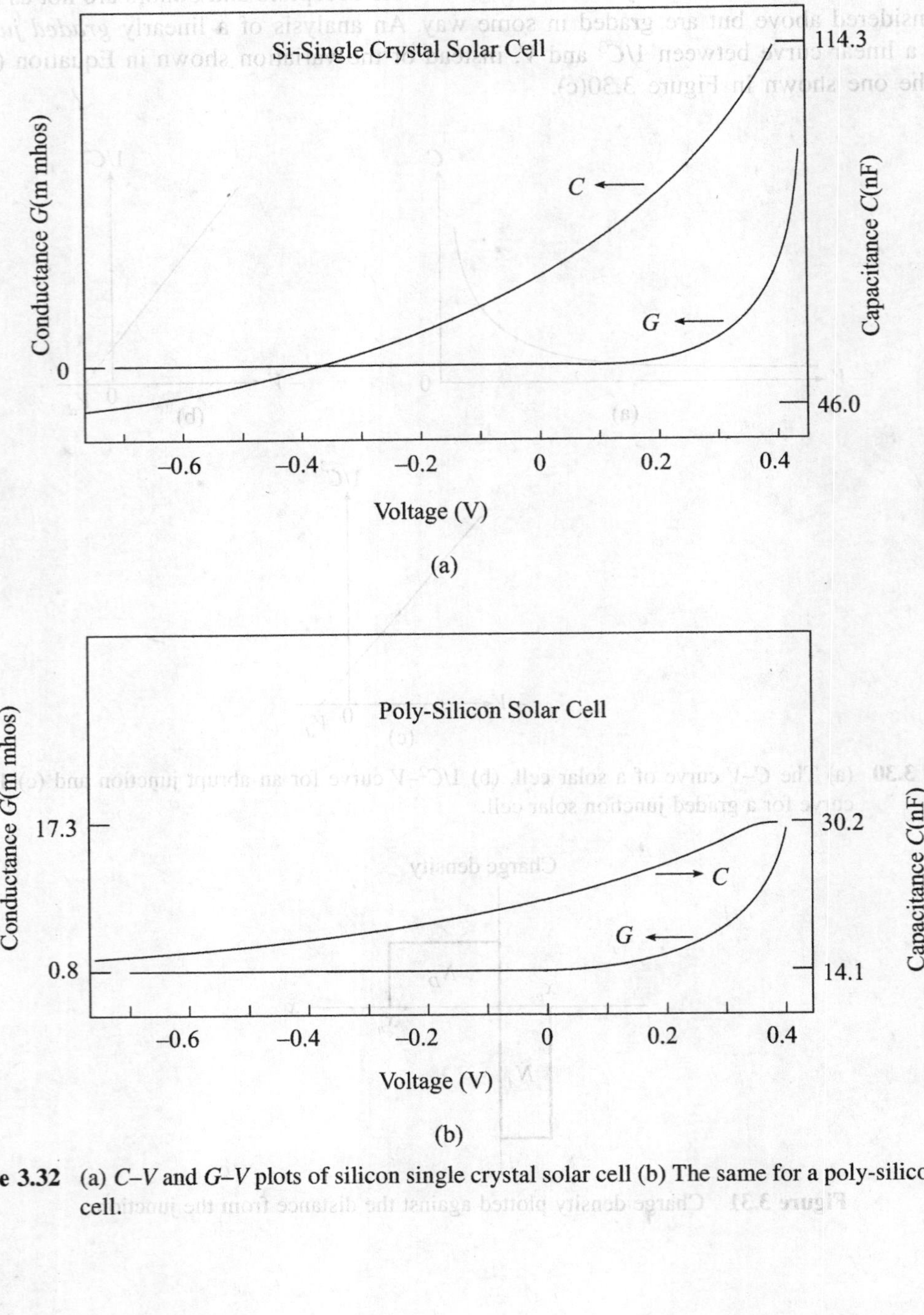

Figure 3.32 (a) C–V and G–V plots of silicon single crystal solar cell (b) The same for a poly-silicon solar cell.

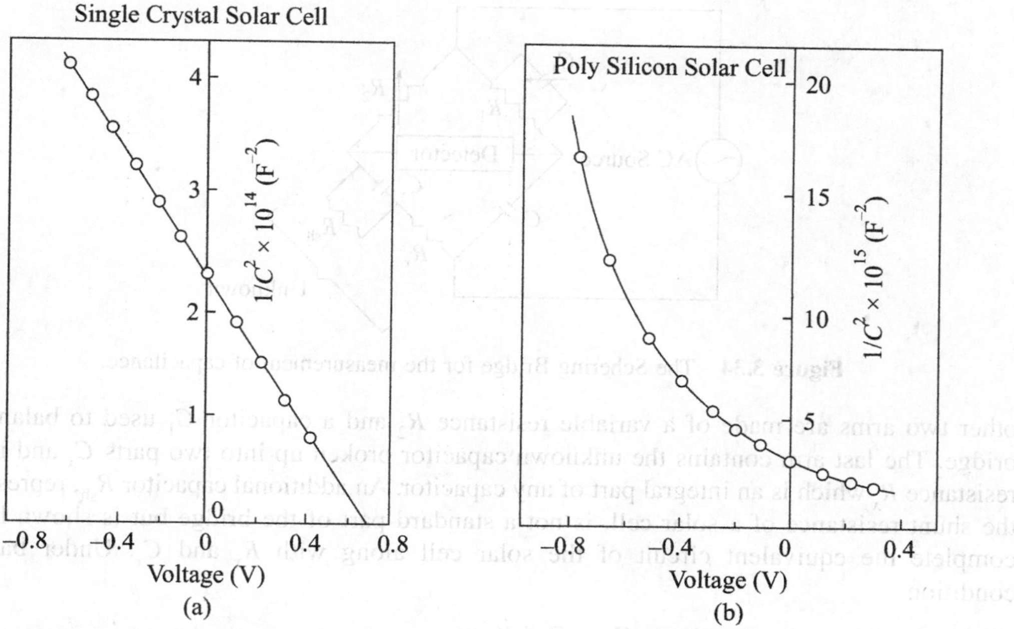

Figure 3.33 (a) $1/C^2$ versus V plot for a single crystal solar cell (b) Same for a poly-silicon solar cell. Both curves are derived from Figure 3.32.

Example 3.2: Calculate the depletion layer width for a silicon single crystal solar cell of 1 cm diameter whose capacitance is 50×10^{-9} F at a reverse bias of –0.2 V as shown in Figure 3.32(a).

Solution: The capacitance is given by the expression

$$C = \varepsilon_s A/W_d$$

For silicon $\varepsilon_s = \varepsilon_{si}\,\varepsilon_0$, where $\varepsilon_0 = 8.854 \times 10^{-14}$ F/cm is the permittivity of vacuum while $\varepsilon_{si} = 11.68$ is the dielectric constant or relative permittivity of silicon. Substituting the values

$$\begin{aligned} W_d &= \varepsilon_s A/C = \varepsilon_{si}\varepsilon_0 A/C \\ &= 8.854 \times 10^{-14} \times 11.68 \times \pi \times (0.5)^2/\ 50 \times 10^{-9} \\ &= 1.62 \times 10^{-5} \text{ cm} \\ &= 1.62 \times 10^{-7} \text{ m} = 0.162\ \mu\text{m} \\ &= 0.162 \text{ microns.} \end{aligned}$$

It is to be noted that the slope of the curve, in Figure 3.33(a), gives $(N_A + N_D)/N_A N_D$. In normal silicon solar cells N_D can be much larger than N_A. Therefore the base acceptor concentration N_A can be obtained from the slope.

Capacitance measurement techniques

The most extensively used classical method for measuring capacitance was the Schering bridge [14] shown in Figure 3.34. A standard capacitor C_s, made of mica forms one arm. The

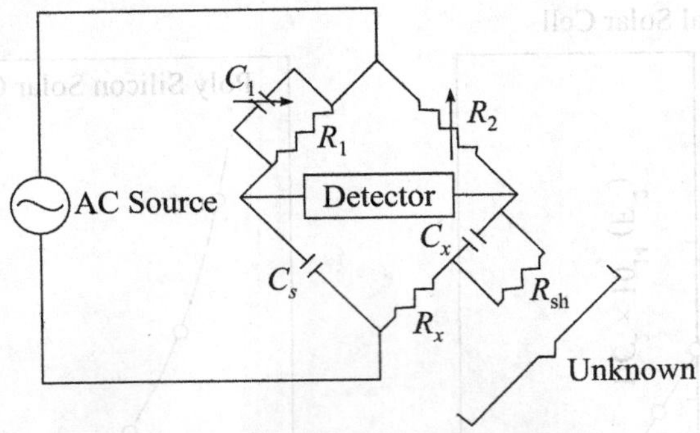

Figure 3.34 The Schering Bridge for the measurement of capacitance.

other two arms are made of a variable resistance R_2 and a capacitor C_1 used to balance the bridge. The last arm contains the unknown capacitor broken up into two parts C_x and its loss resistance R_x which is an integral part of any capacitor. An additional capacitor R_{sh}, representing the shunt resistance of a solar cell, is not a standard part of the bridge but is shown here to complete the equivalent circuit of the solar cell along with R_x and C_x. Under balanced condition

$$Z_x = Z_2 Z_3 Y_1 \tag{3.22}$$

where

Z_x = impedance of the unknown capacitor in the fourth arm
$Z_2 = R_2$, the resistance in the second arm
Z_3 = impedance of the standard capacitor C_s in the third arm, and
Y_1 = admittance of the parallel combination of C_1 and R_1 in the first arm

or

$$R_x - \frac{j}{\omega C_x} = R_2\left(\frac{-j}{\omega C_s}\right)\left(\frac{1}{R_1} + j\omega C_1\right), \quad j = \sqrt{-1}$$

or

$$R_x - \frac{j}{\omega C_x} = \frac{R_2 C_1}{C_s} - \frac{jR_2}{\omega C_s R_1} \tag{3.23}$$

Equating real and imaginary parts, we get

$$\left.\begin{aligned} R_x &= R_2 \frac{C_1}{C_s} \text{ and} \\ C_x &= C_s \frac{R_1}{R_2} \end{aligned}\right\} \tag{3.24}$$

The AC source in normal bridges has a frequency of 100 Hz, for measurement of large capacitances or 1 kHz for smaller ones. However, for semiconductor diodes and solar cells the frequency is kept at 1 MHz or 100 kHz. This signal should not only have equal amplitudes across the detector but also must have the same *phase* for a null condition. If either quantity is different, a current will flow through the detector disturbing the balance of the bridge.

If a solar cell is used in place of the unknown capacitor, its depletion capacitance can also be measured. However, a slight modification is necessary to make the measurements accurate because of the battery included to vary the voltage across the device under test. This modification is shown in Figure 3.35.

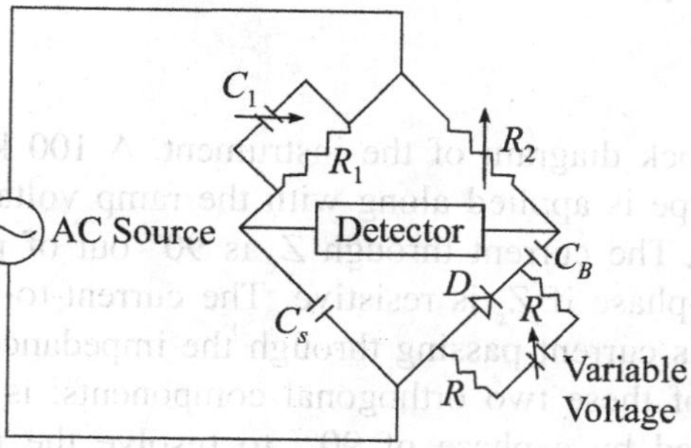

Figure 3.35 *C–V* measurement using the Schering bridges.

The diode or cell is reverse biased by a variable voltage source like a laboratory DC power supply. A blocking capacitor C_B is placed in series with the diode so that the DC bias voltage does not leak into the rest of the bridge arms or the AC source. The value of C_B should be more that 100 times the value of the capacitor associated with the diode otherwise errors will come in since C_B is in series with that of the diode. This means that if the diode capacitance is 50×10^{-9} F the value of C_B should be at least 5000×10^{-9} F or 5 μF. A non polarized capacitor should preferably be used. Another problem arises from the low impedance of the variable voltage supply since the high frequency measurement signal, from the AC source, may pass through it and introduce errors. To isolate the variable DC source from the measurement signal two resistances R are placed such that they offer high impedance to the measurement signal. However, care must be taken to ensure that the total resistance $2R$ is not comparable to the reverse bias resistance of the diode/cell otherwise there will be voltage drop across the resistors and the actual voltage will be less than that recorded at the output of the variable voltage source. To reduce any such voltage drop the resistances may be replaced by inductors of suitable values to provide the necessary isolation. A point by point plot can be obtained by this method; the voltage across the variable source can be plotted against the capacitance of the diode.

In recent years, large area photovoltaic devices (or solar cells), have been the subject of intensive research interest. Naturally, the *C–V* characteristics of these devices have been studied by several workers to gain an understanding of the device performance. However, the capacitance values of these devices generally fall outside the measuring capability of the commercially available plotters especially at a frequency of 1 MHz. The usual practice followed by workers is to cut the devices into small pieces (typically 5 mm × 5 mm or less) to bring the capacitance values within the measurement range of the plotters. This practice is clearly undesirable as it can, and does, lead to extensive damage of the materials comprising the cell, particularly in thin film solar cells. Another serious shortcoming of the commercial plotters is that the voltage signal fed to the X input of the recorder or oscilloscope is taken directly from the ramp generator output instead of across the device under test. The consequences of this are obvious when measuring the *C–V* characteristics of a solar cell under illumination. Under illumination, the solar cell develops a photo-voltage and the voltage across the cell is the algebraic sum of the photo voltage and the externally imposed voltage from the ramp generator. Thus, the *C–V* plotter does not record the true voltage across the device when the cell is illuminated. A new type of *C–V* plotter was designed to overcome these problems and is described below [13].

Principle of operation

Figure 3.36 shows the block diagram of the instrument. A 100 kHz, 20 mV p-p sinusoidal signal from the oscilloscope is applied along with the ramp voltage to the impedance under test (Z_x) through a mixer. The current through Z_x is 90° out of phase with the input signal if Z_x is capacitive and in-phase if Z_x is resistive. The current-to-voltage converter (C–V–C) converts and amplifies this current passing through the impedance to a proportional voltage. This voltage, consisting of these two orthogonal components, is fed to two phase-sensitive detectors (PSDs), separated by a phase of 90°, to resolve the individual components; the output of the two PSDs can be fed to an *X–Y* recorder or an oscilloscope for recording the *C–V* or *G–V* characteristics. The digital panel meter (DPM) reads either the capacitance or the conductance through a selector switch. The curves shown in Figures 3.32 have been obtained with this instrument. A more sophisticated instrument based on this principle is available for these measurements [14]. A pulse can be introduced in the mixer of Figure 3.36 for deep level transient spectroscopy measurements (DLTS) [18].

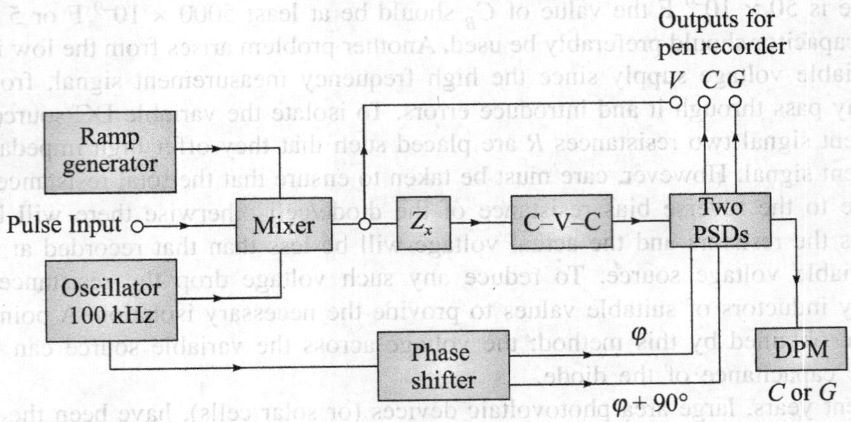

Figure 3.36 Block diagram of the *C–V*/*G–V* plotter.

Conclusion

It can be concluded from these discussions that the capacitance versus voltage plots yields vital information about:

1. The nature of the cell junction, that is, whether the junction is linearly graded or abrupt
2. The width of the depletion layer can be found at any bias condition
3. The average acceptor and donor density N_A or N_D and
4. The conductance gives the shunt resistance for a single crystal silicon solar cell. However, for a multi–diode solar cell analysis is more complicated.

3.3.4 Spectral Response Measurements

Spectral response of a solar cell is the power generated by it at different wavelengths of *same intensity*. The study of spectral response (SR) of a solar cell provides a good method for characterizing a solar cell. As the name suggests the main objective is to find the short circuit current output of the cell at different wavelengths so that a cell can be fabricated ideally to respond to the entire optical spectra. The SR analysis yields such parameters as quantum efficiency (Q), collection efficiency (c) [15], diffusion length of minority carriers (L_n) [16], width of the diffused region (d) [17] and an estimate of barrier height (φ_B) (in Schottky junctions) [18]. Further, information about a variety of spatial defects which reduce the performance of large-area solar cells is also obtained. Two experimental arrangements for SR measurements are given below:

Point by point method

The basic SR measurement consists of a solar simulator and a set of filters. It is shown in Figure 3.37.

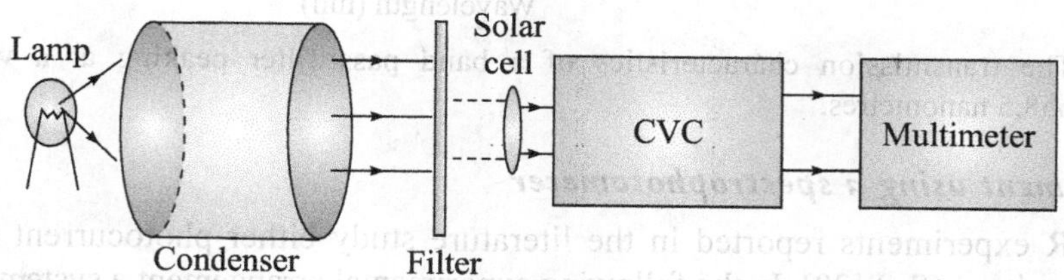

Figure 3.37 Block diagram of the spectral response experiment.

An incandescent or preferably a xenon arc lamp is placed to give an intensity of 100 mW/cm^2 approximately. A condenser lens or a water filter is placed after that to absorb the heat so that the solar cell does not get heated. A narrow band optical band pass filter is placed to pass the desired wavelength of light to fall on the solar cell placed next to it. The solar cell generates a short circuit current corresponding to that radiation which is then amplified by a current to voltage converter (CVC) which also keeps the cell under a short circuit. The output voltage of the CVC is then read by a multimeter. The filter is changed for different wavelengths and the corresponding CVC voltage is registered. It must be remembered that the lamp does not emit all the wavelengths with equal intensity and therefore the SR of the cell thus, measured will be erroneous. To correct this error the measurement must be *normalized.* This process is not tedious but needs a sophisticated radiation measurement device like a thermopile which has the same sensitivity for all the wavelengths. Before or after the SR of the cell the output of the filter is recorded for each of the wavelengths used for measurement by the thermopile. This gives the intensity of the radiation output from the filter. The output voltage of the cell is then divided by the corresponding voltage output of the thermopile at those wavelengths to get the corrected SR values. A *standard solar cell* made of the same material as the device under test is generally used for normalization. The transmission characteristics of a band pass filter peaking at a wavelength of 658.5 nanometres is shown in Figure 3.38. Because of the finite width of the curve light of other wavelengths is also transmitted through it. This results in an error of 6% in the measurement in this case. For this reason spectrophotometers with a diffraction grating or a prism are used to give narrower bandwidths.

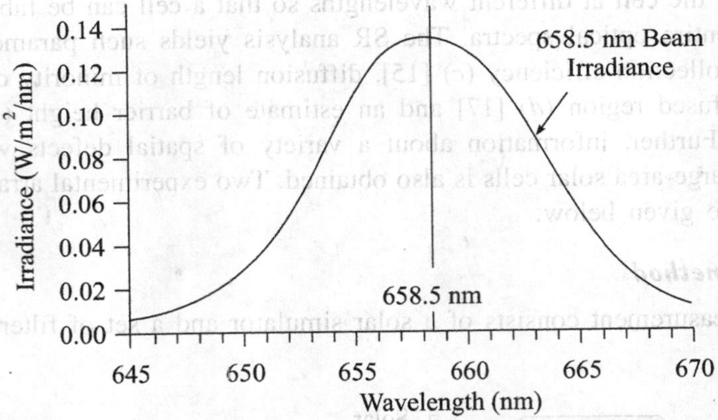

Figure 3.38 The transmission characteristics of a band pass filter peaking at a wavelength of 658.5 nanometres.

SR measurement using a spectrophotometer

Generally, SR experiments reported in the literature study either photocurrent (I_{SC}) [19] or surface photovoltage (S_{pv}) [20]. In the following experimental arrangement a system is described, where the spatial variations of both I_{SC} and S_{pv} can be studied. Polycrystalline silicon, thin film CdS/Cu_2S and SnO_2/SiO_x/n-Si solar cells have been studied with this.

Experimental Details

The system consists of a Carl Zeiss spectrophotometer model VSU-2P, a mobile sample mount, lock-in amplifier, a current-to-voltage converter and a λ–Y recorder as shown in Figure 3.39 [24]; *Copyright (1985) with permission from Elsevier.*

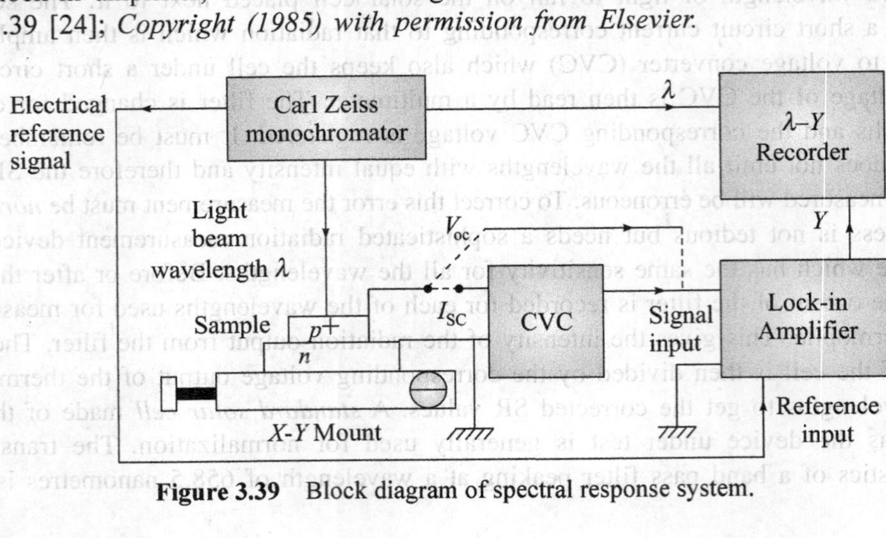

Figure 3.39 Block diagram of spectral response system.

The entire solar spectrum is covered using two sources; a deuterium lamp covers from 0.2 to 0.38 μm and an incandescent lamp covers from 0.3 to 2.7 μm. The lamp compartment has a mechanical chopper of 294 Hz frequency which chops the light from the lamp. An auxiliary single crystal silicon solar cell placed at one end of the chopper, inside the lamp compartment, generates a reference electrical signal for the lock–in amplifier while the light from the other end of the chopper goes to the prism of the monochromator to split up into different wavelengths. The monochromator is a part of a Carl Zeiss (VSU-2P) spectrophotometer, which uses a fused quartz prism as the dispersive element. Rotation of the prism is controlled manually or by a mechanical drive. The spectral slit width (*S*) of the emerging monochromatic radiation is dependent on the exit slit width (*b*) and the range of wavelength at which the monochromator is operating [21], and is given by

$$S(\lambda) = b \times \Delta\lambda$$

Here, $S(\lambda)$ is the spectral slit width in mm, and $\Delta\lambda$ is the band of wavelength between the adjacent graduations on the wavelength scale. The value of the $S(\lambda)$ is kept at 100 Å, by adjusting the exit slit width.

The feeble short circuit current or the photovoltage signal from the solar cell is amplified using either a current to voltage converter (CVC) or a lock-in amplifier respectively. The output from the lock-in amplifier is fed to a λ–*Y* recorder where *Y* is the output of the lock—in amplifier and λ is the wavelength mechanically given out as a drive from the spectrophotometer. The response is normalized with respect to the spectral distribution of the source intensity, measured with a thermopile made by Kipp and Zonen [22] and a nano-voltmeter (Keithley, Model 148) [23]. The sample is mounted on a platform which can be moved both in the *X*- and *Y*-directions using micrometer screws. A 30 W, 5 V lamp is used to bias the solar cell with white light [24].

The following types of cells have been studied: (a) 2 × 2 cm CdS/Cu_2S solar cells of about 4% efficiency fabricated by the spray pyrolysis process, (b) SnO_2/SiO_x/n-Si solar cells of about 2–3% efficiency with an SnO_2 layer deposited by spray pyrolysis on to single crystal silicon wafers (2.5 cm diameter), (c) polycrystalline silicon solar cell (2 cm diameter) of 12% efficiency fabricated by using simultaneous diffusion of boron and phosphorus. Results obtained from the above experiments are discussed below.

The SR of a polycrystalline silicon solar cell with a TiO_2 anti-reflection coating is shown in curve (a) of Figure 3.40. The response peaks at 0.82 μm. The SR of a CdS/Cu_2S of back wall configuration, shown in the curve (b), rises sharply at 0.5 μm (corresponding to the band gap of CdS) and droops slightly till 0.95 μm. The response of the SnO_2/SiO_x/n-Si solar cell, shown in curve (c), rises to a maximum point at 0.57 μm. The improved short wavelength response is due to the fact that the junction in these cells is near the surface.

A detailed description of measurement is given in [25], while calculation of parameters are given in references [26] and [3].

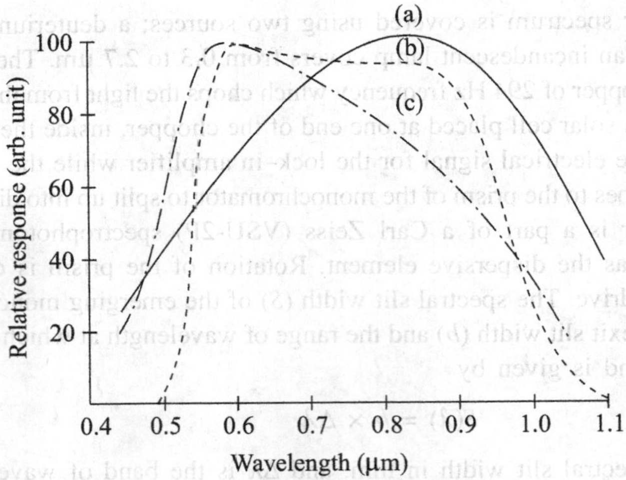

Figure 3.40 Typical spectral response characteristics of three solar cells: (a) Polycrystalline silicon, (b) Thin film CdS/Cu_2S and (c) SnO_2/SiO_x/n-Si.

3.3.5 Light Beam Induced Current (LBIC) Technique for Defect Location

The Helium-Neon laser is a source of a very fine beam of red (monochromatic) light and has been extensively used for LBIC studies. Laser scanning technique has been used to bring out spatial inhomogeneities in semiconductor devices [27–29]. For solar cells in particular, the method is extensively used for studying defects such as cracks, localized shunts, grain boundaries, foreign particle sites, poor metallization regions, and non-ohmic back contact regions [30, 31]. Besides single-crystal silicon solar cells, studies have also been made on polycrystalline Si [32], CdS/Cu_2S [33], SnO_2/Si [34], and liquid-junction solar cells [35]. The basic scanning system used by most authors is reported by Sawyer et al. [36, 37]. A variety of other arrangements have been used by other workers, depending on the nature of their study [29, 35, 38]. We have used a suitably modified version of the galvanometer motor scanning method of Tove et al. [29]. In the method described by Tove et al., two galvanometer motors are driven by two function generators, independent of each other, to scan the cell. It is, therefore, difficult to focus the laser spot on a defective point of the cell. Furtak et al. [35] have kept the light spot stationary and have moved the cell for obtaining a current map. This procedure is essentially slow and a quick evaluation of a large number of cells is not possible. The "logic module" described here enables fast scans of solar cells, where typical times are two-three seconds for routine measurements while permanent records on an *X–Y* recorder can be obtained whenever required [39]. The number of lines scanning the surface of the solar cell can be varied from 16 to 240. Some of the other features of the logic module are listed below:

1. The laser beam with a built in line counter can be moved in both directions which helps to identify defects, that is, regions of low current output of the cell. Line and frame reset switches bring the laser spot back either to the beginning of a line or to the beginning of the raster frame itself.

2. It provides both *Y*- and *Z*-axis modulation signals for recording on a cathode-ray oscilloscope and only *Y* modulation for *X-Y* recorders. Since these modulation signals are derived from the short circuit current of the solar cell under test produced by a laser beam, the pattern produced is a quantitative current profile of the cell.
3. Normally, a current-to-voltage converter (CVC) is sufficient to supply an adequate signal for recording. However, in some very low efficiency devices the current is so small that a noise-free signal is not available from the CVC; therefore, a lock-in amplifier has been incorporated to function with a light chopper at 294 Hz. In case some other operating frequency is desired an external lock-in amplifier can also be used.
4. As mentioned earlier, one scan consists of 16 to 240 lines. This instrument can work in single or multi scan modes.
5. During high-speed operation the *X-Y* recorder is automatically disconnected.

Principle of operation

Figure 3.41 shows the block diagram of the optical scanner. The light beam of approximately 10 μm diameter from a He-Ne laser (λ = 0.6328 μm) is interrupted by a mechanical chopper at the rate of 294 times per second. This chopped light is reflected from mirrors M_1 and M_2 onto the solar cell under study. The solar cell is held vertically by a vacuum metallic chuck which also forms the rear electrical contact; the other contact being a copper wire soldered to the front face of the cell. Mirror M_1 is oscillated horizontally by a galvanometer motor powered by a General Scanning Inc. scanner control model CCX 101. When M_1 returns to its original position after one deflection (called a *line scan*) mirror M_2 is slightly rotated by means similar to that of M_1, such that the beam is deflected downwards. Repetition of this

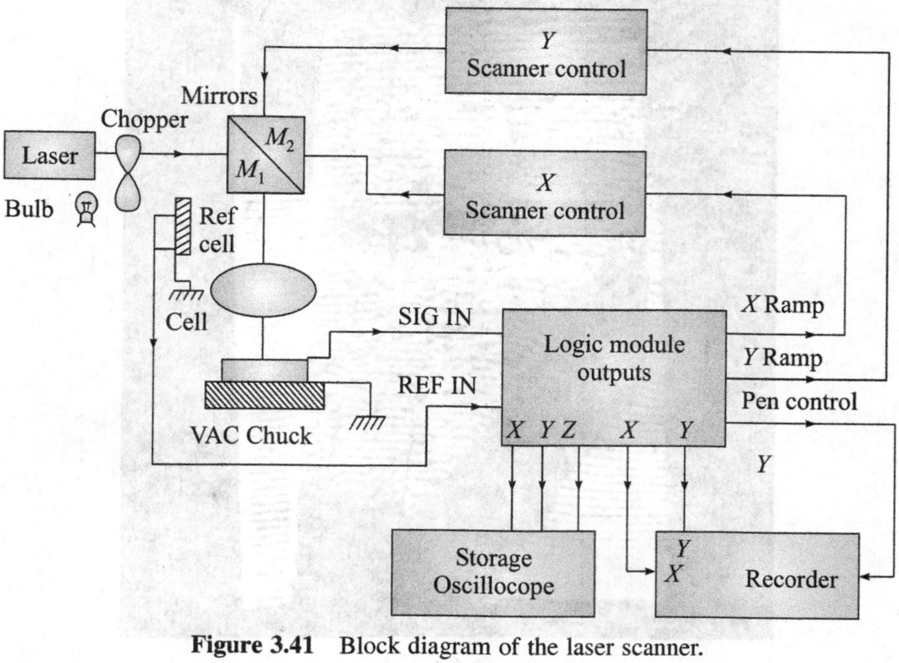

Figure 3.41 Block diagram of the laser scanner.

operation produces a raster similar to that of a television. The logic module generates the signals given to the scanner controls for deflection. The current from the solar cell is modulated on the Y signal of the module and then displayed on the Y channel of a storage cathode ray oscilloscope (CRO) or an X-Y recorder. The pen control signal lifts the pen of the recorder during fly back, and the Z signal carries the blanking pulses and the current signal to the CRO.

The results of the cells studied by the instrument described here are given below. Figure 3.42(a) shows the short circuit current (I_{SC}) map of a polycrystalline silicon solar cell [39]. The oscillogram shows a cell area of 7.5 × 5 mm scanned by 64 lines in a time of about 2 seconds. The fast scan mode of the instrument is useful for quick analysis of a large number of samples. The scanning time, of course, depends on the number of lines in the scan. However, the number of lines, themselves, depends on the resolution of the pattern which is dependent on the oscilloscope beam diameter. The laser beam diameter is about 10 μm. A hard copy taken on an X-Y recorder of the same pattern is shown in Figure 3.42(b). The short circuit current, I_{SC}, plotted here is actually the voltage output from the current to voltage converter (CVC). This map shows grain boundaries (GB) of the grains marked (*a*)–(*d*) in the Figure 3.42(b). And a micro crack across the solar cell. These regions are shown better in Figure 3.42(c) drawn here for comparison with the other two. There is a decrease in the current signal near the GB due to recombination of photo-generated carriers. The effect of the micro crack on the current signal is more serious because it perturbs the lateral flow of charge carriers. Hence the collection of the carriers at the electrode adjacent to this defect is reduced, thus lowering the I_{SC}.

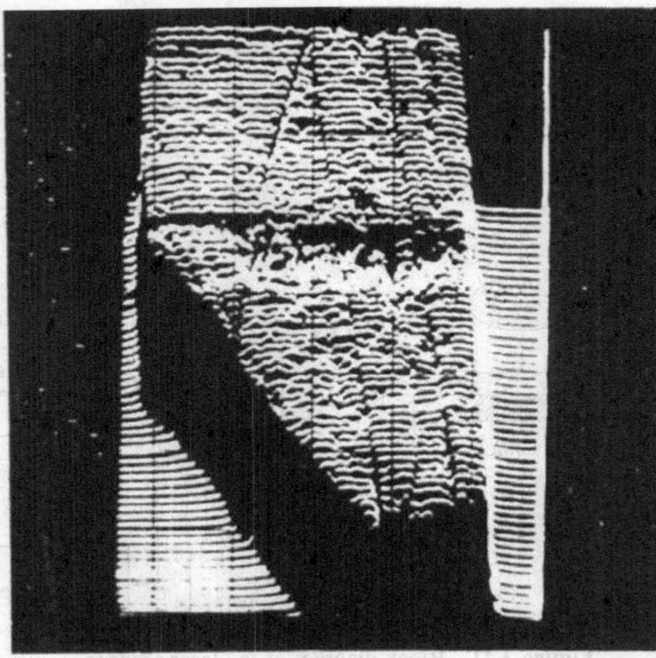

(a)

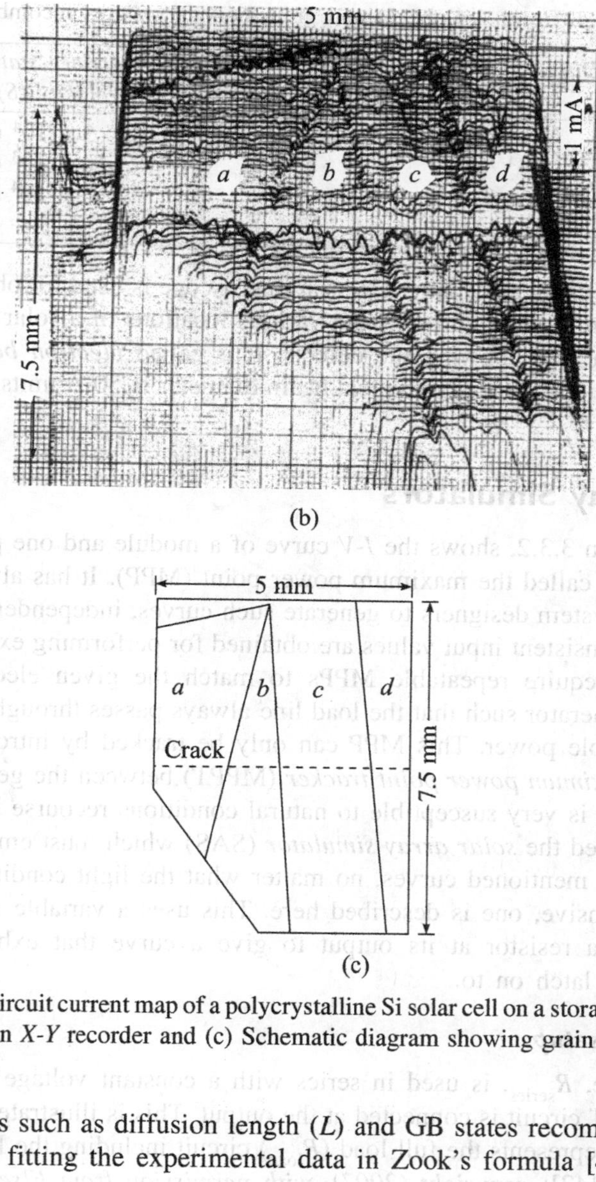

Figure 3.42 (a) Short circuit current map of a polycrystalline Si solar cell on a storage oscilloscope, (b) The same on an *X-Y* recorder and (c) Schematic diagram showing grain boundaries of the solar cell of (a).

Defect parameters such as diffusion length (*L*) and GB states recombination velocity (*S*) can be calculated by fitting the experimental data in Zook's formula [40]:

$$\frac{I(x)}{I(\infty)} = 1 - \frac{2S}{\pi}(1+A)\times\int_0^\infty \frac{\sinh^2(t)\times e^{-(x\cosh t)/L}}{\cosh(t)[S+\cosh(t)][A^2+\sinh^2(t)]}\,dt \qquad (3.25)$$

where

$A = aL$, $S = sL/D_n$; $I(\infty)$ and $I(x)$ are the values of current far from the GB and at a distance from the GB respectively. D_n is the diffusion coefficient of minority carriers. The calculated values of *S* and *L* of different GB's are shown in Table 3.3. References [26] and [41] give further details.

Table 3.3 Values of diffusion length (L) and grain boundary states recombination velocity (S)

Grain	*Diffusion length* (L) (μm)	*Grain boundary states recombination velocity* (S) (cm/s)
a	45	5×10^4 (*a*, *b*)
b	60	8×10^3 (*b*, *c*)
c	48	6×10^4 (*c*, *d*)
d	40	–

A similar method using a scanning electron microscope is used for obtaining finer details. Here an electron beam induces current from various locations of a solar cell, by scanning its surface, to generate a map showing the defects. It is called *electron beam induced current* or simply EBIC technique. Since the typical beam diameter is 50 Å units very high resolution images are obtained.

3.4 Solar Array Simulators

Figure 3.27 of Section 3.3.2, shows the *I-V* curve of a module and one point on it where the power is the highest, called the maximum power point (MPP). It has always been the desire of the photovoltaic system designers to generate such curves, independent of the atmospheric conditions, so that consistent input values are obtained for performing experiments. Interested experimenters also require repeatable MPPs to match the given electrical load, such as pumps, to the PV generator such that the load line always passes through that point to extract the maximum available power. This MPP can only be tracked by introducing an electronic circuit called the *maximum power point tracker* (MPPT) between the generator and the load. Since a solar module is very susceptible to natural conditions recourse has to be made to an electronic circuit called the *solar array simulator* (SAS) which must emulate a solar module, to generate the afore mentioned curves, no matter what the light condition is. The simplest, and the most inexpensive, one is described here. This uses a variable voltage analogue DC power supply with a resistor at its output to give a curve that exhibits a peak for the electronic tracker to latch on to.

Description of the setup

A variable resistance, R_{series}, is used in series with a constant voltage DC regulated power supply and the MPPT circuit is connected at the output. This is illustrated in Figure 3.43. The box labelled MPPT represents the full load (R_{load}) circuit including the DC-DC converter and its control circuitry [42]; *copyright (2007), with permission from Elsevier.*

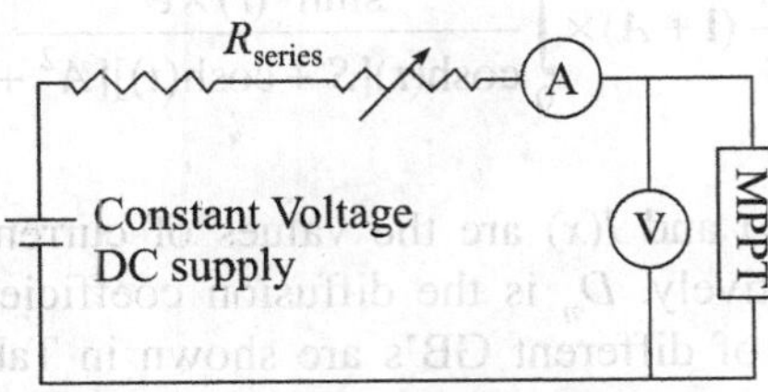

Figure 3.43 Schematic of the entire system.

For a fixed resistance R_{series}, if R_{load} is varied from its minimum to maximum possible value and the voltage across it is plotted against the current in the circuit then a linear curve with a negative resistance is obtained. This is illustrated in Figure 3.44, for two values of R_{series}.

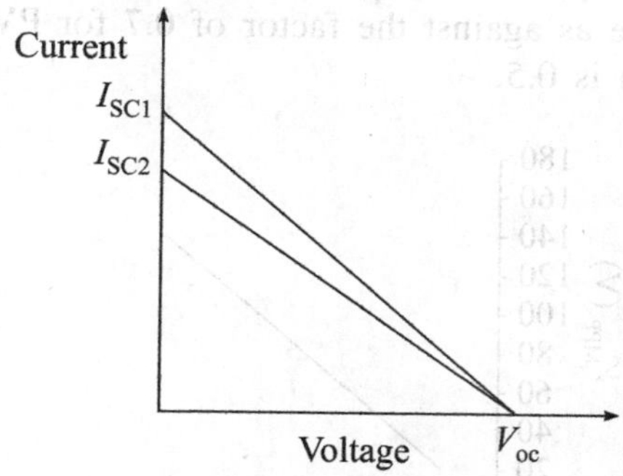

Figure 3.44 Ideal *I–V* curves for fixed V_{oc} and varying series resistance.

The power, $I \times V$, dissipated goes through a maximum when the voltage drop across this resistance is about half of the applied DC voltage as shown in Figure 3.45. This curve passes through a maximum value similar to the *P–V* curve generated by a PV module although in the latter case the peak occurs at about 0.7 of the V_{oc}. The MPPT algorithms can track this MPP.

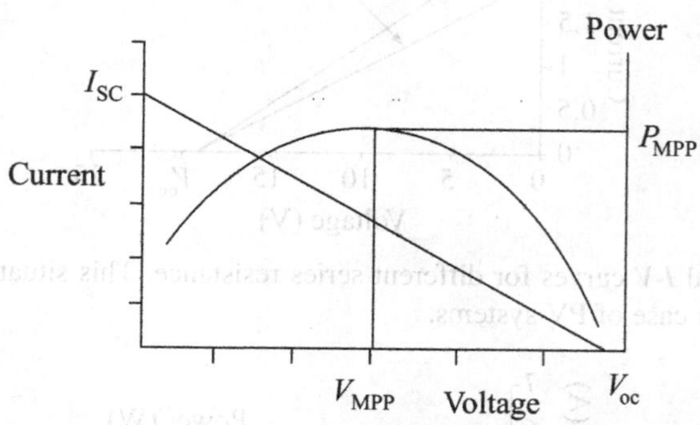

Figure 3.45 Ideal *I-V* and *P-V* curves showing that power peak occurs at half of the input voltage from supply (not to scale).

According to the 'maximum power transfer theorem', the maximum power is transferred to the load when its resistance is equal to that of the internal resistance of the source. Therefore, in the present case, maximum power transfer condition will arise when the voltage drop across the load is half of that supplied by the input DC power supply. If the open circuit voltage in this simulator has to be changed, the input DC supply voltage is simply altered. For increasing or decreasing the short circuit current, the value of R_{series} is changed appropriately. One should ensure that a certain finite resistance is always in series with the power supply to avoid short circuit and thus avoid damage to it.

Experimental Results

Figure 3.46 shows the linearity of tracking of a 2 kW MPPT with the change of the input voltage. The MPP lies midway to the applied voltage. Figure 3.47 shows the experimental voltage versus current curves for a fixed V_{oc} and different series resistances. Figure 3.48 shows experimentally that the peak in the power curve occurs at a voltage that is half of the input voltage (V_{oc}). Therefore as against the factor of 0.7 for PV sources, the ratio of V_{MPP} to V_{oc} for the present system is 0.5.

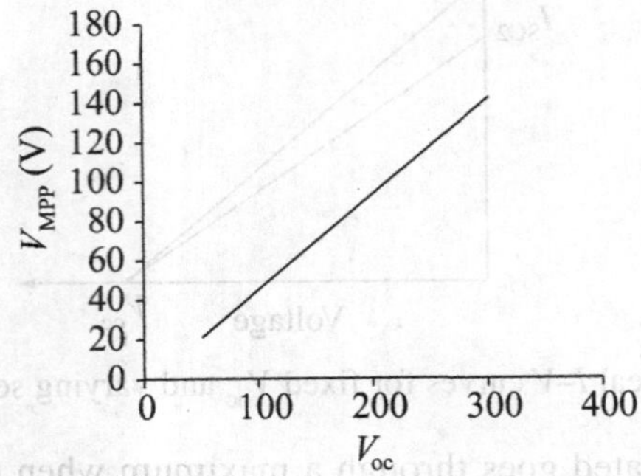

Figure 3.46 V_{MPP} tracked by the MPPT over a wide range of input voltage showing $V_{MPP}/V_{oc} = 0.5$.

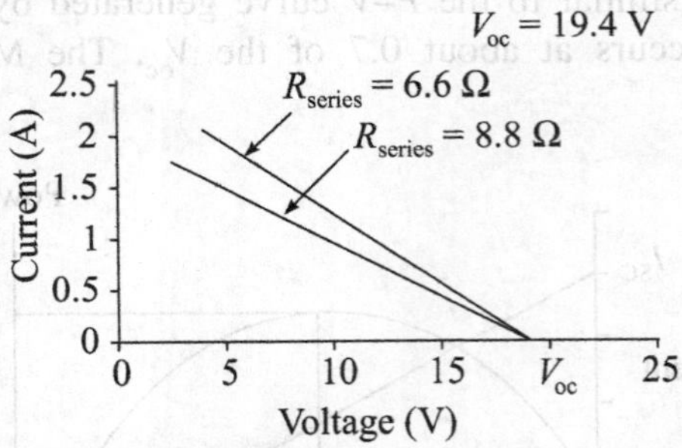

Figure 3.47 Experimental *I*-*V* curves for different series resistance. This situation is similar to changing insolation in case of PV systems.

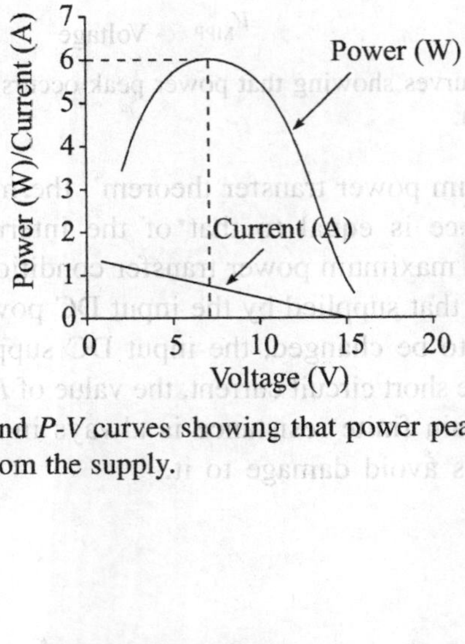

Figure 3.48 Experimental *I*-*V* and *P*-*V* curves showing that power peak occurs at approximately half of the input voltage from the supply.

In photovoltaic sources, when the cell temperature changes, the open circuit voltage changes significantly. This situation is simulated in the present system by simply varying the DC supply voltage keeping the series resistant constant. This was conducted on a 13 W MPPT system and the voltage versus power curves are shown in Figure 3.49. R_{series} is maintained at 8.8 Ω and the V_{oc} is varied through 15.4 V, 17.5 V and 19.2 V. The V_{MPP} point shifts towards the right as the voltage value is increased from the supply.

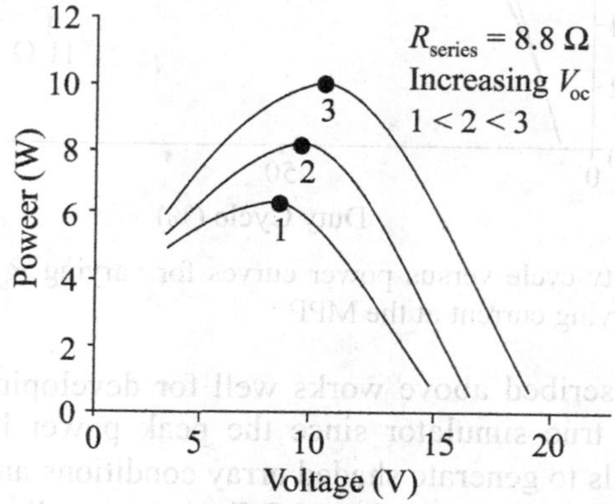

Figure 3.49 Experimental *P-V* curves for varying supply voltages showing the power peak shifting to the right with V_{oc} values. This is similar to changing temperature conditions in PV systems.

For PV sources, when the solar insolation varies, the short circuit current changes linearly. Varying the series resistance of the DC supply keeping the input voltage constant simulates this condition. Figure 3.50 shows the voltage versus power curve for 19.4 V and resistances changing through 6.6 Ω, 8.8 Ω and 11 Ω. Since the V_{oc} is fixed the MPP lies midway in all the three cases. Figure 3.51 shows the duty cycle versus power curve. The MPP is seen to shift towards the right with the decrease in R_{series}, that is, for increasing current. To account for an increasing current the duty cycle must also increase.

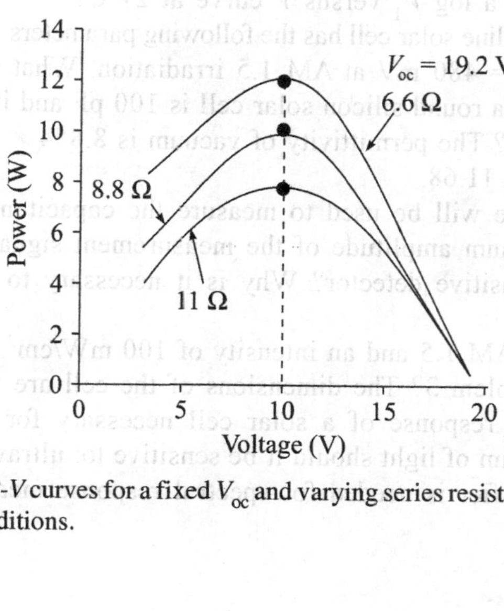

Figure 3.50 Experimental *P-V* curves for a fixed V_{oc} and varying series resistances. This simulates changing irradiation conditions.

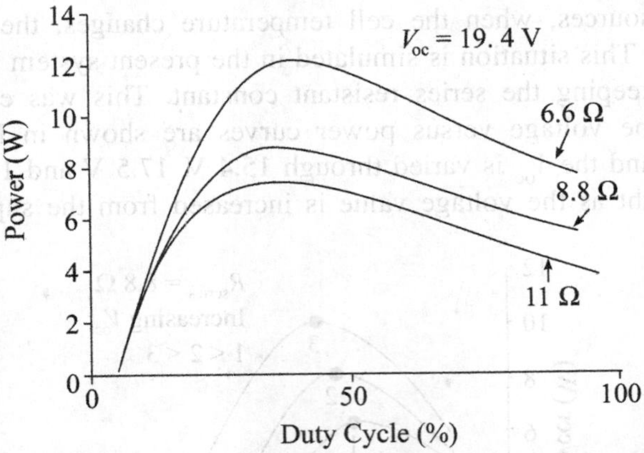

Figure 3.51 Experimental duty cycle versus power curves for varying R_{series} values. Duty cycle value changes with varying current at the MPP.

The solar simulator described above works well for developing MPPT algorithms. This circuit cannot be called a true simulator since the peak power here occurs at 0.5 the V_{oc} instead of $0.7V_{oc}$. Also it fails to generate shaded array conditions and lacks other sophisticated features. For developing such instruments digital DC power supplies are needed with elaborate control software. All these features are available at a cost. The Agilent solar array simulator (SAS), model E4351B, is one such versatile instrument [43] used by workers for study of solar energy systems [44], [45].

Questions

1. Which method should be used to determine whether a *p–n* junction is abrupt or linearly graded?
2. Which portion of a solar cell *I-V* curve is used for making a logarithmic amplifier? What is the slope of a log V_1 versus V curve at 27°C?
3. A square single crystalline solar cell has the following parameters: I_{SC} = 4.5 A, V_{oc} = 600 mV, I_M = 4, 0 A and V_M = 480 mV at AM 1.5 irradiation. What is its fill factor?
4. If the capacitance of a round silicon solar cell is 100 pF and its diameter is 1 cm, what is its depletion width? The permittivity of vacuum is 8.854×10^{-14} F/cm, the dielectric constant of silicon = 11.68.
5. Which type of bridge will be used to measure the capacitance of a solar cell? What should be the maximum amplitude of the measurement signal across the solar cell?
6. What is a phase sensitive detector? Why is it necessary to measure the capacitance with it?
7. At an irradiation of AM 1.5 and an intensity of 100 mW/cm^2, what is the efficiency of the solar cell of problem 3? The dimensions of the cell are 12.5 × 12.5 cm.
8. Why is the spectral response of a solar cell necessary for its development? Which portion of the spectrum of light should it be sensitive to: ultraviolet, visible or infrared?
9. Why a lock-in amplifier is needed for spectral response measurements?

10. What are the test conditions for a solar cell? What is the role of a solar simulator in this? Describe a solar simulator with diagrams.
11. What is the fundamental parameter that is measured in both LBIC and EBIC measurements? (a) Which should be used for very small samples of solar cells? (b) Which is the cheaper of the two methods?
12. Why is the short circuit current I_{SC} used more often for making measurements on a solar cell instead of the open circuit voltage V_{oc}? Give reasons to support your answer.

References

[1] Roger Messenger and Jerry Ventre, *Photovoltaic Systems Engineering*, CRC Press, Boca Raton, 2002.

[2] Alan L Fahrenbruch and Richard H. Bube, *Fundamentals of Solar Cells*, Academic Press, New York, 1983.

[3] Hans S. Rauschenbach, *Solar Cell Array Design Handbook*, Van Nostrand Reinhold Co., New York, 1980.

[4] David F. Stout and Milton Kaufman, *Handbook of Operational Amplifier Circuit Design*, McGraw-Hill, New York, 1976.

[5] G.E. Tobey, J.G. Graeme and L.P. Huelsman, *Operational Amplifiers: Design and Applications*, McGraw-Hill, New York, 1971.

[6] Ramakanth A. Gayakwad, *Op-Amps and Linear Integrated Circuits*, 3rd ed., Prentice–Hall of India, 1988.

[7] A.K. Mukerjee, S.K. Mahajan and K.D. Chaudhuri, Design of a Logarithmic Amplifier for Ultrasonic Pulse Echo Technique, *Indian Journal of Pure and Applied Physics*, vol. **11**, December 1973, pp. 926–928.

[8] M. Purshotham, A.K. Mukerjee and K.L. Chopra, Logarithmic I–V Plotter for Solar Cells, *Indian Journal of Pure and Applied Physics*, vol. **22**, September 1984, pp. 545–549.

[9] http://www.newport.com/file_store/PDFs/tempPDFs/e6388_Oriel-I-V-Test-Station.pdf.

[10] PV Measurements, Inc., www.pvmeas.com/products.html.

[11] Spire Corp., www.spirecorp.com/spire-solar/pv-business/index.php.

[12] William David Cooper, *Electronic Instrumentation and Measurement Techniques*, Prentice-Hall of India, 1976, p. 180.

[13] M. Purshotham, S.R.Das, A.K.Mukerjee and K.L.Chopra, Automatic Capacitance–Voltage (C–V) Plotter for Solar Cells, *Review of Scientific Instruments*, vol. **53**(11), November 1982, pp. 1749–1753.

[14] Boonton Corp., http://www.eseproject.com.hk/rfteam/datasheet/EN/Boonton/7200.pdf.

[15] H.K. Gummel and F.M. Simms, Evaluation of Solar Cells by Spectral Analysis, *Bell System Technical Journal*, vol. **43**, 1964, p. 1103.

[16] E.D. Stokes and T.L. Chu, Diffusion Lengths in Solar Cells, from Short-Circuit Current Measurements, *Applied Physics Letters*, vol. **30**, 1977, p. 425.

[17] A. Silard, R. Marinescu and M. Tazlauanu, The Spectral Response of BSF Silicon Solar Cells Fabricated Through Masked Ion Implantation, *IEEE Electron. Dev. Lett. EDL–4*, 1983, p. 164.

[18] S.M. Sze, *Physics of Semiconductor Devices*, Wiley, New York, 1979, p. 404.

[19] L.M. Kilgren, N.C. Wyeth and W.E. Devaney, The Design and Utilization of a Microprocessor-Controlled Absolute Spectral Response System, *Solar Cells*, vol. **1**, 1979/80, p. 225.

[20] W.E. Phillips, Interpretation of Steady–State Surface Photovoltage Measurements in Epitaxial Semiconductor Layers, *Solid State Electronics*, vol. **15**, 1972, p. 1097.

[21] Operating and Instruction Manual of Carl Zeiss Jena VSU-2P, Spectrophotometer.

[22] www.kippzonen.com.

[23] www.keithley.com/products.

[24] O.S. Sastry, A.K. Mukerjee and K.L. Chopra, Spectral Response Inhomogeneities in Large Area Solar Cells, *Solid State Electronics*, vol. **28**(6), June 1985, pp. 579–585.

[25] J.S. Hartman and M.A. Lind, Spectral Response Measurements for Solar Cells, *Solar Cells*, vol. **7** (1982–83), pp. 147–157.

[26] Kasturi Lal Chopra and Suhit Ranjan Das, *Thin Film Solar Cells*, Plenum Press, New York, 1983, pp. 161–165.

[27] D.E. Sawyer and H.K. Kessler, *IEEE Transaction on Electron Devices* ED–27, 1980, p. 864.

[28] D.E. Sawyer, D.W. Berning and D.C. Lewis, Laser Scanning of Active Integrated Circuits and Discrete Semiconductor Devices, *Solid State Technology*, vol. **20**, 1977, p. 37.

[29] M. Slapa, P.A. Tove and G. Boberg, Characterization of $CdTeHgI_2$ Crystals and Detectors by Light Spot Scanning, *Nuclear Instruments and Methods*, vol. **150**, 1978, p. 55.

[30] D.A. Yates and R.O. Bell, *14th IEEE Photo–Voltaic Specialists Conference*, San Diego, California, 1980, p. 1402.

[31] D.E. Sawyer, *13th IEEE Photo–Voltaic Specialists Conference*, Washington D.C., 1978, p. 1249.

[32] G.M. Storti, S.M. Johnson, H.C. Lin and C.D. Wang, *14th IEEE Photo–Voltaic Specialists Conference*, San Diego, California, 1980, p. 191.

[33] P.G. Lasswell and B.C. Plunkett, *15th IEEE Photo–Voltaic Specialists Conference*, Orlando, Florida, 1981, p. 1021.

[34] J.R. Szedon, T.W. Okeeffe and T.A. Temofonte, *13th IEEE Photo–Voltaic Specialists Conference*, Washington D.C., 1978, p. 774.

[35] T.E. Furtak, D.C. Canfield and B.A. Parkinson, Scanning Light Spot Analysis of the Carrier Collection in Liquid Junction Solar-Energy Converter, *Journal of Applied Physics*, vol. **51**, 1980, p. 6018.

[36] D.E. Sawyer and D.W. Berning, NBS Special Publication, S.D. Cat. No. C 13–10 (U.S. GPO, Washington D.C., 1977), pp. 400–424.

[37] C.N. Potter and D.E. Sawyer, A Flying Spot Scanner, *Reviews of Scientific Instruments*, vol. **39**, 1968, p. 180.

[38] W. Schuler, Simple Light Spot Scanner and its Applications on Semiconductor Nuclear Particle Detector, *Reviews of Scientific Instruments*, vol. **38**, 1967, p. 1374.

[39] V. Khanna, O.S. Sastry, A.K. Mukerjee and K.L. Chopra, Automatic Laser Scanner for Solar Cells, *Reviews of Scientific Instruments*, vol. **55**(10), 1984, p. 1580.

[40] J.D. Zook, Effect of Grain Boundaries in Polycrystalline Solar Cells, *Applied Physics Letters*, vol. **37**, 1980, p. 223.

[41] O.S. Sastry, V. Dutta, A.K. Mukerjee and K.L. Chopra, Defect Analysis in Polycrystalline Silicon Solar Cells, *Journal of Applied Physics*, vol. **57**(12), June 1985.

[42] A.K. Mukerjee and Nivedita Dasgupta, DC Power Supply Used as Photovoltaic Simulator, *Renewable Energy*, vol. **32**, 2007, p. 587.

[43] http://www.home.agilent.com/agilent/product.jspx?nid=-536902317.536880477.00&cc=US&lc=eng.

[44] D.P. Hohm, M.E. Ropp, Comparative Study of Maximum Power Point Tracking Algorithms, *Progr Photovolt Res Appl*, vol. **11**, 2003, p. 47.

[45] Ashish Pandey, Nivedita Dasgupta and Ashok Kumar Mukerjee, High-Performance Algorithms for Drift Avoidance and Fast Tracking in Solar MPPT System, *IEEE Transactions on Energy Conversion*, vol. **23**(2), June 2008.

CHAPTER 4

The PV System

4.1 General Photovoltaic (PV) Systems

All photovoltaic (PV) systems are similar in nature but differ in details from each other. It is therefore necessary to have a general view of the system design. Once the true nature of the system is understood any other individual product design becomes a mere corollary. For example, a solar water pump and a solar lantern both have the inputs from a solar panel or an array of different power, processed through a maximum power point tracker, followed by a load, which is either a pump or a lamp. Figure 4.1 shows the types of solar powered systems currently in use.

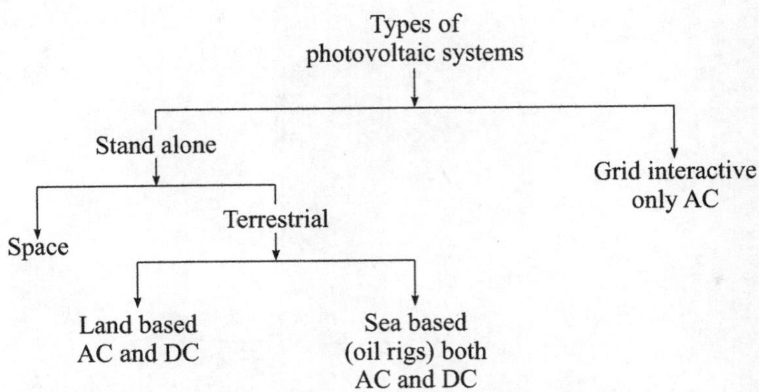

Figure 4.1 Types of photovoltaic systems in use at present.

The terrestrial systems can either operate in a standalone mode for both AC and DC loads or upload power to the grid for supply to other places and is used for AC generation only. Typical DC loads are solar lanterns, DC operated water pumps, fans and general lighting service (GLS) lamps, which are normally low power devices. When connected to an inverter that can generate 220 V AC, the PV source can drive AC fans, motors, water pumps, etc. capable of consuming several kilowatts of power.

A typical PV power supply is shown below in Figure 4.2.

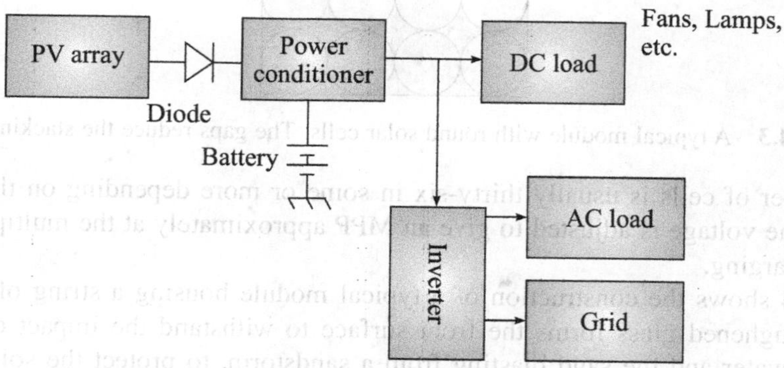

Figure 4.2 Block diagram of a typical photovoltaic system.

The PV array converts the solar energy to DC power, which is directly dependent on the incident solar radiation (Insolation). This is followed by a power conditioner which contains a maximum power point tracker (MPPT) and a battery charge and discharge controller. The MPPT ensures that at all times the maximum power generated by the PV array is extracted while the charge/discharge controller is responsible for preventing overcharging or over-discharging the battery bank required to store the electricity generated by the solar energy for use during sunless times. The life of the batteries will reduce if they are subjected to overcharge or excessive discharge. The DC power thus stored or generated is then fed to the DC load comprising fans, lamps and other DC operated devices. If an inverter is used to convert the DC to normal 220 V AC, then it can operate standard domestic fans, lamps or pumps that are easily available in the market at low prices because of their large scale production. This AC can also be uploaded on to the mains or "grid" to be transmitted to remote places for use. In such cases the inverters acquire the name of "Grid Interactive Inverters".

4.2 The PV Module [1]

Whereas the PV array is made up of several PV modules, each module is assembled from a large number of solar cells which can be made from single, polycrystalline or amorphous silicon or other materials like cadmium sulphide (CdS)/cadmium telluride (CdTe). A module is assembled from a large number of cells to increase its voltage primarily by connecting them in series. Each solar cell produces only a few watts of power, typically 2 W, at approximately 0.5 V DC. Since this power is insufficient to drive most devices the voltage has to be increased with the consequent increase of power. A typical module is shown in Figure 4.3 with sixteen cells in series.

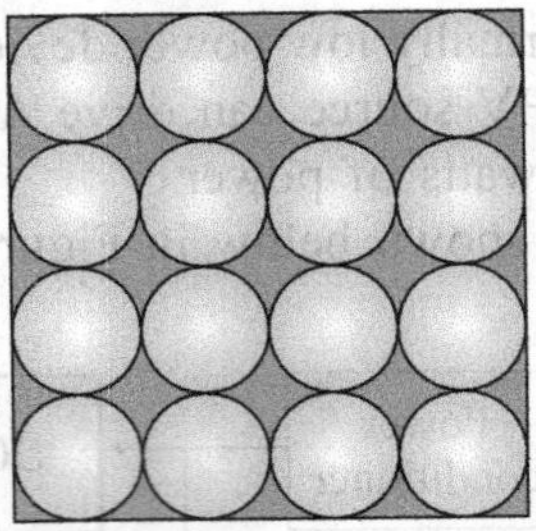

Figure 4.3 A typical module with round solar cells. The gaps reduce the stacking factor.

The number of cells is usually thirty-six in some or more depending on the power level required but the voltage is adjusted to give an MPP approximately at the multiple of six volts for battery charging.

Figure 4.4 shows the construction of a typical module housing a string of solar cells. A transparent toughened glass forms the front surface to withstand the impact of hails during a storm, rain water and the sand blasting from a sandstorm, to protect the solar cells placed inside. A layer of Ethylene Vinyl Acetate (EVA) is then placed on the glass. This is followed by the string of solar cells with their active front faces towards the glass. This string is then covered by a sheet of non-woven glass cloth called *scrim glass* also known by its trade name 'Crane glass'. The scrim glass is covered on top by another layer of EVA. This multiple layer structure is finally covered by a sheet of Tedlar-Polyester-Tedlar combination or Tefzel™. The wires connecting the two ends of the string are then taken out from predetermined holes before lamination of this assembly. EVA has totally replaced polyvinyl butyrate (PVB) as the laminating material since the latter acquired a brown colour on prolonged exposure to sunlight. The back surface of the module was earlier made of Tedlar-Aluminium-Tedlar sheets but this had problems of electrical insulation breakdown and was subsequently replaced by Tefzel.

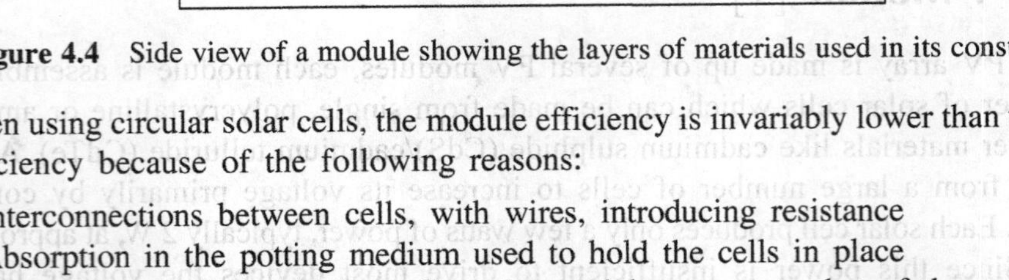

Figure 4.4 Side view of a module showing the layers of materials used in its construction.

When using circular solar cells, the module efficiency is invariably lower than the individual cell efficiency because of the following reasons:

1. Interconnections between cells, with wires, introducing resistance
2. Absorption in the potting medium used to hold the cells in place
3. Solar cell mismatch which confines the module current to that of the weakest cell
4. The transmission of the radiation is reduced because of reflection of the protective glass sheet on top and absorption in it, and

5. The packing factor. That is, the entire area of the module is not covered by the solar cells but there are large gaps between the adjacent cells.

Suppose that the radius of the cells is r. Then the total area required to place four cells is:

$$A_t = (2r + 2r) \times (2r + 2r) = 4r \times 4r = 16r^2 \tag{4.1}$$

However, the actual area covered by the four cells is:

$$4 \times (\pi r^2) = 4\pi r^2.$$

Therefore, the ratio of the cell area to the total area A_t, is:

$$4\pi r^2/16r^2 = \pi/4 = 0.7854.$$

This shows that if the individual cell efficiency is 10%, their efficiency when placed in a module is reduced to 78.5% of their individual value, that is, 7.85% only because of the packing factor.

4.3 The PV Array

The modules are connected in series to increase the voltage and in parallel to increase the current further, to suit any particular requirement, as shown in Figure 4.5. A_1, A_2 and A_3 comprise one string of modules connected in series to increase the voltage required for a desired bus-bar. For instance if a module has a maximum power voltage V_m of 12 V whereas the required voltage to drive a load is 36 V, then the string will have three modules in series. Similarly, the other strings consist of modules B_1, B_2 and B_3, C_1, C_2, and C_3, and also D_1, D_2, and D_3. These other strings will help to increase the current to drive the load. For example, if the A string has a maximum power point current I_m of 2 A, whereas the required current to drive the load is 8 A, then the strings B, C and D will be connected in parallel to achieve this goal.

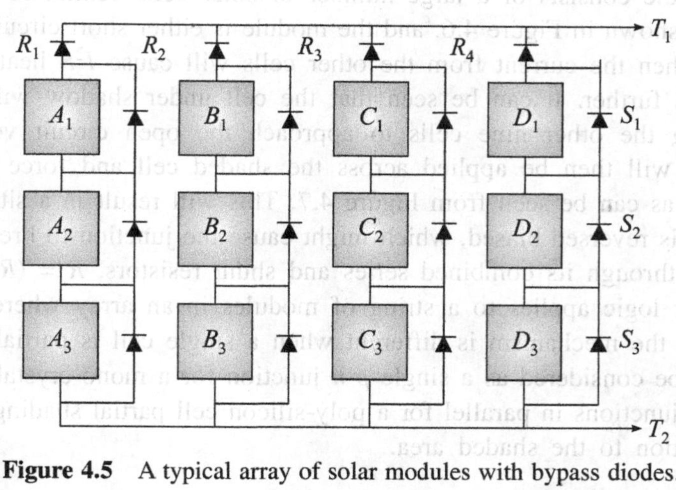

Figure 4.5 A typical array of solar modules with bypass diodes.

Whenever such a large number of power generators, in this case solar modules, are connected in a series-parallel combination they have to be isolated from each other. A large number of rectifiers, R_1 to R_4 and S_1 to S_3 and others, shown in Figure 4.5 do this work. R_1 to R_4, called *blocking diodes*, act as an OR logic so that any of the strings that is operational can contribute to the array current. They also ensure that in the case of a mismatch one string does not force its current through any other. Since the current involved in each string is 2 A the current rating of each rectifier should be ideally 2 A. However, the short circuit current of each module is 2/0.7 = 2.86 A and further the rise in ambient and rectifier temperature generally de-rate the device current. It is advisable to use a current rating of 100% more than the required value. Hence a 4 A, 36 V rectifier will work well for R_1, R_2 and R_3 as well as S_1, S_2 and S_3, which are also called the *bypass diodes*.

4.4 Partial Shading of a Solar Cell and a Module

The function of the bypass diodes is important in any array construction involving multiple modules. If a module in a string fails due to some reason, or comes under the shadow of some object, called *partial shading*, then the current in that module will reduce drastically and will limit the current from the other two which pass through it. In short, the current through a string will reduce. In such a case the bypass diode associated with that module will allow the current to pass through itself. For example, if module D_2 fails then S_2 will bypass the current generated by D_1 and D_3. However, the bus voltage of 36 V required between terminals T_1 and T_2 will be almost equivalent to the addition of the open circuit voltages of D_1 and D_3 but there will be very little contribution of current from the D_1–D_3 string. On the other hand if S_2 is removed from the circuit only a little current will flow out from that string. Hence it is a balancing act between the two situations which has to be determined by the designer.

The other problem, in absence of the bypass diodes, is the formation of hot spots in a module. A module consists of a large number of solar cells connected in series. If one cell is shaded, as shown in Figure 4.6, and the module is either short circuited or connected to a heavy load then the current from the other cells will cause I^2R heat to be generated in it as explained further. It can be seen that the cell under shadow will present a high resistance causing the other nine cells to approach the open circuit voltage V_{oc} of the module. This V_{oc} will then be applied across the shaded cell and force a current in the reverse direction, as can be seen from Figure 4.7. This will result in a situation where not only the junction is reversed biased, which might cause the junction to breakdown, but also force the current through its combined series and shunt resistors, $R = (R_s + R_{sh})$, causing heating. A similar logic applies to a string of modules in an array where one of them is shaded. However, the mechanism is different when a single cell is partially shaded. Since a single cell can be considered as a single *p-n* junction for a mono-crystalline silicon solar cell or many *p-n* junctions in parallel for a poly-silicon cell partial shading will just reduce the I_{SC} in proportion to the shaded area.

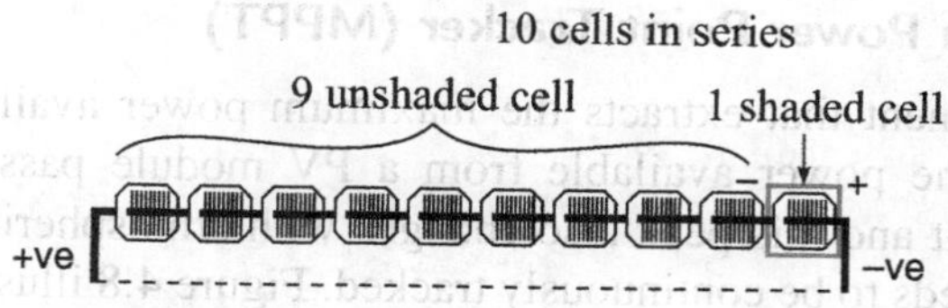

Figure 4.6 If one cell, of several series connected cells, is shaded, it will get heated.

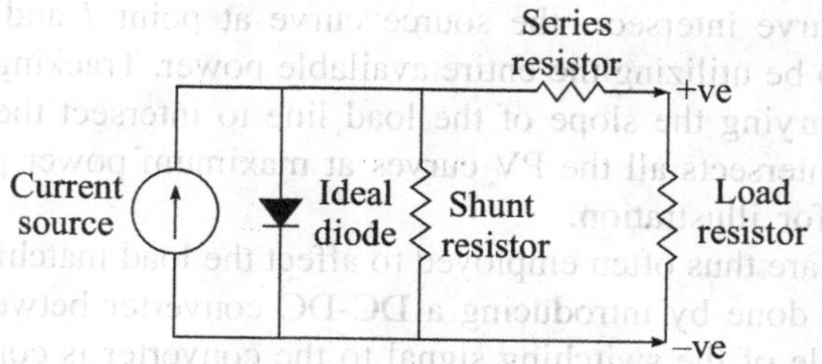

Figure 4.7 Equivalent circuit of a solar cell showing how a current through the resistors can cause heating.

4.5 The Diode

The diode shown between the PV array and the power conditioner, in Figure 4.2, is used to stop the reverse flow of current when the array is connected directly to the battery during sunless hours. In such cases, the voltage output of the array will be low whereas that of the battery will be high, which will drive a current from the battery into the array since all the diodes of the solar cells will be forward biased. A large battery current will certainly destroy the diodes and cause irreversible damage to the array. The forward current rating of the diode must exceed the short circuit current I_{SC} of the array; preferably be $2I_{SC}$. Because of the flow of high current through the diode from the array to the PCU there is considerable heat generated in the diode which has a magnitude of

$$W = I \times V \tag{4.2}$$

where I is the current through the diode, V is voltage drop across the diode which is equal to 0.7 V for a silicon device and W is heat generated in watts.

If, for example, I = 10 A then W = 0.7 × 10 = 7 W. To dissipate this heat the diode must be mounted on a suitable heat sink and cooled with a fan if necessary or the diode will be over heated and damaged. For higher currents, bigger heat sinks are required.

However, in presence of the power conditioning unit the diode may be omitted since it will have its own protections against reverse flow of current.

4.6 The Power Conditioning Unit (PCU)

The power conditioning unit, Figure 4.2, has two parts:

1. The maximum power point tracker, and
2. The battery charge/discharge controller.

4.6.1 The Maximum Power Point Tracker (MPPT)

The MPPT is an instrument that extracts the maximum power available from the PV array at any given instant. The power available from a PV module passes through a peak at a particular operating point and this peak also changes with atmospheric conditions. Hence, the changing peak power needs to be continuously tracked. Figure 4.8 illustrates the load mismatch during direct connection that results in operation at a sub-optimal point. The maximum power (MP) hyperbola intersects the *I-V* curve (at a particular insolation level), at the MP point M_1 and the actual load curve intersects the source curve at point I and the hyperbola at I_{max} where it is supposed to be utilizing the entire available power. Tracking algorithms thus, have the task of virtually varying the slope of the load line to intersect the *I-V* curve at M_1. The ideal load curve that intersects all the PV curves at maximum power points (M_1, M_2, etc.) is also shown in figure for illustration.

Electronic MPPTs are thus often employed to affect the load matching condition described above. This is usually done by introducing a DC-DC converter between the PV source and the load. The duty cycle of the switching signal to the converter is controlled by pulse width modulation to optimize the current drawn from the array. The correct amount of current in turn forces the array to work at the optimum power point. Thus, this converter basically regulates the input voltage to the optimum value continually irrespective of the variations in the cell temperature and the insolation.

Figure 4.8 Typical current and power versus voltage curves of a solar module.

Tracking algorithms are implemented through analog and digital circuitry. Algorithmic deficiencies lead to oscillations around Maximum Power Point (MPP). Efforts to improve tracking occasionally degrade dynamics of MPPT system. With continually changing atmospheric conditions, this trade-off leads to poor utilization of solar cells. Another problem occurs during changing atmospheric condition. Algorithms based on the general gradient following method have an inherent tendency to drift away from the MPP. The control logic governing MPP tracking are usually carried out either through analog circuits or digitally with microcontrollers and digital signal processors (DSPs). The current state-of-the-art tracking methods will be explored and discussed in Chapter 7.

The block diagram of an MPPT is shown in Figure 4.9. The current and the voltage of the array output are fed to the input of the microcontroller, where these two quantities are multiplied to give the power. The power levels at different currents of the array are then compared and a pulse of a varying width is then generated according to the algorithm used. This pulse is then used to switch the MOSFET/IGBT switch of the DC-DC converter.

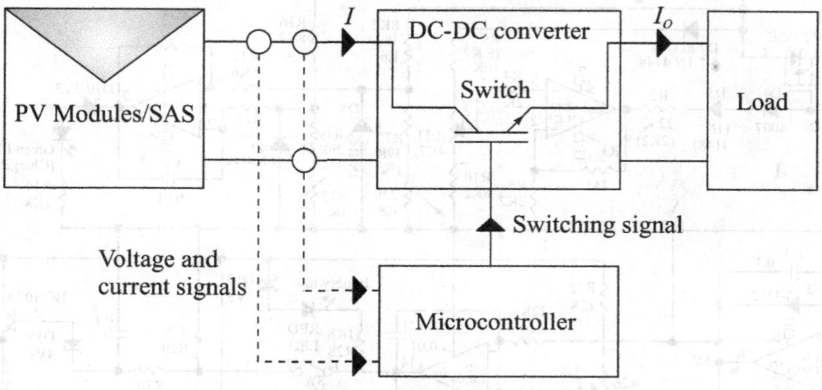

Figure 4.9 Block diagram of an MPPT.

4.6.2 Battery Charge/Discharge Controller

The charge controller ensures that the battery is not overcharged or discharged below a certain limit, and thus prevents the battery from being damaged. An ideal controller will measure the battery voltage and when fully charged will reduce the duty cycle of the MPPT to 10% so that a trickle charge is maintained to compensate for the inherent loses in the battery due to leakage. On the other hand when the battery is close to discharge the MPPT microcontroller will open a switch to disconnect the load from the battery. A simpler circuit is given in Figure 4.10.

The charging of batteries in most PV systems is done in the constant current-constant voltage mode. This is easily accomplished since the PV generator has a constant current region on its *I-V* curve which confines its current to safe values with the maximum limited to the short circuit current I_{SC}. As the battery charges its voltage rises slightly until at full charge the voltage reaches a certain specified value beyond which permanent damage can occur. At this point the voltage input to the battery is *floated* by using electronic circuits. This means that the voltage remains steady at this level. Most of the batteries when left idle over long periods lose charge because of internal losses. This slow leakage of charge lowers the voltage leading to the discharge of the battery which if left unattended may cause irreversible damage. For this reason a small charge should always be allowed to enter the battery to replenish the losses.

This process is called *trickle charging* and can be best achieved by leaving a constant voltage at the terminals of the battery from a *float* charger. For a 12 V battery the float charging voltage, is 14.5 V and the minimum voltage below which the battery should not be discharged is 11.5 V and the temperature coefficient of voltage, that is, the increase of battery voltage with temperature, is 14.56 mV/°C [2]. Table 4.1 gives the details of charging and discharging voltages for a 12 V battery.

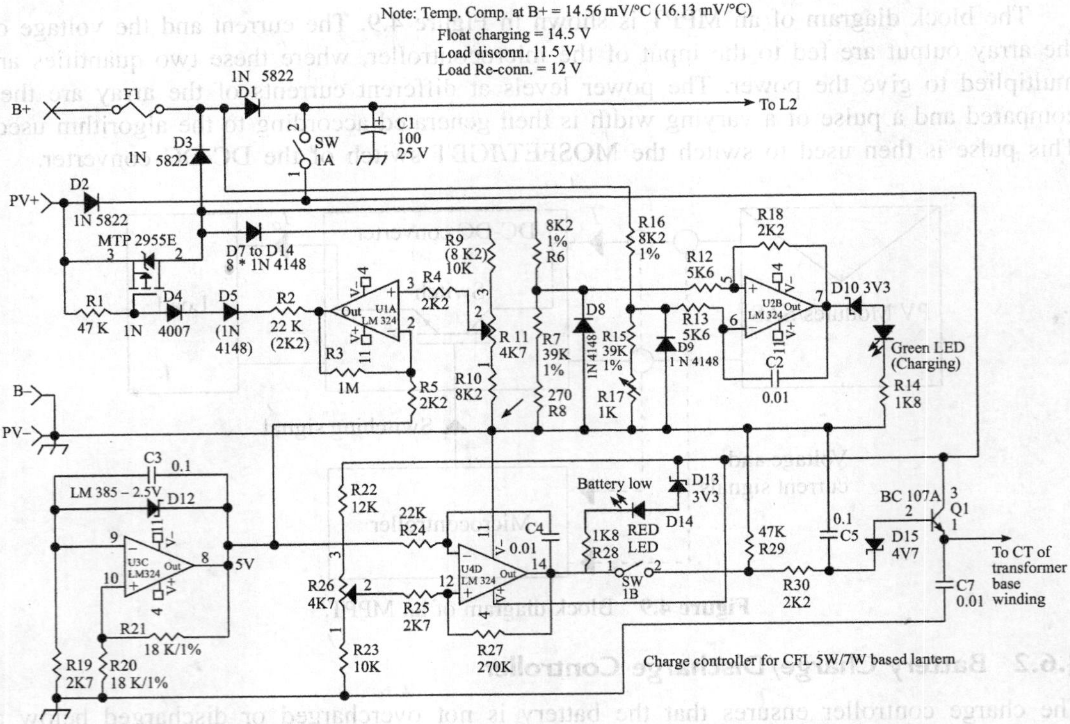

Figure 4.10 Charge-discharge controller of the CFL based solar lantern.

Figure 4.10 shows the circuit diagram of the charge-discharge controller of a CFL based lantern, which can operate either a 4 pin 5 W, or a 7 W, CFL of any standard make [3]; *Copyright (2007) Reprinted with permission from Elsevier.* An LM324, 14 pin quad operational amplifier (four OPAMPS in one envelope), is used to operate the controller. The four sections of the OPAMP are indicated by four triangles with their accompanying components. One section U3C, with its associated components, makes a stable 5 V regulator used as a reference voltage source. U1A uses this voltage to form a 14.5 V feedback regulator with the *P*-channel MOSFET MTP 2955E as the series pass element. Diodes D_1, D_2, and D_3 are low voltage drop Schottky diodes, which protect the circuit against accidental reversal of photovoltaic panel and battery polarities.

Table 4.1 Battery voltages for normal operation

Parameter	*Battery type and range*	
	VRLA (V_{DC})	*Lead–acid* (V_{DC})
SPV disconnection level (Float voltage)	14.1 ± 0.2	14.5 ± 0.2
SPV reconnection level	13.2 ± 0.2	12.5 ± 0.2
Battery low cut-off level	11.5 ± 0.2	11.5 ± 0.2
Battery reconnection level	13.0 ± 0.2	13.0 ± 0.2
Deep discharge level	10.8	10.8
Over-voltage level	14.4	15.2

Diodes D_7 to D_{14} are 1N 4148 silicon diodes in the feedback loop of the regulator, used to provide temperature compensation for the battery during the process of charging. The 12 V lead-acid maintenance-free battery is connected between B+ and the B– terminals to store the charge generated by the PV panel during daytime while the PV panel is connected between the PV+ and PV– terminals. U2B acts as a comparator by detecting the voltage across diode D_3. When the battery is being charged, the terminal 5 of U2B is more positive than the terminal 6 and the green light emitting diode, D_{11}, glows to indicate it.

When the battery voltage goes lower than 11.5 V, the output voltage of operational amplifier U4D goes low and the transistor Q_1 stops conducting. This shuts down the oscillator for driving the CFL while the red LED D_{14} gives an indication of 'Battery Low'. When the battery is charged from a partially discharged condition, the series resistance of the PV panel reduces its output voltage to latch on to that of the battery plus the voltage drop across the MOSFET and diodes that come in series. When the battery is fully charged, the output of the regulator settles down to a fixed value of 14.5 V so that overcharging does not damage the battery. However, the regulator continues to furnish a small current to compensate the loss of charge of the battery, an inherent property of lead-acid cells. This is known as *trickle charging*. A double pole, double throws switch, SW1, turns ON/OFF the lantern manually.

A 10 W solar panel is used for charging the 12 V battery. Although this is the nominal battery voltage, its actual value really depends on its state of charge. This may vary from 10.8 V, for deep discharge, to 14.5 V for a fully charged state for both lead-acid and VRLA types. Therefore, the open circuit voltage of the module should be 12/0.8 = 15 V so that it operates at its *maximum power point*. Since there are some diodes and a MOSFET in the charging path the panel voltage is usually kept between 16 to 17 V to also account for the temperature rise in the tropics.

4.7 Inverter for AC Loads

Batteries give DC voltages at different voltages, in multiples of 6 V, to drive DC loads. However, to operate conventional lamps and fans etc. for domestic use the operating voltage has to be increased to a nominal value of 220 V AC. Since these devices are mass produced they are not only cheaper and have better conversion efficiency but also easily available. To accomplish this conversion an inverter is necessary. Inverters of various power capacities with sinusoidal and, the cheaper, quasi-sinusoidal output waveforms are available to operate for independent, or *stand-alone*, purposes or the more complex ones capable of uploading power to the mains or the grid. The latter are known as *grid interactive inverters*. A detailed discussion on inverters is given in Chapter 6.

4.8 Mounting of Panels for an Array

Solar modules must be appropriately mounted on suitable structures for proper functioning. Needless to say those structures have to be mechanically robust to withstand extreme environmental conditions for a very long time so as to recover the cost during the specified operating period. Some of the desired properties are:

- The structure should be stable under high velocity wind conditions up to 150 km/hr. This is done by grouting it in a firm foundation.
- The materials should be corrosion resistant and should not chemically react with the materials of the modules. Anodized aluminium and painted or plated steel sections are often used but galvanized steel may be avoided for the fear of corrosion when connected to other metal parts. As a matter of fact identical metallic parts should be used to avoid galvanic corrosion in parts that are grounded.
- Enough space must be provided between adjacent modules for easy wind flow.
- A string of modules must not cast any shadow on the one behind it during sunlight.

4.8.1 Fixed Tilted Modules

In most solar energy applications, the modules comprising an array are mounted on stationary holders having a fixed tilt with respect to the horizontal. This tilt angle is adjusted to collect the maximum solar radiation by an adjustment in the structure. The angle of tilt itself depends on the latitude of the site of installation, time of the year and the atmospheric conditions. A side view of a typical fixed tilt mechanism is shown in Figure 4.11. Solar radiation falls on the front side of the module that is held in place by the structure giving a tilt of $\beta°$ to the horizontal surface.

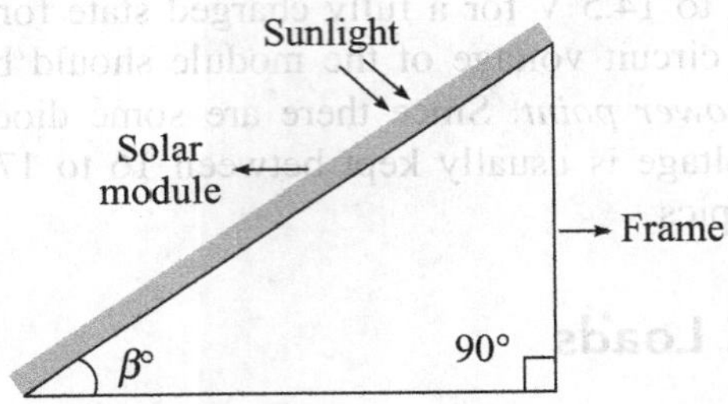

Figure 4.11 A solar module tilted at an angle $\beta°$ to the horizontal surface.

Figure 4.12 shows the situation on earth where a module is located at $S1$, in the northern hemisphere at ϕ degrees latitude. The angle of tilt of the module to the horizontal surface there, shown as the tangent at $S1$, is $\beta°$. Since the sun's apparent motion is from east to west over the day, the module is tilted in the *north-south* direction to collect sunlight. The sun at 'Sun 1' position causes summer in the northern hemisphere and is located directly at 90° to the normal at the point marked *A* called the *Tropic of Cancer*. A similar module is placed at point $S2$ in the southern hemisphere. During the winter in the northern hemisphere the sun occupies the position 'Sun 2' over the point *B* called the *Tropic of Capricorn*, and is responsible for causing summer in the southern hemisphere. These two positions present the limiting values of the declination angles of the sun on either side of the equator. It can be proved that under the ideal solar condition, where there is only direct or beam radiation, maximum annual collection of the solar energy is possible when the tilt of the module β is equal to the latitude ϕ of the site where it is placed [4]. Generally, however, the maximum energy collection over the year is when β is approximately equal to 0.9ϕ.

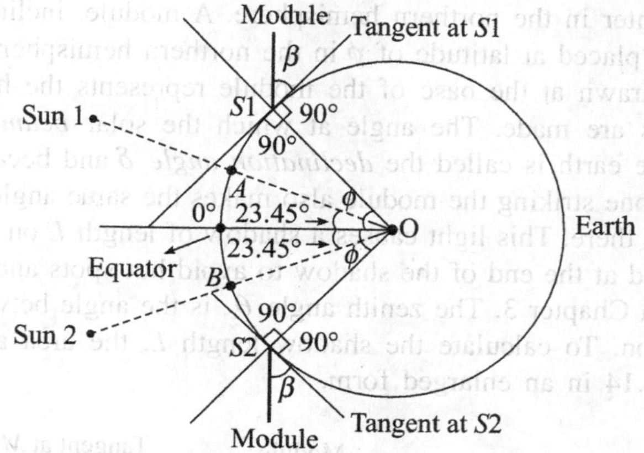

Figure 4.12 The position of the modules at latitude of ϕ from the equator.

For further explanations, see references [1], [5] and [6]. Also some important concepts are given in reference [7]. Some manufacturers have empirically determined the optimum tilt angle and their recommended values are given in Table 4.2.

Table 4.2 Tilt angle corresponding to the site latitude

Site latitude in degrees	*Optimum module tilt angle in degrees*
0–15	15
15–25	Equal to Latitude
25–30	Latitude +5
30–35	Latitude +10
35–40	Latitude +15
More than 40	Latitude +20

The increase in the tilt angle is more for higher latitudes to increase collection of solar energy during the winter months when the days are shorter. Some authors do not agree with the tilt values given above and suggest their outright rejection. However, the data from [6] indicates that the annual total energy collection has a maximum value of 2,131.2 kWhr/m^2 when the tilt angle is 25° for Delhi, at latitude of 28.58°, which is about 0.9 × ϕ (=28.58) as mentioned above. Thus, a lower tilt angle than the latitude in this case harvests more energy. This is due to other factors like temperature and a large diffused component of radiation. For approximate work the values given in Table 4.2 will suffice but for more accurate calculations the exact data for the site may be obtained from [6] or other works [7], [8].

When two or more arrays are placed one behind the other, a sufficient distance must be maintained between successive arrays so that the one in front does not cast its shadow on the array placed at its rear. To calculate this distance, some basic quantities have to be defined with the help of Figure 4.13. The diagram shows a section of the earth in the vertical plane. Since the sun is very far from the earth it can be considered as an extended source that covers the earth with parallel beams of light. In this case the sun is over the southern hemisphere, that is, during the winter in the northern hemisphere. A module, inclined at angle β to the earth surface at M, is placed at latitude of ϕ in the northern hemisphere. The tangent to the surface of the earth drawn at the base of the module represents the horizontal plane from which all calculations are made. The angle at which the solar ***beam radiation*** meets the equatorial plane of the earth is called the ***declination angle*** δ and because all the beams of light are parallel, the one striking the module also makes the same angle δ with a line drawn parallel to the equator there. This light causes a shadow of length L on the surface. The next module must be placed at the end of the shadow to avoid hot spots and lowering of current, as explained earlier in Chapter 3. The zenith angle θ_z, is the angle between the normal at M and the beam radiation. To calculate the shadow length L, the area around the module is presented in Figure 4.14 in an enlarged form.

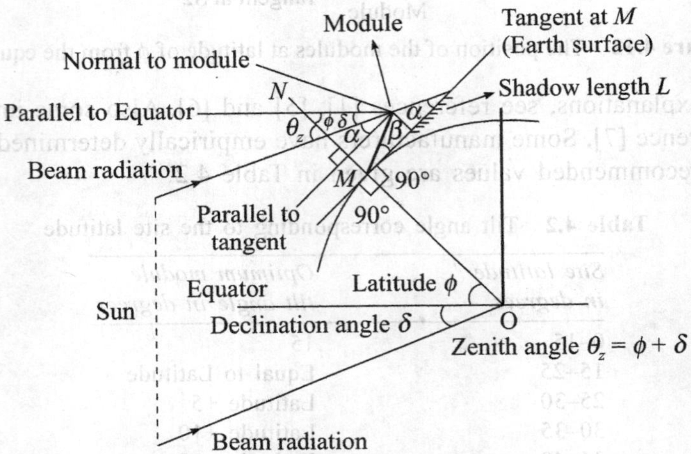

Figure 4.13 The shadow cast by a module is shown with parameters used to calculate its length.

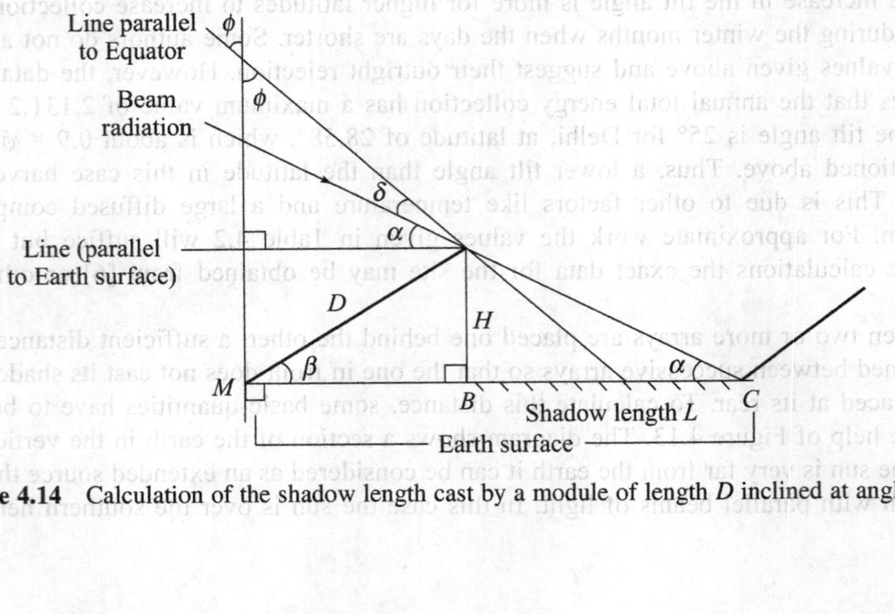

Figure 4.14 Calculation of the shadow length cast by a module of length D inclined at angle β.

A beam of sunlight touches the top of the module of length D and falls on the horizontal surface at an angle α thus producing a shadow length L where the next module cannot be placed. The next module must be placed beyond the point C.

Now $\quad \dfrac{H}{D} = \sin\beta$

or, $\quad H = D\sin\beta \qquad (4.3)$

But $\quad L = H/\tan\alpha = H/\tan(90° - \phi - \delta)$

$\quad = H/\cot(\phi + \delta) = H\tan(\phi + \delta)$

Therefore, $\quad L = D\sin\beta\tan(\phi + \delta) \qquad (4.4)$

Because $\quad \alpha + \delta = 90° - \phi$ from Figure 4.14.

Therefore, $\quad \alpha = 90° - \phi - \delta = 90° - (\phi + \delta)$

When the sun travels to northern hemisphere the situation is altered as shown in Figure 4.15. Now the zenith angle becomes, $\theta_z = \phi - \delta$, which is smaller than that in the earlier case. When the relevant portion of this figure is expanded, it looks like Figure 4.16. As expected the shadow length is smaller than that of shown in Figure 4.14. Also the length is given by

$$H = D\sin\beta \qquad (4.5)$$

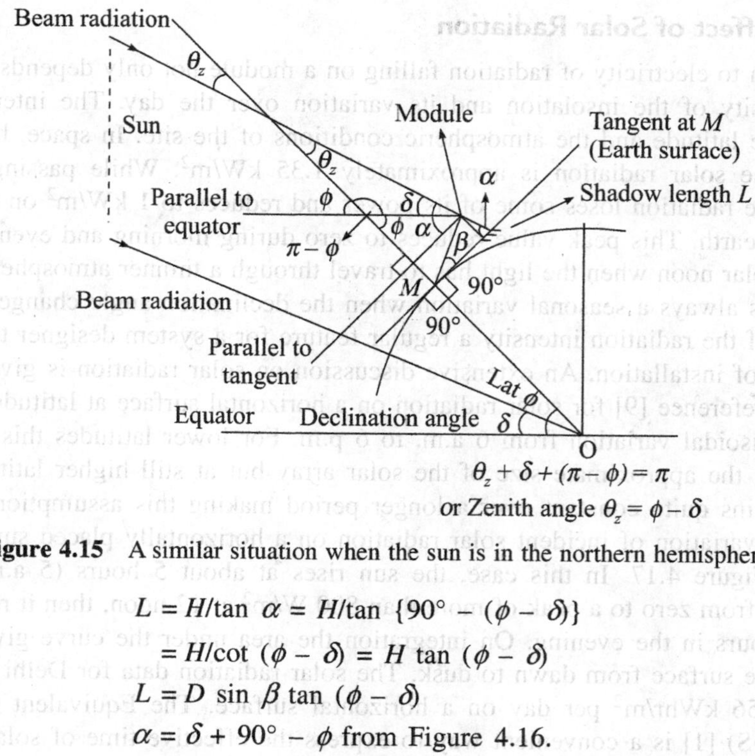

Figure 4.15 A similar situation when the sun is in the northern hemisphere.

Since $\quad L = H/\tan\alpha = H/\tan\{90° - (\phi - \delta)\}$

$\quad = H/\cot(\phi - \delta) = H\tan(\phi - \delta)$

Therefore $\quad L = D\sin\beta\tan(\phi - \delta) \qquad (4.6)$

Because $\quad \alpha = \delta + 90° - \phi$ from Figure 4.16.

or $\quad \alpha = 90° - (\phi - \delta)$

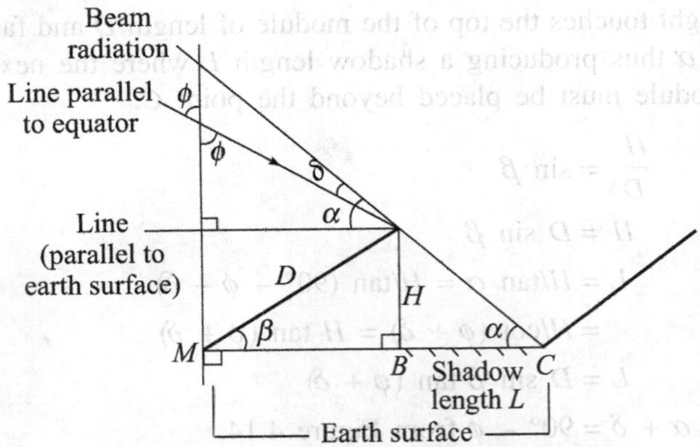

Figure 4.16 The shadow length reduces when the sun is in the northern hemisphere.

Generally, the angle of declination δ is considered negative when the sun is on the southern hemisphere. The numerical value of the angle is therefore $-(-\delta)$ in Equation (4.6), which now becomes $L = D \sin \beta \tan (\phi + \delta)$ showing that the shadow becomes longer. In conclusion, it is noted that the spacing between two modules or arrays must be designed for the worst case of winter when the shadows are longest.

4.8.2 The Effect of Solar Radiation

The conversion to electricity of radiation falling on a module not only depends on its tilt but also the intensity of the insolation and its variation over the day. The intensity, in turn, depends on the latitude and the atmospheric conditions of the site. In space, beyond earth's atmosphere, the solar radiation is approximately 1.35 kW/m^2. While passing through the atmosphere, the radiation loses some of its power and reduces to 1 kW/m^2 on the horizontal surface of the earth. This peak value reduces to zero during morning and evening hours and is highest at solar noon when the light has to travel through a thinner atmosphere. In addition to this, there is always a seasonal variation when the declination angle changes. This makes the variation of the radiation intensity a regular feature for a system designer to worry about for every site of installation. An extensive discussion on solar radiation is given in [9]. The data given in reference [9] for solar radiation on a horizontal surface at latitude 43° suggests an almost sinusoidal variation from 6 a.m. to 6 p.m. For lower latitudes this fact is useful for calculating the approximate size of the solar array but at still higher latitudes the peak radiation remains quite constant over a longer period making this assumption non tenable.

A typical variation of incident solar radiation on a horizontally placed surface on Earth is shown in Figure 4.17. In this case, the sun rises at about 5 hours (5 a.m.), while the intensity rises from zero to a peak of more than 800 W/m^2 at 12 noon, then it reduces to zero again at 20 hours in the evening. On integration the area under the curve gives the energy received on the surface from dawn to dusk. The solar radiation data for Delhi for the month of April is 6.56 kWhr/m^2 per day on a horizontal surface. The Equivalent Hours of Full Sunlight (EHFS) [1] is a convenient way to express the effective time of solar radiation on a unit surface if the solar radiation was effectively constant at its peak value of 1 kW/m^2.

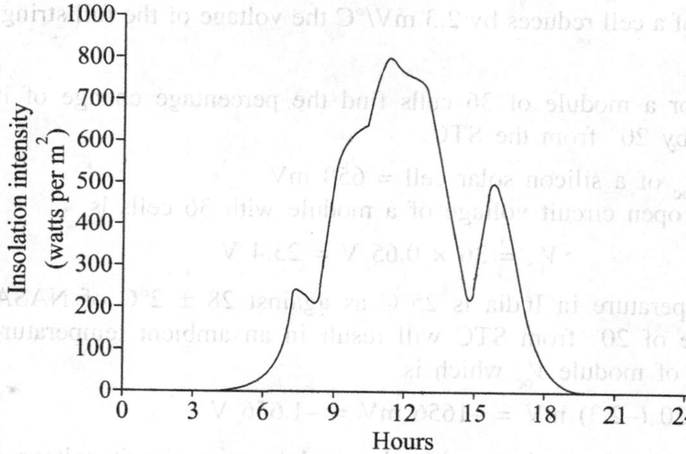

Figure 4.17 Typical variation of daily insolation on a unit area horizontal surface.

In this case, it means that if the sunlight was assumed to be constant at 1 kW/m^2 then the EHFS is 6.5 kWhr/m^2/day divided by 1 kW/m^2 = 6.5 hr.

Example 4.1: What is the energy output of a 36 W solar module placed horizontally in the month of April when the radiation falling on it is 6.5 kWhr/m^2/day?

Solution: PV output energy

= peak module power under standard test conditions (STC) × EHFS

= 36 × 6.5 = 234 Whr

Similarly, if the module is tilted at 25° with respect to the surface of the earth, the EHFS will change slightly. The radiation on a tilted surface is 6.35 kWhr/m^2/day. Therefore the EHFS = 6.35 hours. Hence the

PV output energy = 36 × 6.35 = 228.6 Whr

The EHFS is not to be mistaken for the actual hours of sunlight. The actual hours of sunlight might be different for two places at the same latitude but widely different longitudes. The EHFS values can be computed from the radiation data of desired sites available in literature [6] or in meteorological handbooks.

4.8.3 The Effect of Temperature

Another factor that profoundly affects the module or array output is the temperature of operation. The data given by the manufacturers are at the standard test conditions (STC) but the actual field operation differs in many respects. The ambient temperature, under the sun, in India varies from almost sub zero temperatures at higher altitudes, for example in Leh, to about 50°C in the south of the country. This vastly changes the output of the same module at these two different places. The temperature inside the module is still higher because there is no heat dissipation from the cells encapsulated there. The cell operating temperature (COT) is normally taken to be 25°C above the ambient temperature.

$$\text{COT} = \text{Ambient temperature} + 25°\text{C} \tag{4.7}$$

Since the V_{oc} of a cell reduces by 2.3 mV/°C the voltage of the full string of cells reduces appreciably.

Example 4.2: For a module of 36 cells find the percentage change of its V_{oc} when the temperature rises by 20° from the STC.

Solution: The V_{oc} of a silicon solar cell = 650 mV

Therefore the open circuit voltage of a module with 36 cells is

$$V_{oc} = 36 \times 0.65 \text{ V} = 23.4 \text{ V}$$

The STC temperature in India is 25°C as against 28 ± 2°C of NASA/DOE standard. Hence an increase of 20° from STC will result in an ambient temperature of 45°C. This causes a decrease of module V_{oc} which is

$$36 \times 20\ (-2.3) \text{ mV} = -1656 \text{ mV} = -1.656 \text{ V}$$

Therefore the percentage decrease in the module open circuit voltage is

$$\%\ V_{oc} = (-1.656/23.4) \times 100 = -7\%$$

The V_{oc} of the module at the higher temperature = 23.4 – 1.656 = 21.744 V.

This lowering of V_{oc} causes a substantial reduction of PV output power that must be taken into consideration while designing a PV system. From above the module degradation coefficient comes out to be = –7%/20°C = –0.35% per °C rise in temperature.

However, when power reduction with temperature has to be accounted for the change in V_m is more important than the change in V_{oc}. Since for all practical purposes $V_m = 0.7\ V_{oc}$, the V_m of the module is 23.4 × 0.7 = 16.38 V. However it is customary to slightly enhance it to –0.45%/°C rise in temperature. It will be shown below in Example 4.3 that the reduction of power with temperature is 0.5%/°C. This effect of temperature when coupled with the EHFS results in a more accurate value of the array size for any application. The actual mathematical expression then assumes the form for the array output (AO)

$$\text{AO} = \text{EHFS} \times \{1 - (\%\text{ of degradation coefficient}/100) \times \text{COT} - 25\} \tag{4.8}$$

If the 36 W module of Example 4.1 is exposed to the sunlight its PV energy output will become peak power at STC × AO

= Peak power at STC × EHFS × {1 – (% of degradation coefficient/100) × COT – 25}

$$36 \times 6.5 \times \left[1 - \frac{0.5}{100} \times \{(25 + 25) - 25\}\right] \text{ Whr/day} = 204 \text{ Whr/day}$$

This is quite a reduction in power as compared to 243 Whr/day as found in Example 4.1, where the effect of temperature was neglected.

When the cell temperature increases, the output of the cell is affected in three ways.

(i) The current output increases typically by about 0.1%. This essentially corresponds to the cell's energy conversion efficiency which, in turn, is due to a change in the collection efficiency with temperature. Primarily the short circuit current is a function of the level of illumination, the spectral distribution of the illumination and the spectral response of the solar cell.

(ii) The output voltage of the cell decreases as shown in Figure 4.18. This change is due to the lowering of the conduction voltage of the cell *p-n* junction. The entire *I-V* curve shifts to lower voltages, at the rate of 2.2 to 2.3 mV/°C, with increase in temperature. This shift in voltage is nearly the same for V_m and V_{oc} values of all solar cells, as well as conduction voltage of rectifiers, made from silicon.

(iii) The roundness of the knee region is altered.

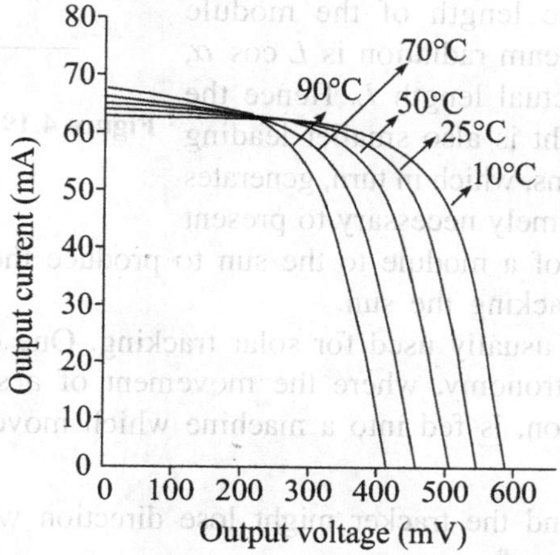

Figure 4.18 The variation of the cell output voltage with temperature.

Example 4.3: For a single solar cell of V_{oc} = 650 mV, find the percentage change in its maximum power when the temperature is increased by 1°C.

Solution:

V_{oc} = 650 mV (given) at a temperature *T*.

∴ $V_m = 0.7 \times 650$ mV = 455 mV

Suppose the current at the maximum power point is I_m and also assume that it does not change with temperature.

The power at temperature *T*,

$$P_T = I_m \times V_m = (I_m \times 455) \text{ mW.}$$

If the temperature rises by 1°C then the voltage V_m reduces to 455 – 2.3 mV = 452.7 mV. Then the power at *T* + 1°C is

$$P_{T+1} = (I_m \times 452.7) \text{ mW.}$$

Hence the % decrease in power = $(P_T - P_{T+1})/P_T$

$$= \frac{I_m\,(455 - 452.7)}{I_m \times 455} \times 100$$

$$= \frac{2.3}{455} \times 100 = 0.5\%$$

Thus, the reduction of power with increase in temperature is 0.5%/°C and is more than the reduction of the V_{oc} as found in Example 4.2.

4.9 Sun Tracking

Figure 4.19 shows a module BC of length L inclined at an angle to earth. A light beam falls on it at an angle α with respect to the normal ON at point O. Therefore, the angle BOA is $\theta = 90° - \alpha$. Since BDO is a right angled triangle, angle DBO is α.

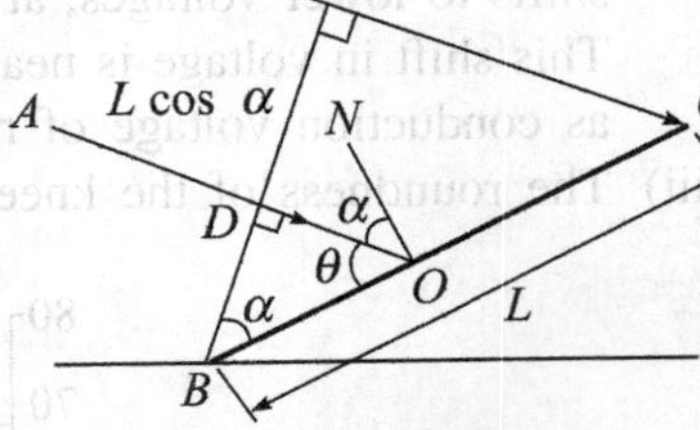

Figure 4.19 The effect of module tilt on the area presented to beam radiation.

Therefore, the effective length of the module presented to the incoming beam radiation is $L \cos \alpha$, which is smaller than its actual length L. Hence the area presented to the sunlight is also smaller leading to a lesser collection of photons, which in turn, generates lower power. It is thus extremely necessary to present the maximum surface area of a module to the sun to produce the largest amount of power. This can be achieved by tracking the sun.

There are two methods usually used for solar tracking. One of them is a computational one extensively used in astronomy, where the movement of a star or the sun, either in a tabulated form or an equation, is fed into a machine which moves the tracker. This method has two disadvantages:

(i) During a strong wind the tracker might lose direction which needs to be corrected manually.

(ii) The table or the data fed into the equation has to be changed with the site of the tracker.

However, it is a time tested method and therefore of unquestioned reliability.

The second method uses a feedback correction to guide the tracker. In case the direction of the tracker is disturbed by natural forces like winds or the object is lost behind clouds, the driving electronics can automatically correct the defect afterwards. This method is used for precision tracking of the sun as some trackers have concentrator solar cells mounted on them. Concentration of sunlight, up to five hundred or more, is achieved normally by Fresnel lens. The only limitation here is that the tracking mechanism needs a strong source of light like the sun or may be the moon. This then excludes the stars or other distant celestial bodies that are of interest to the astronomers. Apart from carrying solar modules trackers are also used to heat oils and water to generate steam for driving turbines.

4.9.1 Single Axis Tracking

Figure 4.20 shows a sensor to generate an error signal for the feedback based dual axis tracker. A twenty five centimetre long tapered rod made of a non-rusting material like a galvanized iron, blackened aluminium or stainless steel is firmly fixed on a plate of a similar material. Four solar cells A, B, C and D, closely matched for the short circuit current, are placed around the rod such that A, the rod and C are in a straight line. Similarly B, D and the rod should also be placed in a straight line. The line

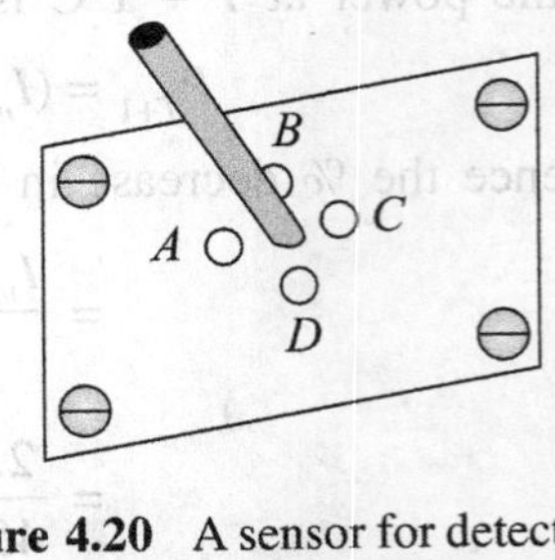

Figure 4.20 A sensor for detecting the movement of the sun.

connecting A and C should be perpendicular to the one connecting B and D so that when AC is placed along the east-west, then BD must be in the north-south direction.

The tracking error signal is generated by the shadow of the rod on a solar cell. Consider a position where the rod is pointed straight towards the sun. Since the surface of the tracker carrying the solar modules is perpendicular to the rod, the entire areas of the modules get the beam radiation normally. Also it is worth noticing that the rod does not produce any shadow since it is tapered at the top. If the sun now moves west, the shadow of the rod will fall on the solar cell C reducing its short circuit current I_{SC}. This current when compared with that of A there will give a difference which can be treated as an error signal. It should be noted that initially when both A and C were equally illuminated with sunlight the two cells were generating the same I_{SC} so that there was no error signal. Let the control electronics now move the tracker to the west using this error signal as the driving indicator. In such a case the shadow will reduce and eventually move away from C which will now generate the initial I_{SC} and reduce the error signal to zero. This means that the tracker has been able to track the sun. After a while the sun will move westwards again and the same procedure will be repeated. A solar tracker is described here for carrying either a solar array or a concentrating collector.

Concentrating collectors are required to produce high temperatures since it is not possible to obtain such temperatures with flat plate collectors. All the concentrating collectors, whether they are the linear or the point focusing type, require partial or complete tracking. A fully tracked (dual axis) flat plate photovoltaic (PV) array gives about 40–43% more annual energy output than a fixed flat (PV) array [11]. Although the cost of the whole system increases still tracking systems have great importance and several designs of such systems are given in references [12–14]. Most heliostat systems, large concentrating paraboloids and some PV systems require computer based control systems for tracking the mirrors, concentrators or arrays. In this case, very precise tracking is required and the tracking errors are minimized by the mechanical design and the software of the computer. However, in the sensor controlled feedback systems, the tracking error is determined by the design of the sensor, the error signal produced by the sensors and the gain of the amplifiers etc. Teague et al. [15] have given an experimental procedure to find out the tracking errors in azimuth and latitude angles using an auto collimating method. In this method they have used a theodolite and an auto collimating eye piece capable of measuring up to one arc second.

In this chapter, an indirect method is given to determine the tracking errors for a horizontal east-west (E-W) tracking system. The tracking errors are determined by measuring the average duration between the two consecutive pulses generated at the Schmitt trigger output for driving the motor and finding the angle travelled by the sun in that duration. The values so obtained have been compared with the calculated values obtained by considering the design of the sensors, differential amplifier gain etc.

Description of the solar tracker

A. Tracking system

A single axis E-W tracking system was fabricated to hold a parabolic concentrator of aperture area 3.75 m^2 or a photovoltaic array of the same size. It has the facility to tilt the collector to improve the energy collection in winter when the declination δ angle of the sun rays is small. The tracking system, shown in Figure 4.21, consists of a strong rectangular frame of 3 m × 1.5 m, made of mild steel angle iron, which can be fixed to a cemented base. Another similar frame movable at one end was build to provide the required tilt in the north-south direction. A steel shaft, 2 m long and 0.05 m in diameter, was driven in the middle by a three phase motor through a pulley and rope arrangement. The speed of the 1500 rpm motor was reduced with gears. This motor has a high initial torque and load capacity. The total electrical energy consumed by the motor in a day is about 0.3 Whr. To minimize the load on the motor, counter weights were used to bring the centre of gravity of the moving part to the middle of the shaft.

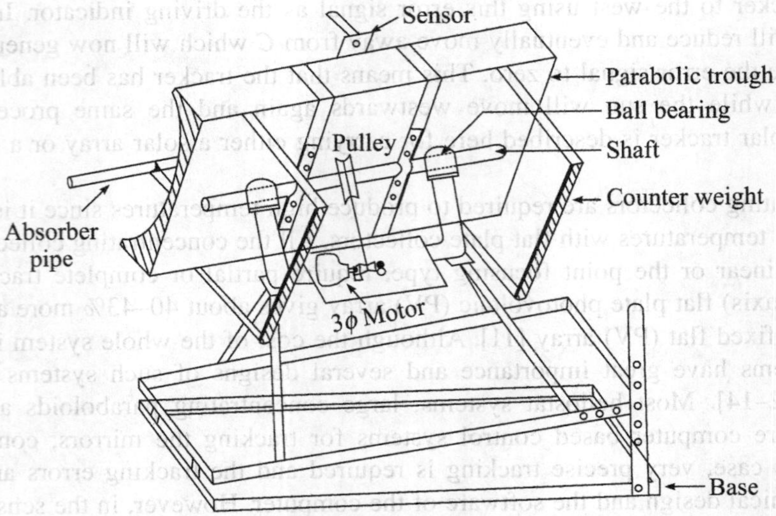

Figure 4.21 Schematic of a single axis solar tracker.

B. Sensor assembly and control unit

The tracking of the system is controlled electrically with the help of sensors. The block diagram of the control unit is shown in Figure 4.22. The sensor consists of two matched silicon solar cells placed on either side of a rectangular bar. This is similar to the one, shown in Figure 4.20. When any one of the cells is shadowed by the bar, an error signal is produced which is amplified to activate the Schmitt triggers which, in turn, drive the motor in the required direction. When the solar intensity is below 250 W/m^2 and automatic tracking is not possible, a cloud signal is generated which activates the motor after every 10 min for 0.4 s.

Three phase pulse width modulated (PWM) signals are produced by the generator circuit to drive the motor. The PWM signal has an advantage of generating a high torque to start even low speed motors easily. The logic unit contains circuits for (a) manual and automatic motion of the tracker (b) generation of a cloud signal in case of low solar radiation intensity and (c) returning the tracker to the 'morning position' after sunset so that it is ready to receive sunlight the next day.

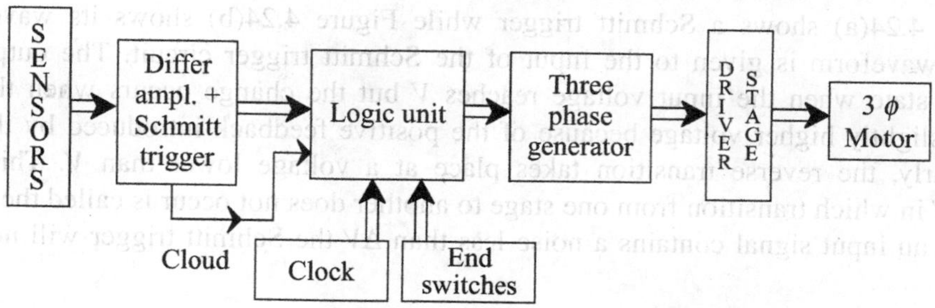

Figure 4.22 Block diagram of the control unit.

Figure 4.23 shows the circuit diagram of a differential amplifier with the outputs from the two solar cells. Solar cell S_1 is connected to an operational amplifier A_1 which works in a current to voltage converter mode. The current generated by S_1 flows through the feedback resistor R_f to produce a voltage V_1. Similarly, solar cell S_2 produces a voltage V_2. When light of equal intensity falls on both S_1 and S_2, they produce almost the same current, since they are matched, and the output voltage V_1 equals V_2. These two voltages are fed into the two inputs of the differential amplifier consisting of operational amplifier LM 741 along with resistors R_1, R_2, R_3 and R_4. A differential amplifier is an analogue circuit for subtracting one voltage from the other. When V_2 is more than V_1, o/p is positive but when V_2 is less than V_1, o/p is negative. The magnitude of the voltage at o/p depends on the differential gain of the amplifier A_D. For $R_1 = R_2$ and $R_3 = R_4$, A_D is unity. The gain can be increased to enhance the tracking accuracy but up to certain limit beyond which the whole system becomes unstable due to a large amount of noise being introduced. To overcome the unwanted affect of noise interfering with the system performance a Schmitt trigger is introduced. This circuit has a built in hysteresis that eliminates the random triggering by noise to some extent. A Schmitt trigger is a comparator circuit that incorporates positive feedback. When the input is higher than a certain chosen threshold, the output is high; when the input is below another (lower) chosen threshold, the output is low; when the input is between the two, the output retains its value. The trigger is so named because the output retains its value until the input changes sufficiently to trigger a change. This dual threshold action is called *hysteresis*, which implies that the Schmitt trigger retains its memory.

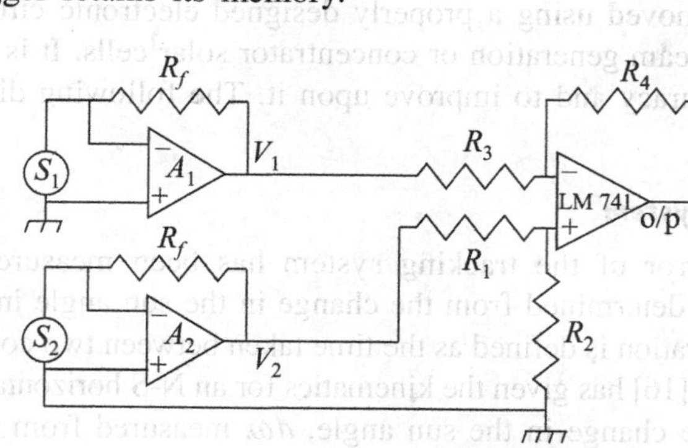

Figure 4.23 A differential amplifier.

Figure 4.24(a) shows a Schmitt trigger while Figure 4.24(b) shows its waveforms. A triangular waveform is given to the input of the Schmitt trigger circuit. The output should change its state when the input voltage reaches V but the change occurs when the voltage reaches a slightly higher voltage because of the positive feedback introduced by the resistor R_f. Similarly, the reverse transition takes place at a voltage lower than V. This band of voltage ΔV in which transition from one stage to another does not occur is called the *hysteresis voltage*. If an input signal contains a noise less than ΔV the Schmitt trigger will not respond to it.

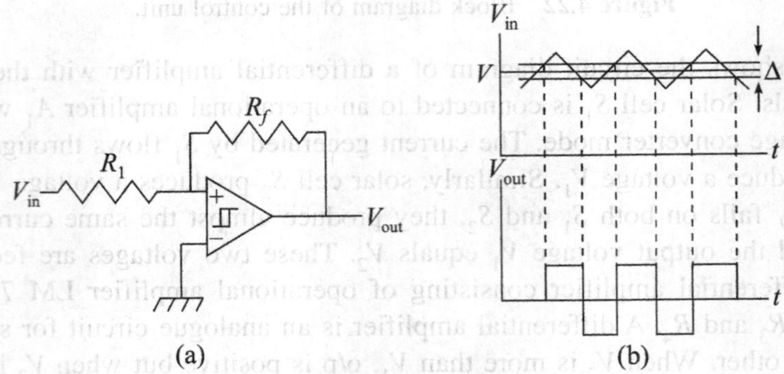

Figure 4.24 (a) A Schmitt trigger circuit and (b) Its waveforms.

The "cloud signal" is actually an indication for a cloudy sky when the insolation drops to below 250 W/m^2. This signal comes from the either V_1 or V_2 and fed into the logic unit, bypassing the Schmitt trigger.

The "clock" is a precision quartz controlled oscillator used for moving the tracker at the rate of 0.4 second after every ten minutes among other functions. The tracker moves approximately 15° per hour.

The "end switches" save the tracker from mechanical damage should the motor continue to run after the trough reaches its extreme position of tilt while tracking the sun. These two switches are placed at the two positions at the end of the rotational arc, physically disconnect the motor from the electricity source so that the entire driving electronics is disabled.

The tracker can be moved using a properly designed electronic circuitry but this is not sufficient for its use in steam generation or concentrator solar cells. It is extremely important to know its tracking accuracy and to improve upon it. The following discussion is aimed at this only.

Tracking error of the system

The average tracking error of the tracking system has been measured experimentally at different intensities. It is determined from the change in the sun angle in an average tracking duration. The tracking duration is defined as the time taken between two consecutive movements of the tracker. Vasilevskis [16] has given the kinematics for an N-S horizontal axis (E-W tracking) collector, from which the change in the sun angle, $d\omega$, measured from the southern horizon is determined from Figure 4.25. According to Vasilevskis, the sun angle ω is related to hour angle H by the Equation (4.9)

$$\tan \omega = \frac{\sin H}{\sin \phi \tan \delta + \cos \phi \cos H} \tag{4.9}$$

The rate of change of ω with respect to hour angle is obtained by differentiating Equation (4.9) and substituting

$$\tan q = \frac{\sin H}{\cot \phi \cos \delta + \sin \delta \cos H} \tag{4.10}$$

and
$$\cos n = \cos \phi \sin \delta - \sin \phi \cos \delta \cos H \tag{4.11}$$

The result so obtained is given by

$$\frac{d\omega}{dH} = \frac{\cos \delta \cos q}{\sin n} \tag{4.12}$$

where

ϕ, the latitude angle of the place and δ is the declination angle.

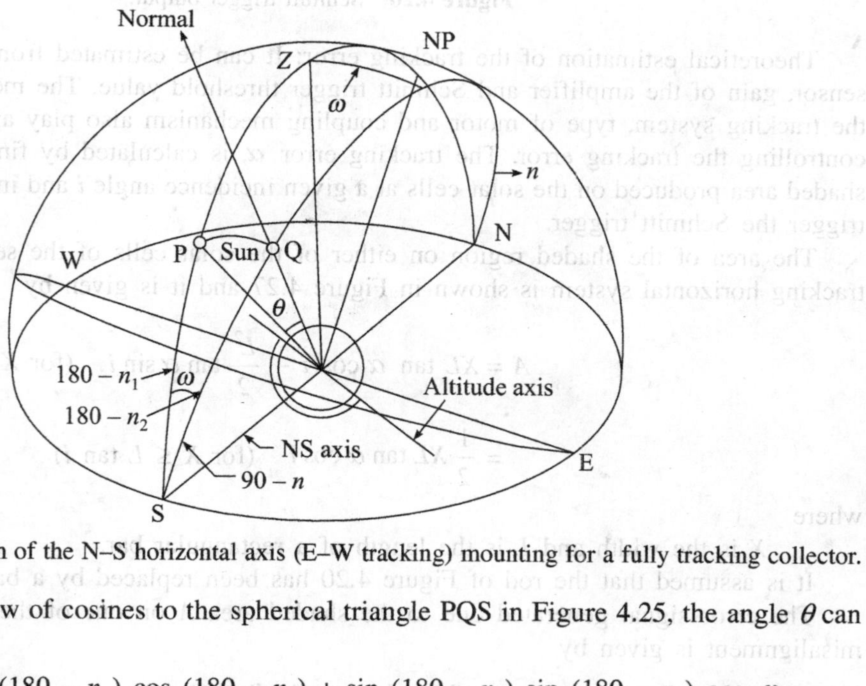

Figure 4.25 Diagram of the N–S horizontal axis (E–W tracking) mounting for a fully tracking collector.

Applying the law of cosines to the spherical triangle PQS in Figure 4.25, the angle θ can be determine as

$$\cos \theta = \cos (180 - n_1) \cos (180 - n_2) + \sin (180 - n_1) \sin (180 - n_2) \cos \omega$$

$$= \cos n_1 \cos n_2 + \sin n_1 \sin n_2 \cos \omega \tag{4.13}$$

where θ is the angle subtended by two different positions of the sun at the centre of collector and n_1, and n_2, are the angles subtended by the sun with North Pole at the centre of collector at different sun positions. The values of ω and n are calculated for every solar hour and substituted in Equation (4.13) to get the angle θ travelled by the sun in one hour. The tracking error is calculated from experimentally determined average duration between two consecutive pulses generated at the output of Schmitt trigger for every solar hour as shown in Figure 4.26. This duration is equal to the tracking duration. The minimum tracking error is the angle moved by the sun in the tracking duration time.

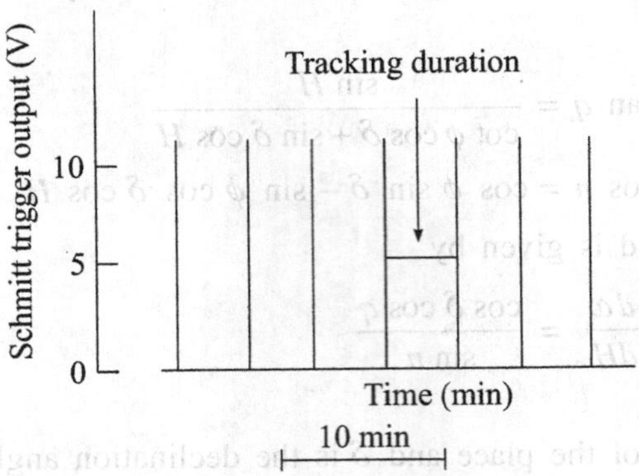

Figure 4.26 Schmitt trigger output.

Theoretical estimation of the tracking error: It can be estimated from the design of the sensor, gain of the amplifier and Schmitt trigger threshold value. The mechanical design of the tracking system, type of motor and coupling mechanism also play an important role in controlling the tracking error. The tracking error α, is calculated by finding the minimum shaded area produced on the solar cells at a given incidence angle i and intensity I which can trigger the Schmitt trigger.

The area of the shaded region on either of the solar cells of the sensors for the E-W tracking horizontal system is shown in Figure 4.27 and it is given by

$$A = XL \tan \alpha \cos i - \frac{L^2}{2} \tan \alpha \sin i \quad (\text{for } X > L \tan i)$$

$$= \frac{1}{2} XL \tan \alpha \cos i \quad (\text{for } X \le L \tan i) \tag{4.14}$$

where

X is the width and L is the length of a rectangular bar.

It is assumed that the rod of Figure 4.20 has been replaced by a bar.

The error signal generated due to the shaded area A on one of the solar cells, during misalignment is given by

$$(J_L \times a)\, R_L - J_L\, (a - A)\, R_L = J_L A R_L \tag{4.15}$$

where J_L is light current density at a given intensity, a is the area of solar cell ($a = X^2$) and R_L is the load. Here it is assumed that both the solar cells have identical I-V characteristics and equal areas are exposed during alignment.

The error signal generated due to misalignment is enlarged by the amplifier gain factor G, and the output so obtained should be greater or equal to the threshold voltage of the Schmitt trigger to activate the motor. So

$$(J_L A)R_L\ G \geq V_{\text{Th}}$$

or

$$A \geq V_{\text{Th}}/(J_L R_L G) \tag{4.16}$$

From the Equations (4.14) and (4.16), we get

$$XL \tan\alpha \cos i - \frac{L^2}{2}\tan\alpha \sin i \geq \frac{V_{\text{Th}}}{(J_L R_L)G} \tag{4.17}$$

The minimum value of α, can be determined from the Equation (4.17) at a given incidence angle and the light current density since the other terms V_{Th}, R_L and G are constants.

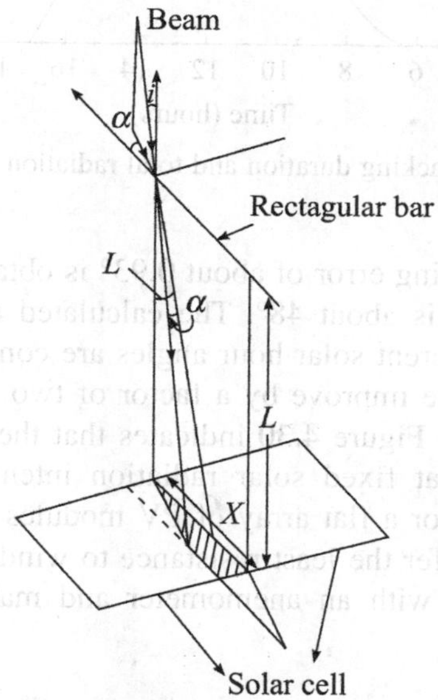

Figure 4.27 Shadow area on the solar cells due to misalignment.

Performance of the tracking system

The tracking performance of the system was studied with a composite parabola of aperture size 3 m × 1.20 m [17]; *Copyright (1987)Reprinted with permission from Elsevier.* A black coated copper tube, of 0.088 m diameter, was used as the receiver and 0.025 m wide mirror strips were used to form a concave reflector which was used to focus sunlight on the receiver. The diurnal variations of tracking duration and solar radiations on the N-S horizontal surface around winter solstice at New Delhi (latitude = 28.63°N, longitude 77.33°E) are shown in Figure 4.28. The tracking duration is smallest at solar noon when the solar intensity on a horizontal surface is strongest and it increases as the solar radiation decreases. When the solar intensity becomes low and the sensor is unable to produce an automatic tracking, a cloud signal is generated which moves the tracker after fixed intervals of time. Therefore, the tracking duration is constant for about three hours before and after solar noon. The tracking error variation of N-S horizontal axis (E-W tracking) system with hour angles is

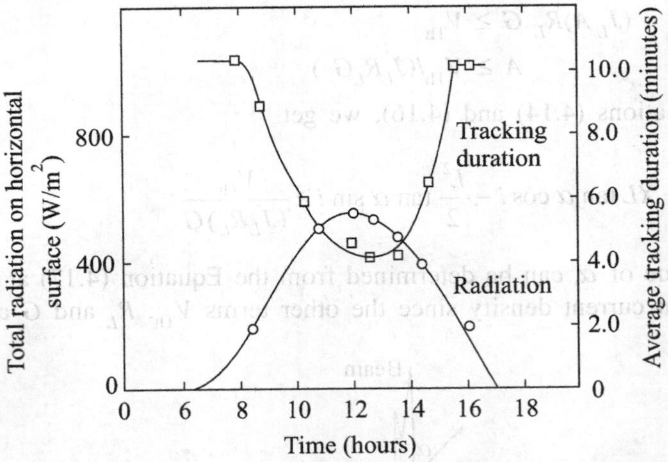

Figure 4.28 Diurnal variation of tracking duration and total radiation on a horizontal surface at New Delhi in winter.

shown in Figure 4.29. The tracking error of about 0.93° is obtained at solar noon in the worst case when the incidence angle is about 48°. The calculated and experimentally determined values of tracking errors at different solar hour angles are comparable. The calculated values of tracking error show that these improve by a factor of two at solar noon in summer when the incidence angle is about 5°. Figure 4.30 indicates that the tracking duration increases as the incidence angle increases at fixed solar radiation intensity. Care should be taken to compensate for strong winds. For a flat array of PV modules the logic of the tracker should be so designed that it should offer the least resistance to wind. An acceptable solution would be to sense the wind velocity with an anemometer and make the array horizontal to the ground.

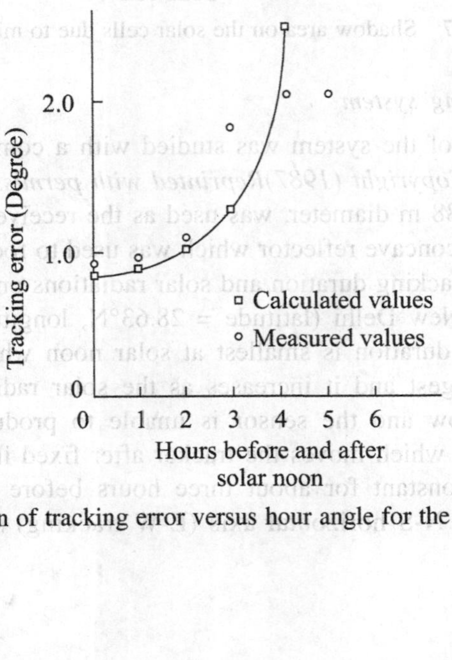

Figure 4.29 Variation of tracking error versus hour angle for the E-W tracking system.

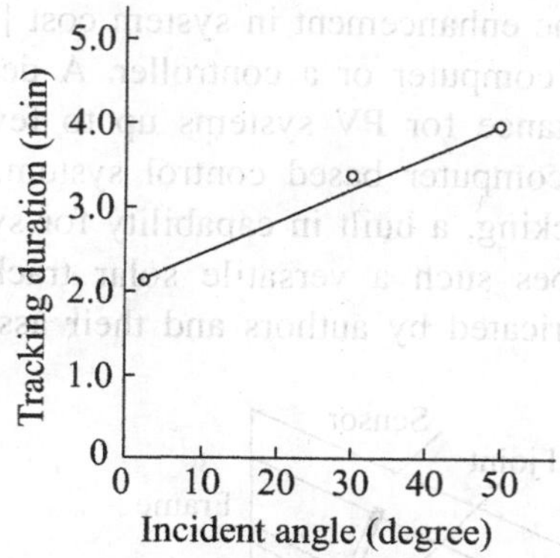

Figure 4.30 Variation of tracking duration versus incidence angle at solar noon.

This tracker was used to generate steam for water distillation. For such applications the tracking accuracy of 0.93° is acceptable. However, for concentrator solar cells of more than a concentration factor of fifty a smaller error must be achieved and a N-S tracking should also be introduced. This leads to a dual axis tracker.

4.9.2 Dual Axis Tracking

A dual axis tracker requires movements in both E-W (*azimuth*) and N-S (*altitude*) directions and therefore two motors are needed along with two sets of sensors and driving electronic circuitry. This is indeed a replication of the single axis tracker arrangement for movement in two directions.

A simple tracker capable of moving in two directions is shown in Figure 4.31. A section of a circular arc makes a track on which a shaft called the *azimuth shaft* can move with the help of a 3-phase electric motor. When this shaft moves from S_4 to S_3 the frame carrying the solar modules moves from east to west. This shaft is firmly attached to the central pillar that bears the load of the solar array, through a ball bearing mechanism for easy movement. When the sun moves from the tropic of cancer to the tropic of capricorn, the other motor designated Motor_1 moves the Altitude Shaft to increase or decrease the tilt of the frame. The central pillar is grouted to a concrete base to enable it to resist windy conditions. The operation is similar to the single axis tracker, as the sensor used here, is similar to the one shown in Figure 4.20. In case, the control electronics moves the tracker beyond the permissible limits of the azimuth angle, the end switches S_3 and S_4 disconnect the electricity supply to the electronics to stop any further movement. Similarly, S_1 and S_2 stop the change in the tilt angle. These are however extreme measures which are not encountered during normal operation.

It is clear from the above discussion, that solar tracking is essential for improving system performance [18] for many solar energy based power generation systems. The change in the sun's position is monitored, and the array is maintained normal to the incident radiation for optimum performance. Accurate tracking is essential for concentrator systems because the spot of light produced by a concentrating lens might not fall on the small area solar cell because of improper alignment. However, even in flat-plate PV system, consisting of modules, an increase in power output by 30–60% (depending on the location) has been observed by tracking as compared to fix tilt [19]. The increase is significant enough to make tracking a viable proposition in spite of the enhancement in system cost [20]. It is possible to align the tracking heliostats by using a computer or a controller. A dedicated microprocessor based controller has immense importance for PV systems up to few kilowatts of power. Larger systems do require elaborate computer based control system. Such a dedicated controller should have, in addition to tracking, a built in capability for system control and monitoring. The following section describes such a versatile solar tracking controller based on the microcontroller Intel 8085 fabricated by authors and their associates.

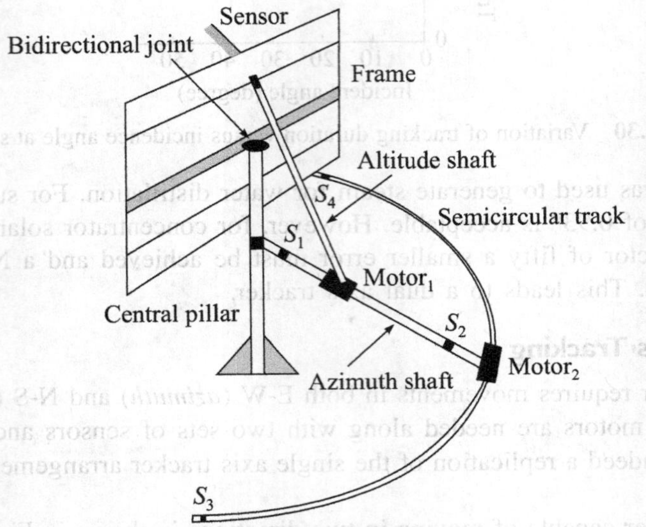

Figure 4.31 Schematic diagram of a dual axis tracker.

Design of a tracking controller

PV systems have found main application as decentralized power generation system for remote areas [24]; *Copyright (1990) IEEE.* Keeping this in mind, the controller has been designed with following criteria:

- It should be totally automatic and simple to operate. Operator interference should be restricted to only during emergencies.
- The design should be flexible to operate different types of tracking systems. For this 6 PWM (Pulse Width Modulated) signals for 3-phase, 50 Hz AC motors or 4 step signals for stepper motors are provided. The program for signal generation is stored in an EPROM (Electrically Programmable Read Only Memory) which can be changed to suit the tracker requirement. The software is designed in a modular fashion for ease in modifications. Also implemented are both open loop and closed loop strategies for tracking.
- Facility is provided for control of batteries/load. The state of charge of batteries and load requirement should be checked to prevent any excessive charging or discharging of batteries. The state of charge of batteries is sensed through the voltage of the battery bank (configured for 24 V, 48 V DC applications). The software senses the voltage

signals during the day. If the batteries get overcharged, an OVS (Over voltage signal) is generated to cut-off the battery charging. On the other hand, if batteries are undercharged, a UVS (Under voltage signal) is generated to cut off the load. In addition to this, load management can also be provided.

- A system monitoring facility is included in the design. The PV parameters (I_{SC}, V_{oc}, P_m etc.) and meteorological parameters need to be studied to evaluate the PV system's performance in the field condition. Therefore, the controller should have a data acquisition and storage facility. In order to keep the design simple voltage to frequency conversion (VFC) for data acquisition has been utilized. Though this is a slower process, it does not severely affect the data acquisition since most of the parameters are slowly varying.

The VFC signals can be easily acquired by the processor and can also be simultaneously recorded in a cassette player. Alternately these signals can be stored in a RAM (Random Access Memory) during the day and can be transferred to the cassette recorder at the end of the day. The data thus collected can be processed at a central location. This obviates the need for providing expensive computer facilities at remote locations. Such a design is affected using different cards for each feature. Figure 4.32 shows the block diagram of the tracking controller. It consists of:

(i) *Converter Card.* This provides +5 V and –5 V signals using DC-to-DC converter from a 48 volts input.

(ii) *Microprocessor card.* It contains an Intel 8085 central processing unit (CPU), two type 8255 (#1 and #2) Programmable Peripheral Interfaces, a type 2716 (or 2732) EPROM for program storage and a type 6116 RAM for data storage. The ports in 8255 #1 are used exclusively for tracking and the ports in 8255 #2 are used for data acquisition and control. Additional chips can be used for display and any other functions that may be desired.

(iii) *Driver card.* The PWM or step signals from the microprocessor card are connected to the driver stages comprised of BDX 62 and BDX 63 power transistors in complementary push-pull configuration through a buffer. Using tri-state buffers (74 LS 367), these signals can be connected to either of the driver stages by providing a gate signal. This is elaborated later on.

(iv) *Data acquisition card.* It consists of a Voltage to Frequency Converter (VFC) to which the signals are provided for a fixed interval of time. Different voltages signal are sequentially selected by a relay selector which is started by a STart Data acquisition (STD) signal from the processor. The VFC signal is serially given to the input on SID pin and the number of step are counted. A timer provides timing pulses of 4 second interval which then increments a 4-bit binary counter at each pulse. The 4-bits of the counter provide address lines for a demultiplexer to select one of the 16 channel outputs. Since the demultiplexer cannot provide the current required to actuate relays, a bus driver is used. The 'Data' are acquired in one minute (time required to scan all the channels) every 10 minutes. At the end of the day, the data are transferred to the cassette player by providing a 'STart Recording' signal (STR) to start the cassette and an SOD signal for the Data.

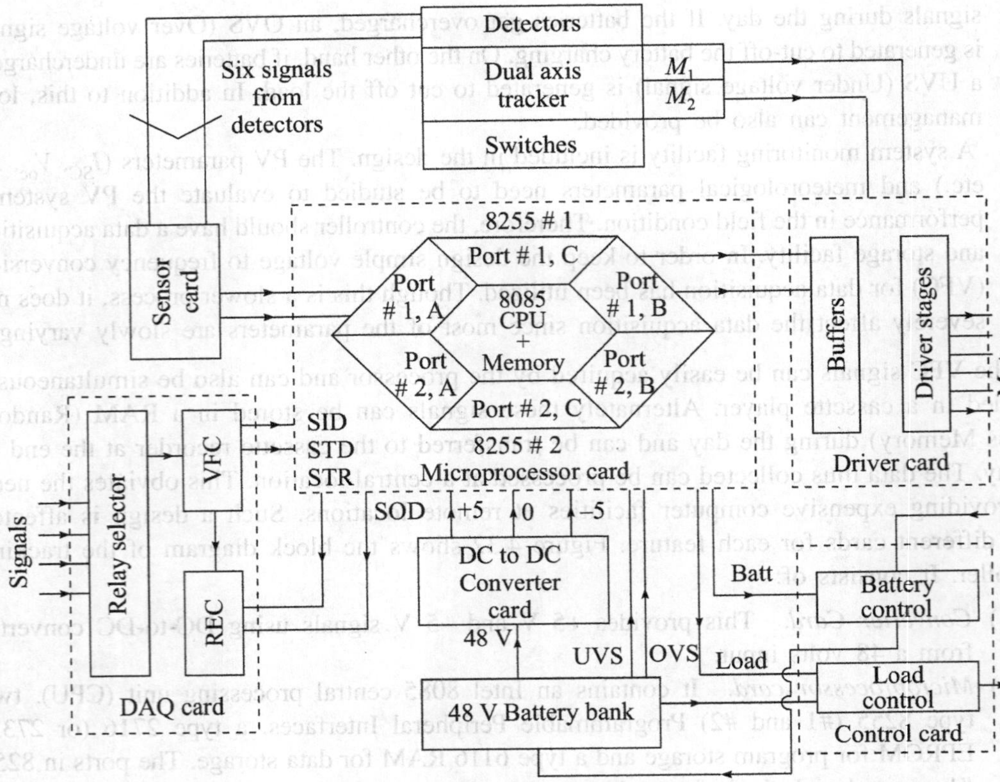

Figure 4.32 Block diagram of the tracking controller showing different cards and signals that were used.

(v) *Control card.* The state of charge of batteries is detected by measuring their voltage. If the voltage is below 10 V (for a 12 V battery), a Schmitt trigger is actuated to give a HIGH on UVS. If the voltage is above 13.2 V, a HIGH is given on OVS. The two signals are indicated from the battery bank in Figure 4.32. Then LOAD or BATT signal from the processor are used by the control card to actuate relays to cut-off the load or input PV power, respectively. If the loads, for example, tube lights at different places, are to be switched off at different times additional signals can be provided to the control card. Each time a signal is received, the controller card can switch on or off a relay to connect or disconnect the load.

(vi) *Sensor card.* The detailed circuitry of the sensor card is given in Figure 4.35. In closed loop mode the sensor card is given six signals (WIND, CLOUD, ALTITUDE REVERSE, ALTITUDE FORWARD, AZIMUTH REVERSE, and AZIMUTH FORWARD) from the detectors mounted on the tracker for AUTO motion. MANUAL motion is done using front panel switches. In open loop mode, the sensor card is eliminated, since the tracker is brought to a pre-calculated position depending on the month and time of the day. The signals from front-panel switches are directly connected to the microprocessor card.

Design of controller software

The details of controller software for AC motors in closed loop mode are given in [21]. Here the software for stepper motors for both closed loop and open loop modes are described.

The Ports A, B and C in 8255 #1 are used by the controller for monitoring the tracking system. Port A and Port C (upper nibble) are initialized as Input and Port B and Port C (lower nibble) are Output ports. The error signals from the tracking system for azimuth and altitude motion are converted to TTL signals by means of the sensor card. The interpretation for different bits is indicated in Figure 4.33.

In closed loop mode these bits are generated as error signals from the detector/manual switches. However, in open loop mode, these bits are generated by the front panel switches for manual operation only. The end positions of the tracker in azimuth and altitude directions are given to the input on Port C to invoke different programs when any of the bits goes HIGH. The stepper motors require four steps signals (90° out of phase). These are generated at Port B to run the motors through driver cards. The upper two bits (GAL and GAZ) are used to provide the step signals to either the azimuth or the altitude motor. This way several trackers can be driven simultaneously by adding more driver cards in the controller. Additional gate signals can be provided on Port C (lower nibble).

Port #2, B is identical to Port #1, B and can be used to drive additional trackers. In fact by providing PWM signals on one of the B Ports and step signals in the other Port, trackers having different types of motors can also be operated.

Port #2, A and Port 2#, C are also interpreted as input ports. Port A is used in open loop mode to provide month (MN) and Time (*T*) signals in the manual motion. For instance if the tracker is stopped for maintenance and is to be restarted, MANL bit of Port #1, A is set HIGH. The software then acquires on Port #2, A the starting month and time to determine which of the DELAY WORDS is to be used. Then the motion is started under AUTO mode by making MANL low. Port #2, C is used for Data acquisition and control. The OVS and UVS from the battery bank are provided to detect the state of charge of batteries. If OVS is HIGH, BATT is set to cut-off the PV array from the battery bank. On the other hand, if UVS is high, LOAD is set to cut-off the load to prevent draining of the batteries. A STart Data acquisition bit is provided to the data acquisition card to start the timer. The VFC signal is given to the input on SID pin and the count number is stored in the RAM. A STart Recording (STR) bit starts the cassette recorder for data storage at the end of the day. The data stored in the RAM is serially sent to the output on the SOD pin for recording.

Figure 4.34 shows the memory map of the software consisting of different modules. The entire software is stored in a 2KB EPROM. The program area can be increased to 4KB by changing a jumper for 2732 in the microprocessor card. INIT initializes the ports and memory. STPGEN is used for step generation for running the stepper motors. The speed of the motors can be changed by changing the delay between the steps. In closed loop mode, INSIG is used to acquire the tracker status and then program for azimuth forward/reverse or altitude forward/reverse motion is executed accordingly. In open loop mode, OPLOOP checks if the motion is to be done in AUTO (Automatic) or MANL (Manual) mode.

Port # 2, C:

OVS	UVS	X	X	STR	STD	LOAD	BATT

Port # 2, B:

GAL	GAZ	X	X	ST4	ST3	ST2	ST1

Port # 2, A:

MN4	MN3	MN2	MN1	T4	T3	T2	T1

Port # 1, C:

VERT	HORZ	DAY END	HOME	X	X	X	X

Port # 1, B:

GAL	GAZ	X	X	ST4	ST3	ST2	ST1

Port # 1, A:

WIND	X	MANL	X	ALTR	ALTF	AZR	AZF

Figure 4.33 Bit assignment in different ports in 8255 # 1 and #2.

(B) OPEN LOOP

Address	Module
0800H	
	DATAQ
0700H	
	CONTROL
0600H	
	DLYWAD
0100H	
	OPLOOP STPGEN INIT
0000H	

Address	Module
0800H	
	DATAQ
0700H	
	CONTROL
0600H	
	UNUSED
	INSIG STPGEN INIT
0000H	

Figure 4.34 Memory map showing different modules in the software.

The pre-calculated azimuth and altitude values for every 10 minutes on a particular day for all months [22] are stored in the memory. This avoids the complicated software for calculation of actual position without drastically affecting the tracker performance. The tracker is started from HOME position in the morning. It is brought to the next position by moving the tracker for a fixed time. The tracker waits at new position for an interval decided by corresponding DLYWRD. While waiting at the new position, the new data are acquired and the battery is checked.

Discussion

The controller has several features which make it versatile for tracking, system control and monitoring applications. The controller is totally automatic and does not require any operator intervention unless needed. It has been successfully tested using tracking PV system at Leh (freezing conditions) and Jodhpur (desert conditions) in closed loop mode. In closed loop mode, the tracker starts from the HOME position at about 5 a.m. and moves it under CLOUD motion till the sun is out. Thereafter the tracker is moved to follow the sun throughout the day in both azimuth and altitude directions in the closed loop or feedback mode. In the evening, the DAYEND switch is pressed which stops further forward motion. The tracker is slowly brought back to HOME position at night, where it "sleeps" at this position till the motion is started in the morning. In open loop mode, the motion is done as described before. Since the variation in sun position on the days other than the pre-selected day of the month is within 10°, the output power from the PV system is not drastically affected. Using a maximum power point tracker in Delhi, the data for the photovoltaic parameters and meteorological parameters were acquired every 10 minutes. The data are transferred to the audio cassette at the end of the day.

The complete circuit diagram of the sensor card for the azimuth movement of the tracker is shown in Figure 4.35. There is also a similar card for the altitude variation. The signals from one pair of cells placed horizontally on either side of the rod of the sensor are fed to a differential amplifier IC-1 (LM308). The output of IC-1 is given to the Schmitt trigger IC-4 (74LS132) through an inverting amplifier IC-2. The same signal is fed to the other half of the Schmitt trigger without phase inversion through a voltage follower IC-3. If the tracker is rightly positioned, the error signals are zero and none of the Schmitt trigger is actuated. Otherwise one of them will show logic '1' to indicate misalignment in the direction of motion required (azimuth forward/reverse). The output from the other pair is similarly used to generate two more TTL signals to indicate elevation forward/reverse motion.

Apart from "Auto" mode "Manual" motion can also be achieved by means of five toggle switches, namely (MAN, ALTRM, ALTFM, AZRM and AZFM) on the front panel of the controller unit. The four signals for "Auto" and "Manual" motions are given to Exclusive OR gates (74LS86) through two tri-state buffers IC-5 and IC-6 (74LS367), respectively. Depending on the motion required ("Auto" or "Manual" as set by the front panel MAN switch); the gate signal of one of the tri-state buffers is made '0' in order to apply all the four signals to 74LS86. The four outputs of the Exclusive OR gates, for both the azimuth and the altitude, are connected to port A of 8255 of microprocessor card described earlier. To understand this better, the data sheets of Intel 8085 and Intel 8255 should be consulted.

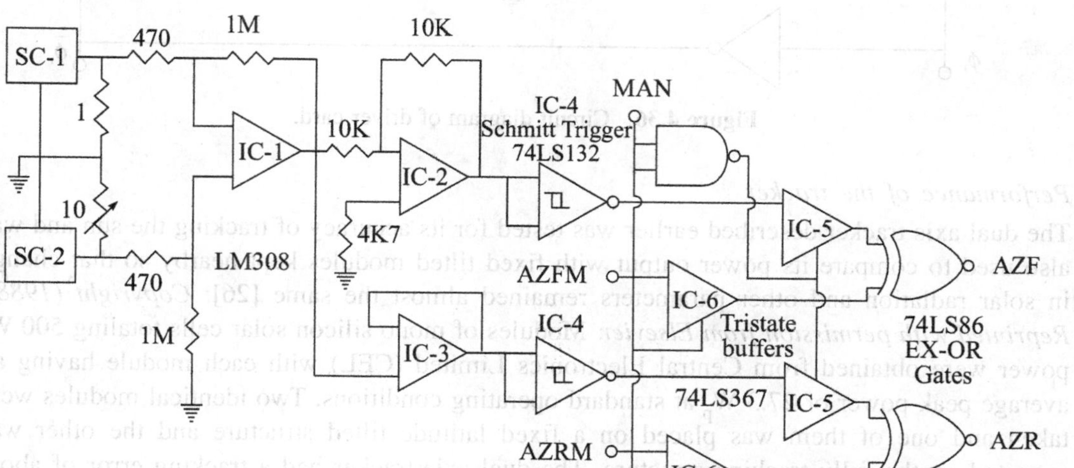

Figure 4.35 Circuit diagram of sensor card (Azimuth movement).

Figure 4.36 shows the circuit diagram of one of the three driver cards needed to drive the three phases of the motor for the azimuth movement. Three more such cards will be required to power the three phase motor for the altitude variation. This is based on the standard full bridge configuration used for pulse width modulated sinusoidal signals [23]. The Darlington pairs ECP055/BD140 and ECN055/BD139 give considerable current gain to the input signals ϕ_1 and $\bar{\phi}_1$ from the microcontroller 8085. The two pairs conduct diagonally and alternately to drive the phase winding connected between R and R-. The diodes labelled as D protect the transistors from high reverse voltage build up due to switching on the inductive load of the motor windings since snubber circuits are not provided.

Detailed descriptions of the software and hardware can be found in references [24] and [25].

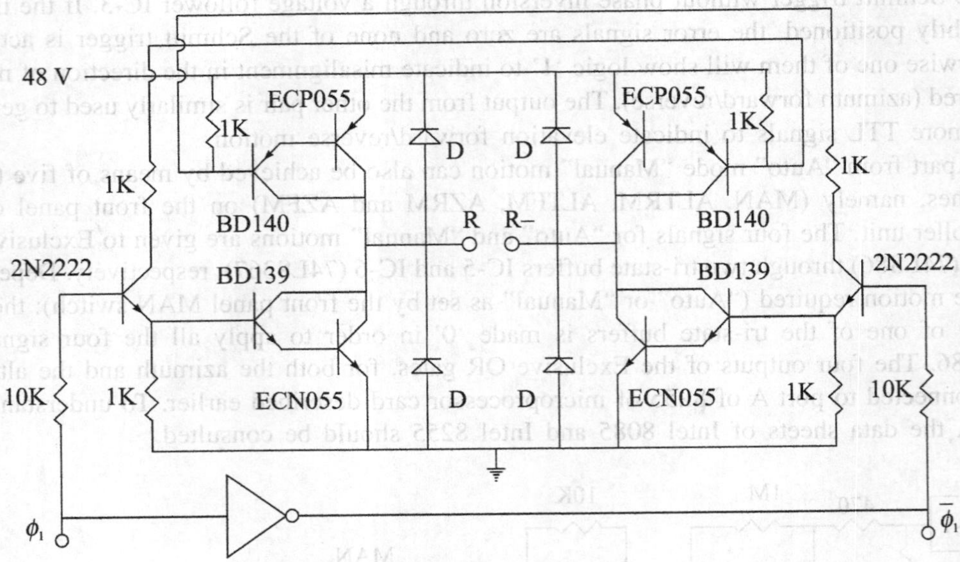

Figure 4.36 Circuit diagram of driver card.

Performance of the tracker

The dual axis tracker described earlier was tested for its accuracy of tracking the sun and was also used to compare its power output with fixed tilted modules kept nearby so that change in solar radiation and other parameters remained almost the same [26]; *Copyright (1988), Reprinted with permission from Elsevier.* Modules of mono silicon solar cells totaling 500 W_p power were obtained from Central Electronics Limited (CEL) with each module having an average peak power of 27.3 W_p at standard operating conditions. Two identical modules were taken and one of them was placed on a fixed latitude tilted structure and the other was mounted on the fully tracking structure. The dual axis tracker had a tracking error of about 0.5°. The photovoltaic energy outputs from both the modules were recorded on a chart recorder at the maximum power point for a few average days of every month. The integrated value of the daily energy output from these modules was calculated. The global solar radiation data on a horizontal surface and both direct and global radiation on the tracking surface were also recorded. The hourly values of the short circuit current, open circuit voltage and current at maximum power point were measured for both the modules. The cell temperature was estimated from the change in open circuit voltage (V_{oc}) and also from the glass temperature of the modules.

The results of these studies are shown in Figure 4.37. The histogram shows that dual axis tracking PV modules produce about 40% more energy in winter as compared to fixed latitude tilted modules. In summer, the global radiation is more than that in winter, still the PV energy output from each module and performance advantage of tracking are less. The major causes for low power output in summer are higher cell temperature and more scattered radiations.

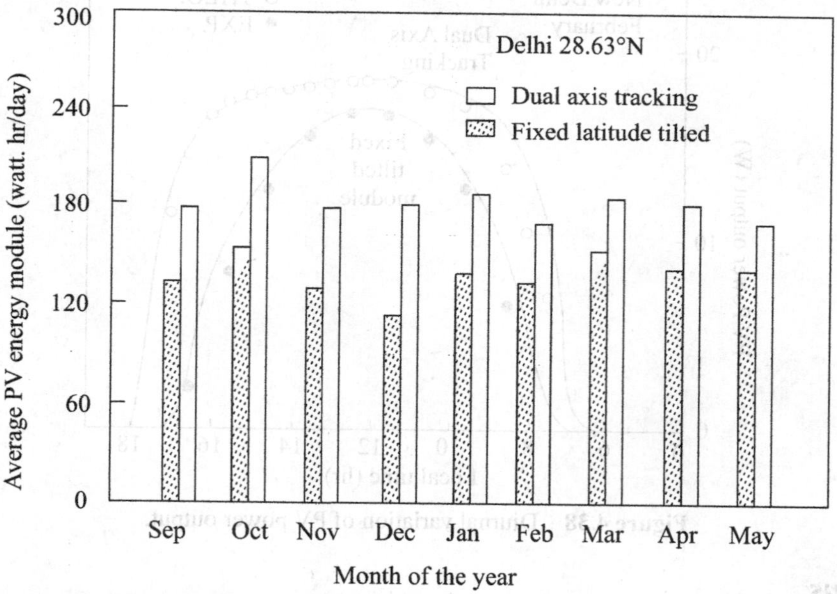

Figure 4.37 Average daily PV energy outputs from tracking and fixed latitude tilted systems.

Figure 4.38 shows the diurnal or daily variation of PV power output from a module of 27.3 W_p rated power. The dots and the circles in the figure indicate theoretical values calculated in reference [26]. The experimental curve and theoretical values of PV power outputs for tracking and fixed latitude tilted modules show some deviation in early and late hours of the day, whereas during midday the results are comparable. The deviation in the experimental and calculated values are due to more scattered light at large incidence angles which is difficult to account for in terms of PV power output. Table 4.3 shows the comparison of experimentally observed and calculated values of daily PV energy outputs averaged over the month from modules mounted on tracking and fixed tilted structures. It has been found that the average annual increase in energy for a tracking module over its fixed tilted version lies between 33 to 37% depending on the site.

Table 4.3 Average daily PV energy output for a module, in watt–hours, of 27.3 W_p power at New Delhi (latitude = 28.63°)

Month (1986)	*Observed*		*Calculated*	
	Fixed latitude Tilted	*Dual axis tracking*	*Fixed latitude Tilted*	*Dual axis tracking*
February	104	141	111	144
March	112	152	120	166
April	108	160	110	164

Similar improvements have been reported by others [27] and [28].

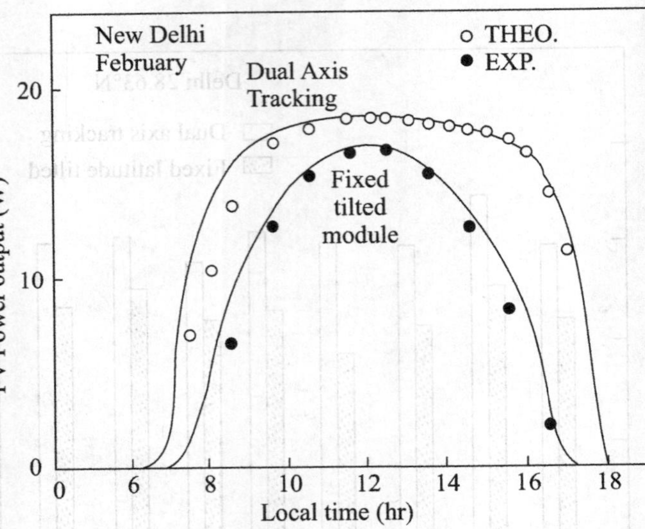

Figure 4.38 Diurnal variation of PV power output.

Questions

1. Draw a diagram like Figure 4.13 when the module is kept at a latitude ϕ in the southern hemisphere.
2. Calculate the shadow length when the module is placed in the southern hemisphere for $\phi = 20°$, $D = 2$ metres and $\beta = 20°$, when the sun is directly over the equator.
3. What should be the optimum tilt angle of the module when there is no diffused radiation? Neglect all reflected light.
4. How can the stacking factor of a module, which has round cells, be improved? For a 100% stacking factor what will be the maximum improvement in power output?
5. Obtain an expression for the distance between two modules, one placed behind the other, when both the modules are raised 50 cm above the ground.
6. What will be the inter-module distance, if the rear module is raised by 50 cm? The other parameters remain the same as in problem 2.
7. Define 'tracking error' of a solar and show its dependence on the amplifier gain.

8. Draw a diagram, similar to Figure 4.25, of North-South tracking for a dual axis tracker.
9. What should be the angle of tilt for a single axis tracker placed in Delhi when the sun is on the tropic of cancer? Take the latitude of Delhi to be 28.7°N.
10. Why is a 3-phase motor used for the trackers? Can a stepper motor replace it?
11. What will happen to the tracking accuracy if a Schmitt trigger is replaced by a comparator without hysteresis?

References

[1] Tapan Bhattacharya, *Terrestrial Solar Photovoltaics*, Narosa Publishing House, New Delhi, 1998, p.76.

[2] A.K. Mukerjee and Vinay Gupta, Communication System using Photovoltaic Power, *Energy Conversion and Management*, vol. **41**, 2000, pp. 551–557.

[3] A.K. Mukerjee, "Comparison of CFL-Based and LED-Based Solar Lanterns", *Energy for Sustainable Development*, vol. **11**, No. 3, September 2007, pp. 24–32.

[4] H.P. Garg and Sat Narain Garg, Model Evaluation and Optimum Collector Slope for a Tropical Country, *Energy Conversion and Management*, vol. **21**, 1981, pp. 299–312.

[5] S.P. Sukhatme, *Solar Energy–Principles of Thermal Collection and Storage*, second ed., Tata McGraw Hill, New Delhi, 1997.

[6] Anna Mani, *Handbook of Solar Radiation: Data for India 1980*, Allied Publishers Limited, New Delhi, 1981.

[7] Roger Messenger and Jerry Ventre, *Photovoltaic Systems Engineering*, CRC Press, New York, 2000.

[8] *Stand-Alone Photovoltaic Systems: A Handbook of Recommended Design Practices*, Sandia National Laboratory, Albuquerque NM, USA, 1995.

[9] John A. Duffie and William A. Beckman, *Solar Engineering of Thermal Processes*, John Wiley, New York, 1991.

[10] http://tel.archives-ouvertes.fr/docs/00/18/13/94/PDF/2721.pdf.

[11] C.F. Gay, J.W. Yerkes and J.H. Wilson, Performance Advantages of Two Axis Tracking for Large Plate PV Energy Systems, *Proceedings of 16th IEEE Photovoltaic Specialists' Conference*, San Diego, California, U.S.A., 1982.

[12] D. Esteve, G. Vialaret, K. Achaibou and D. Follea, Design of a PV Power Array with Concentrated Sunlight, *Proceedings of First E.C. Photovoltaic Solar Energy Conference*, Luxembourg, 1977.

[13] P.O. Carden, Steering a Field of Mirror using a Shared Computer Based Controller, *Solar Energy*, vol. **20**, 1978, p. 343.

[14] T. Kerner, Sun Tracker Control Systems, *Int. Jet Propulsion Lab's Space Program*, Summary 37–34, vol. **4**, 1965, p. 52.

[15] H.L. Teague, C.T. Brown, C.J. Swafford and G.C. Lewis, Identification of Alignment and Tracking Errors in the Open Loop Time Based Heliostat System of 400 kW advanced Component Test Facility, *Proceedings of ISES, Silver Jubilee Congress*, Georgia, U.S.A., vol. **II**, 1979.

[16] S. Vasilevskis, Large Telescope Design Criteria, in *Telescope Design*, Academic Press, New York, 1966.

[17] M.C. Bhatnagar, J.C. Joshi and A.K. Mukerjee, Determination of Tracking Error in an Automatic Sun Tracking System, *Solar and Wind Technology*, vol. **4**(3), 1987, pp. 399–403.

[18] R.E.I. Tolbert and J.C. Arnett, Design, Installation and Performance of ARCO Solar Photovoltaic Plant, *Proceedings 17th IEEE PV Specialist Conference*, 1984, pp. 1149–1152.

[19] Ashok Kumar Saxena and V. Dutta, Performance Study of Solar Cell Modules at Leh, Delhi and Jodhpur, *Renewable Energy for Rural Development Proceedings 89*, Tata McGraw–Hill, New Delhi, 1985, pp. 479–484.

[20] J.C. Joshi, M.C. Bhatnagar, A.K. Saxena and V. Dutta, Study of Power Generation by PV Tracking System in *Recent Trend in Electrical Energy System* (J. Nanda and D.P. Kothari, Eds.) Chapter 6, Prentice-Hall of India, New Delhi, 1988, pp. 241–246.

[21] V. Dutta and Ashok Kumar Saxena, A Microprocessor Based Two Axis Tracking Controller, *Advances in Solar Energy Technology*, vol. **1**, (Edited by W.H. Bloss and F. Pfisterer), Pergamon Press Oxford, 1987.

[22] R. Walraven, Calculating the Position of the Sun, *Solar Energy*, **20**, 1978, pp. 393–397.

[23] Ned Mohan, Tore M. Undeland and Willam P. Robbins, *Power Electronics*, 3rd ed., John Wiley and Sons, INC., 2004, p. 212.

[24] Ashok Kumar Saxena and V. Dutta, A Versatile Microprocessor based Controller for Solar Tracking, *Conference Record of the 21st IEEE Photovoltaic Specialist's Conference*, 1990, pp. 1105–1109.

[25] Ashok Kumar Saxena, Design and Performance Study of Microprocessor Controlled Autonomous Tracking Photovoltaic System, Ph.D. Thesis, Centre for Energy Studies, Indian Institute of Technology, New Delhi 1991.

[26] Mukesh Chander, K.L. Chopra, J.C. Joshi and A.K. Mukerjee, Comparative Study of Different Orientations of a Photovoltaic System, *Solar and Wind Technology*, vol. **5**(3), 1988, pp. 329–334.

[27] Tapan Bhattacharya, op. cit., p. 106.

[28] Roger Messenger and Jerry Ventre, op. cit., p. 34.

CHAPTER 5

Balance of System (BOS)

Batteries

The primary purpose of a battery is to generate electricity by reaction of chemicals stored inside it. The need for powering instruments from toys to satellites is so strong that till the mid nineties their production rose to 25 billion annually. Now it is estimated that the battery market is approximately \$50 billion annually with a growth rate of 6% per annum. For the photovoltaic industry the availability of efficient, light and cheap batteries is crucial for the storage of charge, generated during the day, for use during sunless hours. Somehow, the present high cost of charge storage is inhibiting the exploitation of the PV power at night. For large charge storage the cost of the battery bank itself is comparable to that of the PV array so much, so that the power produced at night is economically unviable.

5.1 Cells and Batteries

The fundamental unit of electricity generation by chemical reactions is called a *cell.* When two or more cells are connected in series the combination is called a *battery.* The cells are usually connected in series, not in parallel. When cells are connected in series, the worst case scenario would be that the battery would go dead, if one cell in the string fails but on the other hand a parallel combination can discharge all the cells under certain conditions. The following explanation will clarify this point further. Consider Figure 5.1(a), where a string of four cells is connected to give the requisite voltage to the load. This series combination which gives a current I at a certain voltage will continue to do so until one or more of the cells are discharged. At this stage the voltage will be lower than the starting value. In case one of them is damaged the voltage will reduce substantially with considerable lowering of the current.

When two strings of cells are connected in parallel, as depicted in Figure 5.1(b), then the situation becomes complex. If the cells in the first string, that is C_{11} etc., are discharged more than the ones in the second string, the total string voltage of C_{11}–C_{41} will decrease with the result that the current I_2 generated by the second string will be divided with one portion I'_2

going through C_{11} to C_{41}. This way only a fraction of the current generated by the second string will actually go to the load while the string itself will discharge faster. The same mechanism will occur if the second string comprising C_{12} to C_{42} has a lower voltage than the C_{11} to C_{41} string. To stop the currents flowing through each other two diodes D_1 and D_2 of the required current capacity are placed in series with each string as shown in Figure 5.1(c). Since finite currents will flow through the diodes some power will be dissipated, but the protection the diodes provide against discharge of the cells will be worth it. For higher current it is advisable to employ batteries of higher charge capacity, usually measured in ampere-hours (Ahr), instead of using the protection diodes.

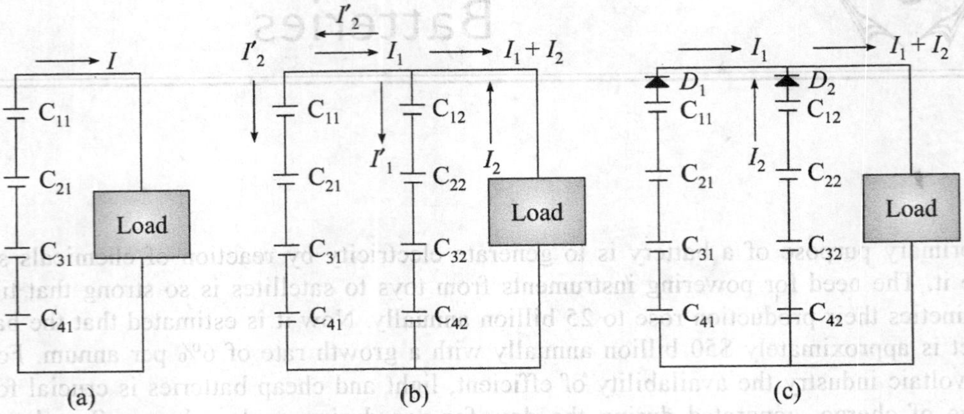

Figure 5.1 Cells put in series and parallel.

5.2 Primary and Secondary Cells

Primary cells are the ones that cannot be normally charged but are discarded after use. In a way, they are wasteful but are used in large numbers in electric torches, toys, remote controls of televisions sets and other portable equipment. Some advantages of these devices are their reasonably long shelf life, a reasonable cost, are disposable after one time use, easy to carry and do not need any maintenance. *Secondary cells* on the other hand have the distinct advantage of being rechargeable so that they can be put to multiple uses and thus, avoid wastage in terms of material. Some of the uses they are put to are: in automobiles, in inverters for standby power during grid power failures, in photovoltaic systems for terrestrial and space applications, in consumer electronics such as computers and a large number of other applications including toys.

The practice of recharging primary cells/batteries must be discouraged as they are not designed for that purpose. Although some manufacturers print a warning note in this regard on the cells, enthusiasts are tempted to do so. The batteries are properly sealed during production so that the chemicals inside do not leak out. If any recharging is attempted an inevitable pressure will build up inside the battery, from the gases that are usually produced, leading to bursting of the outer casing and causing severe personal injury or damage to equipment.

Technically some primary batteries like zinc/alkaline/manganese dioxide system are designed for recharging. There are some other batteries too that can be charged for a few cycles under controlled conditions at very low charging rates but these are unable to deliver back a large amount of the same charge. Also these might not retain the charge for a reasonably long period so that they cease to be of any practical use. Such observations lead to the conclusion that for reuse only secondary batteries should be considered.

The most common primary cell in use for the last one hundred years is the Leclanche cell. This cell, with its modifications, is used in most of the portable instruments like the flash lights, portable radios and similar products where recharging is not needed. It has a large energy density, is light and can be discarded after use. Figure 5.2 shows a vertical section of such a cell. A carbon rod with a metal cap, that acts as the cathode, is surrounded by manganese dioxide mixed with carbon powder and is separated by a jacket from the electrolyte ammonium chloride. The whole system is then encased in a zinc cylinder which acts as the negative electrode or anode of the cell. The load, like a bulb for instance, is connected between the anode and the cathode for use. Chemical reactions inside the cell are complicated and several equations are cited for different rates of discharge. Only the one for prolonged discharge is given here.

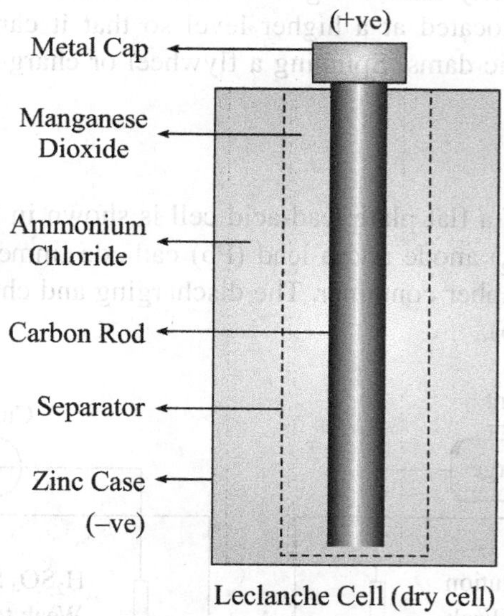

Figure 5.2 Vertical cutaway view of a Leclanche cell.

$$Zn + 6MnOOH \longrightarrow 2Mn_3O_4 + ZnO + 3H_2O \quad (5.1)$$

It is to be noted that this is a non-reversible equation and hence the cell cannot be recharged. Leclanche cell is now facing a stiff competition from another primary cell which has replaced the former from many applications but still it cannot be a total replacement because of its comparatively higher cost. This is the *Alkaline–Manganese Dioxide* cell. Apart from the alkaline cell there are other primary cells: Magnesium and Aluminium cells, Mercuric oxide cells, silver oxide cells, Zinc/Air cells, Lithium cells and solid–Electrolyte cells. A selection of cells exists, each with its merits and demerits, for a particular use which the user has to find out. A comprehensive discussion on batteries is given in David Linden's handbook [1]. Although the primary cells have their area of use they are not suitable for use in PV systems where storage is of primary concern. For multiple use recharging capability is mandatory and therefore only the secondary cells are worthy of consideration. These cells are covered by the Bureau of Indian Standards (BIS) standard IS 586 : 1976. Leclanche type dry batteries for telecommunications, signalling and general purposes.

5.3 Lead Acid Cell

Only three types of cells will be considered here and compared for their cost, charge to weight ratio and shelf life. They are the nickel-cadmium/metal hydride, lithium cells and the lead-acid cells. The most common storage cell in the secondary category is the lead-acid cell which owes its mass production largely to the automobile industry that includes heavy traction machinery. The inverter industry, including uninterruptible power supplies (UPS) for computers, also has a large demand for this product. Because of their high cost other sources of storage are being explored and used. Hydrogen is produced by electrolysis of water and stored in tanks for generating electricity later using fuel cells. Another way is to operate a pump to store water in a reservoir located at a higher level so that it can then be used to generate electricity as in hydroelectric dams. Spinning a flywheel or charging large chemical cells are other available options.

5.3.1 Construction

The general construction of a flat plate lead-acid cell is shown in Figure 5.3. A fully charged cell has a lead oxide (PbO_2) anode and a lead (Pb) cathode immersed in a bath of sulphuric acid stored in a hardened rubber container. The discharging and charging processes are shown in Figures 5.3(a) and 5.3(b).

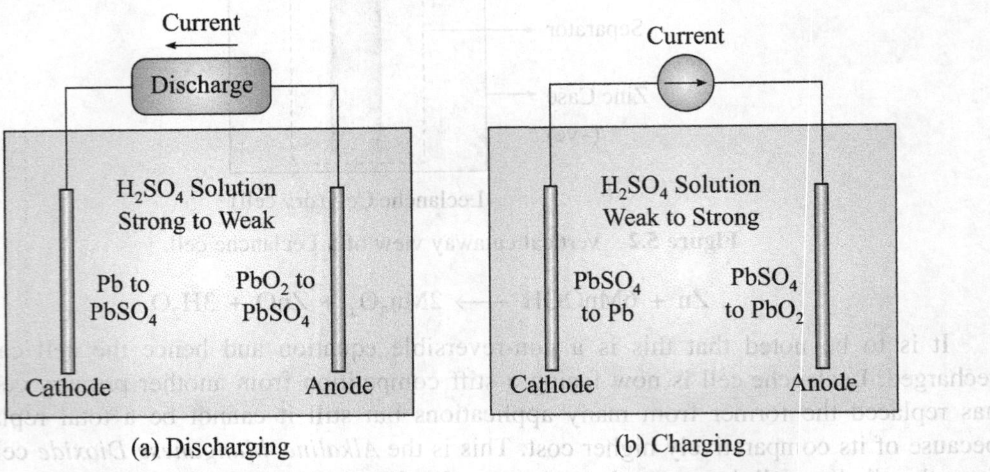

Figure 5.3 Discharging and the charging processes of a lead-acid cell.

If a fully charged cell is taken and discharged through a load, the chemical reaction is the following:

For discharging

at cathode: $$Pb + SO_4^{2-} \longrightarrow PbSO_4 + 2e^- \quad (5.2)$$

at anode: $$PbO_2 + 4H^+ + SO_4^{2-} + 2e^- \longrightarrow PbSO_4 + 2H_2O \quad (5.3)$$

The equations show that lead oxide at the anode turns to lead sulfate with water generated as a byproduct which dilutes the sulphuric acid solution. Similarly, the cathode also produces lead sulfate from the lead while releasing two electrons. These electrons pass through the load and reach the anode to complete the reaction there. Since the direction of current flow is opposite to that of the electrons the current appears to flow from anode to cathode. Clearly, the process of discharge not only produces water to make the electrolyte less acidic but also coats both, the anode and the cathode, with lead sulfate to hamper further reaction. Consequently a discharged cell does not give enough current output leading to a drastic voltage drop on being connected to a load and is termed a *dead cell.* On remaining sulfated over prolonged periods the plates are permanently damaged and no amount of charging thereafter can reverse the damage. For this reason a cell should always remain fully charged by being charged after regular intervals to remove the layer of lead sulphate from the plates. Also on remaining disused over extended periods the cells have a tendency to self discharge internally leading to sulphate formation. Following this discussion, it is only proper to look into the chemistry of charging.

A DC power source, like a PV generator or a laboratory power supply, is connected to the terminals of the cell to drive a current into the anode as indicated in Figure 5.3(b). The current induces chemical reactions inside the cell to reverse the chemical balance to the former state that is before discharging. The *lead sulphide* at the anode changes to lead oxide while that at the cathode becomes lead again and at the same time the strength of the sulphuric acid also increases.

For charging

at anode: $$PbSO_4 + 2H_2O \longrightarrow PbO_2 + 4H^+ + SO_4^{2-} + 2e^- \quad (5.4)$$

at cathode: $$PbSO_4 + 2e^- \longrightarrow Pb + SO_4^{2-} \quad (5.5)$$

These reversible equations can also be written for both the mechanisms as:

at anode: $$PbO_2 + 4H^+ + SO_4^{2-} + 2e^- \rightleftharpoons PbSO_4 + 2H_2O \quad (5.6)$$

and at cathode: $$Pb + SO_4^{2-} \rightleftharpoons PbSO_4 + 2e^- \quad (5.7)$$

Of course H^+ and SO_4^{2-} ions are supplied by the sulphuric acid electrolyte. Because of natural evaporation or gas formation the electrolyte level reduces exposing some of the electrode areas thereby reducing their effective area of reaction. This enhances the chemical reaction of the immersed part of the electrode leaving the exposed part dormant causing the former to weaken by excessive reaction. To guard against such situations distilled water is added to the cell at regular intervals to top up the level of the electrolyte.

The potential difference between the two terminals of a fully charged cell is 2.12 V. When a large number of cells are added in series to give a higher voltage the battery voltage increases in the multiples of this value. Normally three or six cells are serially connected to give 6 or 12 V at the terminals. The area of the electrodes determine the current capacity of a cell and although a large number of cells may be added in a chain to increase the output voltage, the current rating of the battery remains the same as that of an individual cell.

Figure 5.4 gives a cutaway view of a lead-acid battery made by Power-Sonic Corporation. Grids are cast from a lead-calcium alloy free of antimony. The small amount of calcium and tin in the grid alloy imparts strength to the plate and guarantees durability even in extensive cycle service. Lead dioxide paste is added to the grid to form the electrically active material. In the charged state, the negative plate paste is pure lead and that of the positive lead dioxide. Both of these are in a porous or spongy form to optimize surface area and thereby maximize capacity. The heavy duty lead calcium alloy grids provide an extra margin of performance and life in both cyclic and float applications and give unparalleled recovery from deep discharge.

Separators isolate one electrode from another. These are made of non-woven glass fibre cloth with high heat and oxidation resistance. The material further offers superior electrolyte absorption and retaining ability, as well as excellent ion conductivity. The case sealing, depending on the model, is ultrasonic, epoxy or heat seal. The container case and lid material is ABS, high impact, resin with high resistance to chemicals and flammability. Case and cover are made of non-conductive ABS plastic to UL94-HB or UL94 V-O specifications. This case has moulded-in dividers for each 2 V cell.

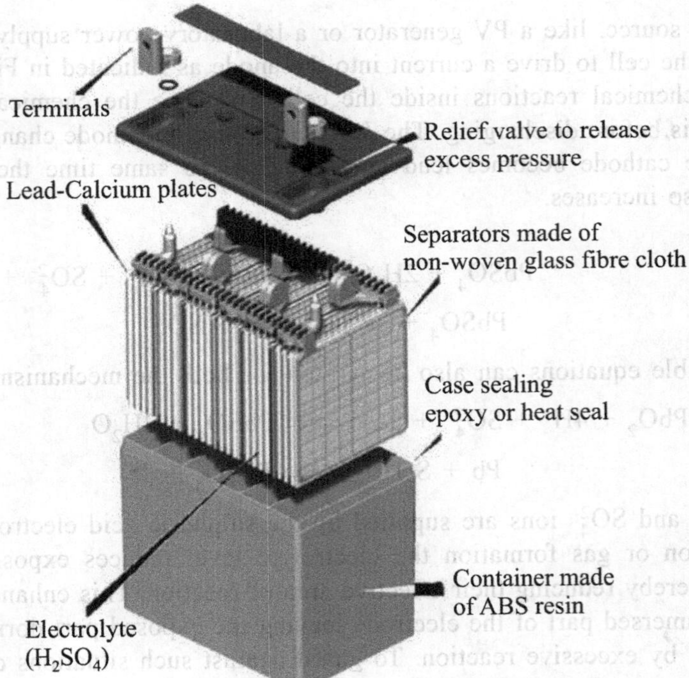

Courtesy: Power Sonic Corporation [11]

Figure 5.4 Internal arrangement of a lead-acid battery [11].

5.3.2 BIS Standards on Lead-Acid Batteries

Battery charging process is not arbitrary but is regulated by certain norms across the world. In India the BIS has laid down the specifications for these batteries in its document numbers:

1. IS 1145 : 1980 Lead-acid storage batteries for motor cycles, auto-rickshaws and similar vehicles.
2. IS 8320 : 2000 General Requirements and Methods of Tests for Lead-acid Storage Batteries.
3. IS 6848 : 1979 Specification for Lead-acid Batteries for Train Lighting and Air conditioning Services.
4. IS 7372 : 1995 Lead-acid storage batteries for motor vehicles (first revision of IS 7372)–Specification.
5. IS 7660 : 1988 Lead-acid batteries for electric locomotives and electrical multiple units.
6. IS 7624 : 1990 Lead-acid starter batteries for diesel locomotives and rail cars.
7. IS 9814 : 1981 Lead-acid storage batteries for marine use.
8. IS 12013 : 1987 Lead-acid batteries for 110 V train lighting system with monobloc containers.
9. IS 12950 : 1990 Battery hydrometer portable syringe type for lead-acid batteries.
10. IS 13568 : 1992 Lead Acid Light Weight Storage Batteries for Motor Cycles and Similar Vehicles Fitted with AC Circuitry–Specification.
11. IS 14782 : 2000 Code of Practice for Maintenance and Testing of Large Lead Acid Batteries for Generating Stations and Substations.
12. IS 15549 : 2005 Stationary Valve Regulated Lead Acid Batteries–Specification.
13. IS 13369 : 1992 Stationary Lead Acid Batteries (with tubular positive plates) in monobloc containers.

It is clear from the list given above that for different applications separate standards must be followed. The last standard (13) is old and therefore withdrawn but it gives the fundamental requirements of a battery. Among other features the material of the container, the concentration and preparation of the sulphuric acid-water electrolyte, the ampere-hour rating, constructional requirements and dimensions are specified. Interestingly the recommended discharging rate is $C/10$ and the final voltage of discharge of an individual cell is 1.8 V.

When a cell is charged beyond its maximum capacity, gasification occurs. During charging some of the hydrogen ions combine with the free electrons to form hydrogen gas. When the process of charging is over, no more lead sulphate is available for conversion to lead at the cathode. If further charging continues the electrons continue to stay in the electrolyte and combine with hydrogen ions to form gaseous hydrogen. To avoid production of the gas, charging must be stopped or reduced to float charging levels. Gas formation is harmful to batteries in general but particularly so in the case of sealed maintenance free types which might explode.

The charge stored in a cell is measured in ampere-hours and is a direct indicator of the energy. The energy stored in a cell is the product of the charge and the cell voltage referred to in watt-hours. For example, the energy stored in a 7 Ahr, 12 V battery is simply 84 Whr.

The charge capacity of a battery is usually referred to as C. Now, if the battery is discharged fully in time t hours, the discharge rate is defined as C/t. Thus, if a battery is fully discharged in 10 hours, as indicated above, the discharge rate is $C/10$. Similarly, for charging purposes $C/10$ rate means that the battery is fully charged in 10 hr.

Example 5.1: What should be the current setting in a battery charger for 12 V, 7 Ahr battery used in an uninterruptible power supply for a computer?

Solution: Since the stipulated charging time is 10 hr, the charging current is 7 Ahr/10 hr = 0.7 A. That means the calculated current value will give the battery a longer lifetime.

A shorter charging time is often needed in practice but there are hidden costs for such a gain in time. It must be remembered that the reaction inside a cell can be construed as an internal resistance R called Thevenin's resistance in electrical engineering and shown in Figure 5.5. Although its value is small there is an appreciable voltage drop across it when the current is large. Also the heat generated by such a resistance is I^2R W, which is instrumental in raising the temperature of the cell. In effect the entire charge (current) driven into the cell is not stored but a part of it is wasted as heat. The only benefit that can be obtained in the process is slightly higher charge storage because of increased temperature.

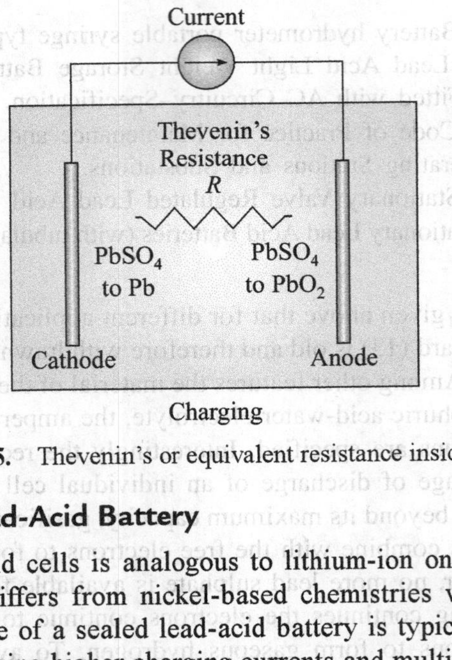

Figure 5.5. Thevenin's or equivalent resistance inside a cell.

5.3.3 Charging the Lead-Acid Battery

Charging mode for lead-acid cells is analogous to lithium-ion ones, where voltage is fixed after a certain stage, but differs from nickel-based chemistries which use current limiting throughout. The charge time of a sealed lead-acid battery is typically ten hours or more for higher capacity batteries. Using higher charging currents and multi-stage charge methods, the charge time can be reduced to less than ten hours. Lead-acid system has a drawback that it cannot be fully charged as quickly as the nickel or lithium-based batteries. Another disadvantage is that a lead-acid system takes about five times as long to recharge, to the same level as it does to discharge, than its lithium or nickel counterparts. The other two systems have either 1:1 ratio, for nickel-based batteries, or roughly 1:2 for lithium-ion. This advantage has put the lithium batteries in the forefront for use in mobile phones and hand held video cameras.

A charge controller gives a constant current charge to the lead-acid battery until its terminal voltage reaches about 0.7 times its maximum rated voltage that is preset for a certain number of cells in the battery (Section 1 in Figure 5.6). The time taken for this process should be set at about five hours by adjusting the current for a *C*/10 charging rate. From this point, the rate of rise of the voltage slows down, with the consequent decrease in current, since the voltage difference between the output of the charger and the input to the battery reduces. The time taken to reach from 0.7 to full voltage, the voltage corresponding to the fully charged condition, takes another five hours as indicated in Section 2. This action of topping the charge is essential for the well being of the battery. If omitted, the battery would eventually lose the ability to accept a full charge. The battery is supposed to be fully charged when the current drops to 3% of the rated current. Section 3 is where the incoming 'trickle' charge merely compensates the charge lost during self-discharge and is called the *float charge region*. In this case the initial charging current was set at 2.5 A which was reduced to a trickle charge value of 75 mA. Two sets of curves are shown: (*A*) for mains and (*B*) for PV charging. In the first case, the voltage is fixed at a higher value in Section 2 for quicker charging while in the latter case, when PV is used for charging, the voltage is kept at 2.16 V per cell because of ease of implementation and also to avoid gasification which may occur if 2.4 V per cell is preset in Section 2.

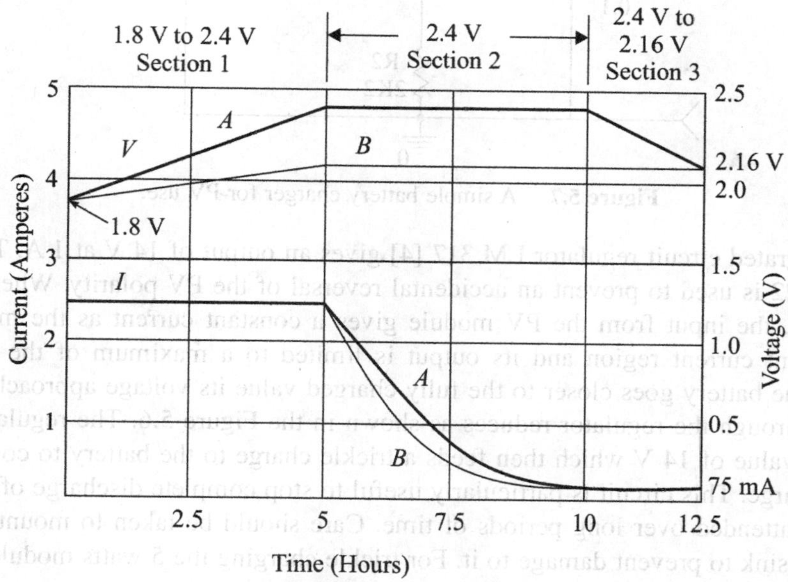

Figure 5.6 Cell charging current and voltage.

In PV systems the *C*/10 rate is not always feasible since the time for charging changes from summer to winter when the full charge of the battery must be replenished within five hours giving a *C*/5 pace. However, if the size of the battery is designed according to conventional practice that ensures half the charge is left inside after the discharge process is over, then the battery is to be charged from the *C*/2 level to full value. Hence for *C*/5 charging rate the ultimate charging rate that is achieved remains at $C/10$. Finally, the entire battery voltage at 2.16 V per cell must lie near the maximum power point of the solar module at least for small systems like the solar lantern. A higher temperature requires slightly lower terminal voltages and vice versa as the voltage limits shift with temperature. Chargers that are exposed to large temperature fluctuations should be equipped with temperature sensors to adjust the charge voltage for optimum charge. A suitable instrument, which can cover a wide range of voltage from 4.5 to 40 V [2], can be made using an integrated circuit UC 3906 specially designed for charging a lead-acid battery. The data sheet of this IC gives the relevant details. Instead of three charging zones shown in Figure 5.6, there are many more which makes the process more elaborate.

The simplest form of a battery charger from PV sources is given in Figure 5.7.

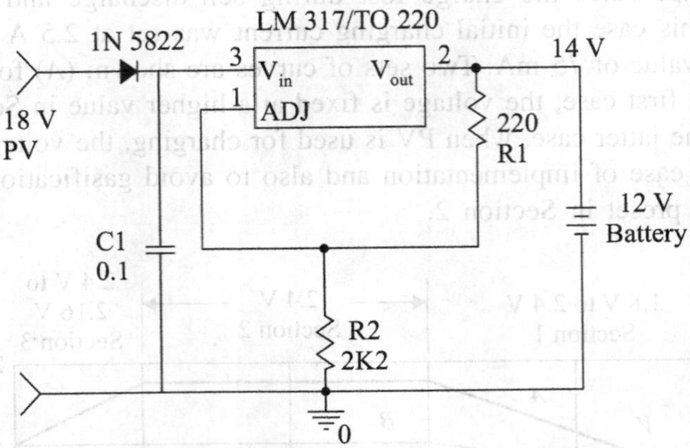

Figure 5.7 A simple battery charger for PV use.

The integrated circuit regulator LM 317 [4] gives an output of 14 V at 1 A. The Schottky diode 1N 5822 is used to prevent an accidental reversal of the PV polarity. When the battery is discharged, the input from the PV module gives a constant current as the module works in the constant current region and its output is limited to a maximum of the short circuit current. As the battery goes closer to the fully charged value its voltage approaches 14 V and the current through the regulator reduces as shown in the Figure 5.6. The regulator then sets a maximum value of 14 V which then feeds a trickle charge to the battery to compensate for its self discharge. This circuit is particularly useful to stop complete discharge of car batteries when left unattended over long periods of time. Care should be taken to mount the IC on a suitable heat sink to prevent damage to it. For trickle charging the 5 watts module is adequate and that can do with a less powerful IC of the same type.

A slightly more advanced version is given in Figure 5.8. In solar systems power has to be conserved wherever possible. The circuit comprising LM 317 consumes somewhat more standby power although it is inexpensive. Since MOSFET based regulators consume less power but are not easily available one has to be constructed with discreet components incorporating the additional feature of temperature compensation so that proper charging can be maintained in all seasons.

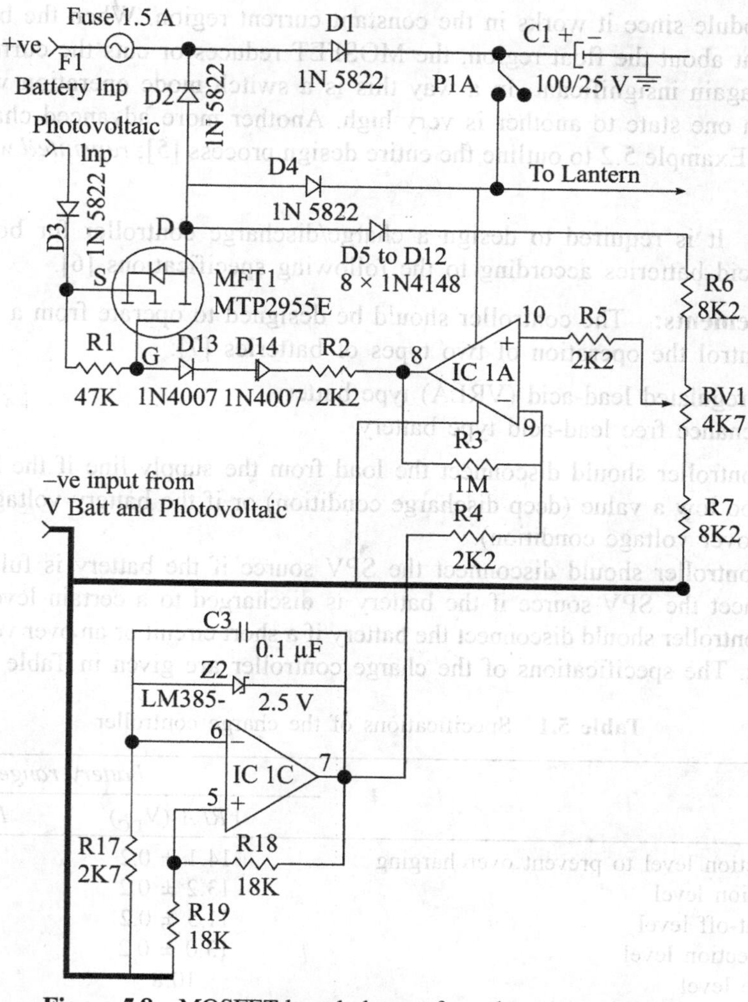

Figure 5.8 MOSFET based charger for solar operated lights.

Circuit diagram of a charger shown here is primarily a conventional feedback regulator with a *p*-channel MOSFET as the series element. Output from the drain *D* is fedback to the inverting amplifier IC 1 A. The reference voltage for this amplifier is given to pin #9 from a stable 5 V supply generated by IC 1C. Both these are parts of a low power quad operational amplifier LM 324. The reference diode is a band gap regulator LM 385–2.5 V situated in the feedback loop of IC 1C. The output voltage of the entire stabilizer is set to float at 14.5 V by the potentiometer RV1 so that the battery receives a float charge voltage of 14.2 V after a drop of nearly 0.3 V across the diode D_2. Eight silicon diodes, D_5 to D_{12}, in the feedback loop of IC 1A offer the necessary thermal correction of 14.6 mV/°C at the battery terminal. Some apprehension might be expressed regarding the power loss due to the linear nature of the circuit. Whereas it is true that switch mode designs waste less power but the situation in this case is different. To start with the MOSFET is in a fully conducting mode similar to that of switching circuit so that there is minimal power loss. The only drop occurs in the solar module since it works in the constant current region. When the battery is close to full charge at about the float region, the MOSFET reduces or cuts the current so that the power loss is again insignificant. In a way this is a switch mode operation where the time to change from one state to another is very high. Another more advanced charge controller is discussed in Example 5.2 to outline the entire design process [5]; *reprinted with permission from Elsevier.*

Example 5.2: It is required to design a charge/discharge controller for both VRLA and flooded lead-acid batteries according to the following specifications [6].

Design requirements: The controller should be designed to operate from a 12 V_{DC} source and should control the operation of two types of batteries [7]:

1. Valve regulated lead-acid (VRLA) type battery.
2. Maintenance free lead-acid type battery.

(a) The controller should disconnect the load from the supply line if the battery voltage falls too low a value (deep discharge condition) or if the battery voltage becomes too high (over voltage condition).

(b) The controller should disconnect the SPV source if the battery is fully charged and reconnect the SPV source if the battery is discharged to a certain level.

(c) The controller should disconnect the battery if a short circuit or an over voltage condition occurs. The specifications of the charge controller are given in Table 5.1.

Table 5.1 Specifications of the charge controller

Parameter	*Battery range*	
	VRLA (V_{DC})	*Lead-acid* (V_{DC})
SPV disconnection level to prevent overcharging	14.1 + 0.2	14.5 ± 0.2
SPV reconnection level	13.2 ± 0.2	12.5 ± 0.2
Battery low cut-off level	11.5 ± 0.2	11.5 ± 0.2
Battery reconnection level	13.0 ± 0.2	13.0 ± 0.2
Deep discharge level	10.8	10.8
Over voltage level	14.4	15.2

All the connections are, of course, automatic.

The required indicators on the controller panel are:

1. Green light emitting diode (LED) to indicate charging from SPV panel.
2. Yellow LED to show battery over charge.
3. Red LED as a warning for low battery voltage.
4. Red LED for battery reverse polarity indicator.

Protections required:

1. Protection against reverse battery polarity to the control unit as well as the load.
2. Reverse current flow protection from the battery to the SPV module.
3. Short circuit protection by a 6 A miniature circuit breaker switch.
4. Battery protection from over charge and deep discharge.
5. Deep discharge and over load conditions are to be set at 10.8 and 15.2 V, respectively.

Principle of operation

The fundamental principle of operation is based on the non-inverting Schmitt trigger circuit [8] shown in Figure 5.9. Assume that the output is negatively saturated. Then, the feedback voltage is negative. This feedback voltage will hold the output in negative saturation until the input voltage becomes positive enough to make the error voltage positive with respect to V_R. To switch the output states, the input voltage has to become negative enough to make the error voltage.

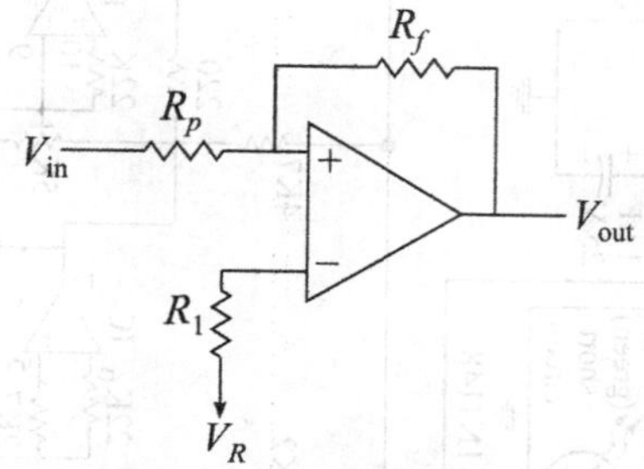

Figure 5.9 Basic circuit of the Schmitt trigger.

Circuit design

The operation of the actual circuit is simple, however, the protections make the entire system elaborate. The four comparators shown in Figure 5.10 are made from a single quad OP-AMP, IC LM 324. To use this for switching from a single power supply, a stable reference is created by the integrated circuit 78L05. The total design concentrates on minimizing power losses since PV power is highly expensive. Power from a solar photovoltaic (SPV) module is given to the input relay RL_2. The coil of this relay switch, C_2, is placed in series with the contact of the reed relay, RRL_1, which is operated by the transistor, T_1. This transistor is operated by the non-inverting Schmitt trigger circuit comprising IC_1 and its associated components. The level is adjusted by a 2K2 Ω preset potentiometer, whereas its hysteresis is controlled by the adjoining preset resistor of 1K Ω. Double poles switch S_1 alters both the switching levels and the hysteresis level of the Schmitt trigger. For charging the Valve Regulated Lead Acid (VRLA) battery, S_{1A} and S_{1B} are kept open, whereas they are closed for charging the lead acid battery.

The non-inverting type circuit has been preferred over the inverting type because of the latter's susceptibility to change of hysteresis levels with the change of switching levels. The green LED glows when the battery is being charged. Once the battery is fully charged, the Schmitt trigger reverses state and the relay switch RL_2 opens. Diode, D_1, prevents reverse polarity of the charge-discharge controller, while diode, D_2, prevents the battery voltage from damaging the SPV module at low intensities of insolation.

The 12 V battery connected after diode D_2 through relay switch RL_4, which is in a normally open state. Switch, S_3, when pressed, during the installation of the system, connects the 12 V battery to the controller part through the diode, D_1. It also energizes coil, C_4, of RL_4 so that its contact closes. The battery supplies power to the load through the polarity reversal prevention diode, D_3, and a short circuit protection Miniature Circuit Breaker (MCB) switch.

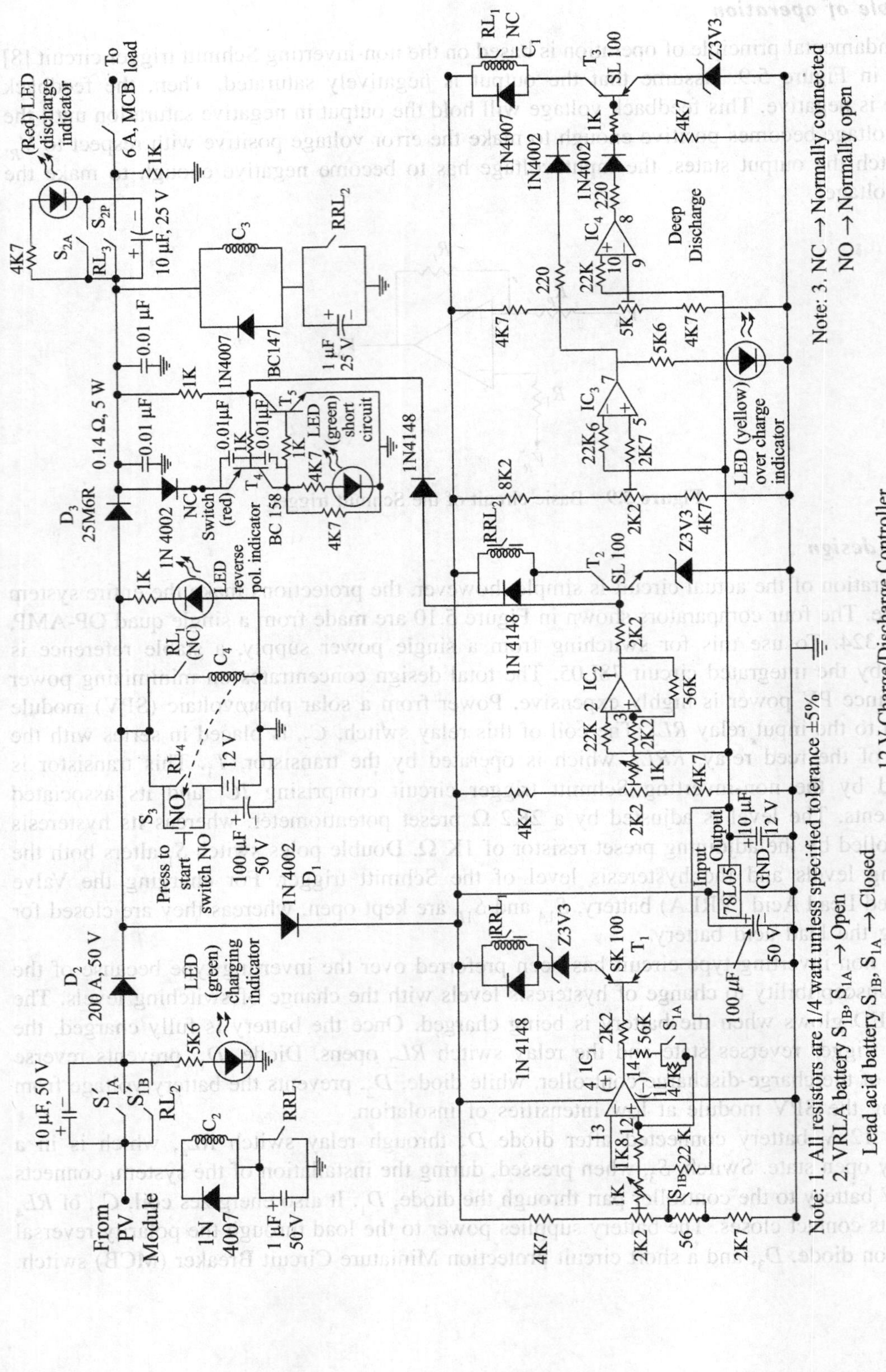

Figure 5.10 Circuit diagram of the battery charge controller.

The discharge level of the battery is determined by the non-inverting Schmitt trigger, connected to IC_2 and its associated circuitry. When the battery voltage falls below the prescribed level, reed relay, RRL_2, opens, thus DC-energizing coil, C_3, of relay switch, RL_3, to isolate the load from the battery.

IC_3 lights the yellow over charge indicator by completely isolating the battery from the circuit or the load by operating relay switch, RL_1, through the transistor, T_3. Similarly, when the battery goes into deep discharge because of some reason or other, IC_3 acts to disconnect the battery.

An electronic short circuit protection prevention circuit is made of transistors, T_4 and T_5, as shown in the circuit diagram. This uses the voltage drop across a 0.14 Ω resistor to operate the relay switch, RL_3, with a simultaneous optical indication from a green LED.

Performance

Since the controller was primarily designed to charge batteries from photovoltaic sources, a solar array was used to test it. The solar array size used in this experiment was made of 25 units of single crystal solar cell modules each of 35 W_P power, made by Central Electronics Limited, India. The specifications of a single module are given below:

1. I_{SC} = 2.5 A
2. V_{oc} = 20.5 V
3. I_{max} = 2.1 A
4. V_{max} = 16.5 V
5. P_{max} = 35 W_p

These were connected in parallel with a Schottky diode in series with each module. A 2 Ω, 100 W, resistor was used as a dummy load. The batteries were fully charged to start with, i.e. before being connected to the PV array. To simulate sunless days, the array was covered until the charging current was negligible. The system worked satisfactorily for about one month as per its design. The battery charging efficiency, defined as voltage × current, ($V \times I$), across the battery terminals divided by $V \times I$ before the relay RL_2, was 85.2%. However, the overall efficiency, when the voltage drop across the Schottky diodes with each module was taken into account, was slightly lower at 82%. A lot of power was lost in the cables connecting the array to the charge controller and in D_2. The efficiency of the battery discharge through the load was more than 90%, but if the battery efficiency is taken into account, the figure deteriorates sharply. The actual ampere hours that flow out of the battery divided by those used to charge the bank yields an efficiency less than 80% (79.4%).

Example 5.3: It is required to make a battery charger to charge a bank of lead-acid batteries from a photovoltaic source. The specifications are:

1. A string of 10 batteries has to be charged from a PV array to a float voltage of 14.5 V for each battery.
2. The circuit must be capable of handling a maximum power of 3 kW from the array, that is, at its maximum power point (MPP).
3. Input voltage: 150 to 210 V_{DC}. That is for V_M = 150 V_{DC}, $V_{oc} \approx 214$ V_{DC}
4. Reverse polarity of input PV must be provided.
5. The circuit should operate only after the cooling fans start working.

6. The output of the charger must be protected against an accidental connection of the battery with wrong polarity.
7. The other specifications remain same as those of Table 5.1.

The circuit diagram of the battery charger is shown in Figure 5.11.

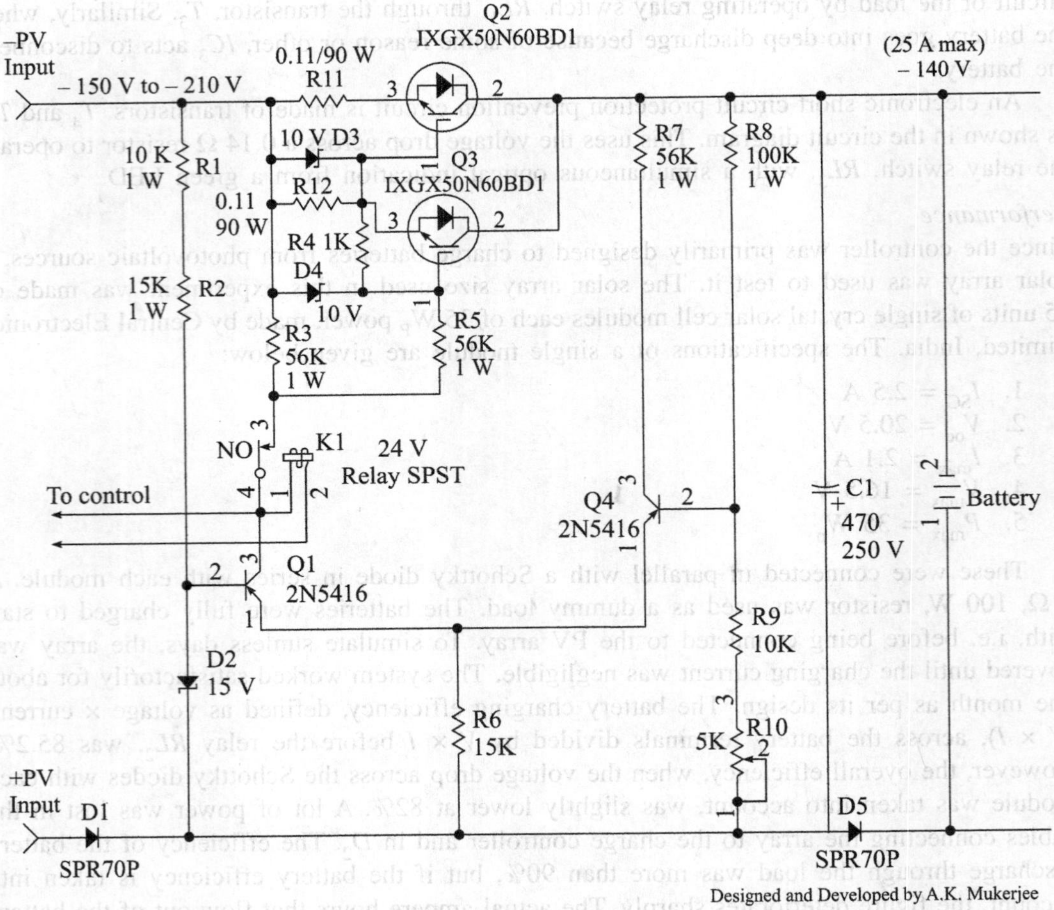

Figure 5.11 3 kW battery charging circuit using a photovoltaic source.

Design procedure

Each battery has to be charged to a maximum voltage of 14.5 V_{DC}, its float value. Therefore, 10 batteries will need a float charge voltage of 145 V_{DC}. Thus, the maximum output of the charger is:

$$V_{max} = 10 \times 14.5 = 145\ V_{DC}$$

When a power of 3 kW is fed to the battery at 145 V, the current that passes through the IGBT series element is

$$I = 3000/145 = 20.689 \approx 20.7\ \text{A}$$

In case of a short circuit, the maximum current of the array can rise to an approximate value of

$$I_{SC} = 20.7/0.7 = 29.6 \text{ A}$$

This means that the IGBTs must be capable of carrying more current than this. The IXGX50N60BD1 device with diode carries a current of 50 A_{DC} at a case temperature of 90°C.

Actually, this is a fast transistor admirably suitable for operation at high frequencies and is therefore expensive. A cheaper version that can work at DC or low frequencies, but with the same current capacity, will do equally well at a lower cost. In addition, this transistor can work up to 600 V across its collector and emitter, which is adequate for the present application since the highest input voltage is 300 V. Two such devices in parallel can deliver a current of 100 A to the battery at 90°C and thus, exceed the ratings by a large amount. As a matter of fact this circuit can be used to charge batteries of double this capacity. In electronic circuits it is always better to work with a 100% over rating as the semiconductor devices are susceptible to damage on slight mishandling.

As in the case of any solar module, this array, used to charge the batteries, goes into an almost constant current mode when the batteries are discharged and so there is no need for any external current control. However, when the batteries are fully charged the voltage of the array should not be allowed to go beyond the stipulated 145 V_{DC}. It can be noticed that the rest of the circuit is nothing more than a simple electronic regulator [9], [10] whose output voltage is controlled by the potentiometer R10 to suit the types of batteries given in Table 5.1. Two 70 amperes rectifiers D_1 and D_5 provide protection against accidental reversal of polarities of the PV array and the batteries respectively. To further protect the circuit against overheating fans are provided to cool the IGBTs and the two rectifiers that are of course mounted on suitable heat sinks. To ensure that the circuit operates only when the fans are operational the fan power supply is made to operate the 24 V SPST relay K1. The relay is normally OFF but the fan power supply energizes it to ON position after the air cooling begins.

The relay K2 in Figure 5.12 can be used instead of K1 of Figure 5.11 since only one of them is sufficient for operation.

The fan power supply is based on the TOP 243 Y switch mode integrated circuit. It is a universal power supply that works both on AC and DC and has an AC input from 90 V_{rms} to 270 V_{rms}. A bridge rectifier following an EMI filter converts it into DC. Alternately a DC source, that can give 125 V to 370 V, can be attached to the input without any problem to feed the IC. After initial filtering by capacitor C_2 the DC is fed to pin #7 of the IC through a high frequency transformer T_3. The snubber elements C_3, R_1 and D_3 protect the IC against overvoltage spikes at the drain of the MOSFET at pin #7. The secondary winding of T_3 gives a train of pulses to the fast recovery rectifier MUR 860 for rectification and subsequent filtration by capacitor C_8 to produce a constant voltage of 24 V at the output. 24 V is maintained by a feedback mechanism involving an opto-coupler 4N35 which also isolates the output from the input which is at a high voltage. The use of the transformer also ensures galvanic separation between the two sides.

After the required voltage level is attained at the output, the relay K3 is activated which in turn sends power to the fans and the coil of the relay K2. On closure of the K3 switch

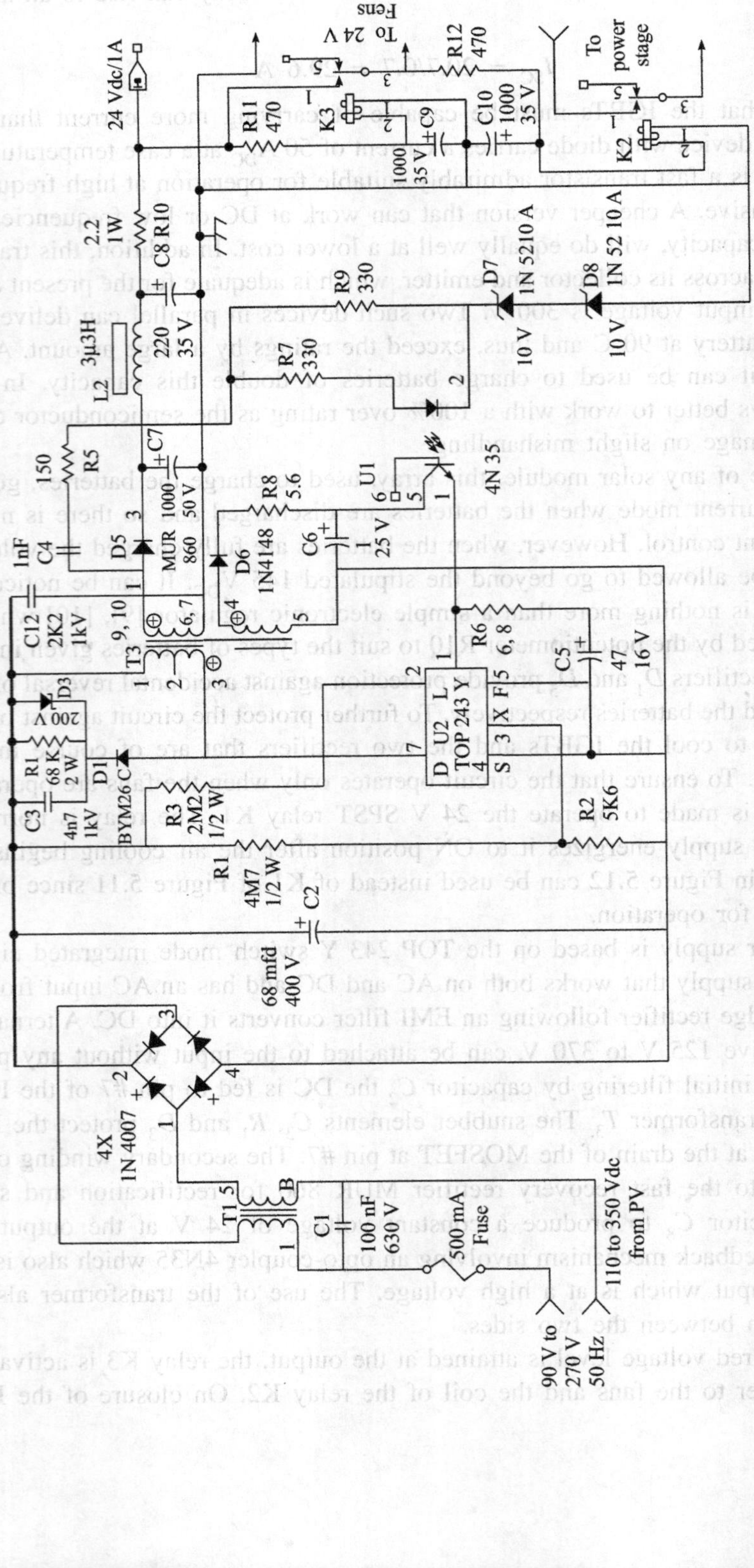

Figure 5.12 24 volts power supply to operate cooling fans.

the battery charger is rendered operational. This interlocking of the two circuits makes sure that only when the fans start cooling the battery charger begins to work without being in the danger of getting overheated.

5.3.4 Properties of Lead-Acid Batteries

The charging and discharging efficiencies of a battery are not 100% as would be expected of a new device. Instead both may be about 95% resulting in an overall cycle efficiency of approximately 90%. The efficiency of charging is defined by the following equation. For a new battery it is 95%.

$$\eta = \frac{\text{AH Discharged after fully charged}}{\text{AH Delivered to battery during full charge}} \tag{5.8}$$

It was mentioned earlier that a Thevenin's resistance is always present which can be considered the internal resistance of a battery. Both the charging and the discharging current produce heat whose magnitude depends on the actual value of the current I. The heat, and therefore the power, loss is I^2R, R represents the internal resistance. The higher the current, the more will be the losses resulting in lower values of efficiency. Thus, neither the entire charging current is stored in the battery nor the entire stored charge is delivered to the load during discharge. Figure 5.13 shows the effect of the rate of discharge on the battery voltage as well as the time taken for the process. At higher rates of discharge, for instance at 3C, the time taken by the battery terminal voltage to reduce to unacceptable limits is far shorter than at lower rates of 0.1C, which is the most advocated value.

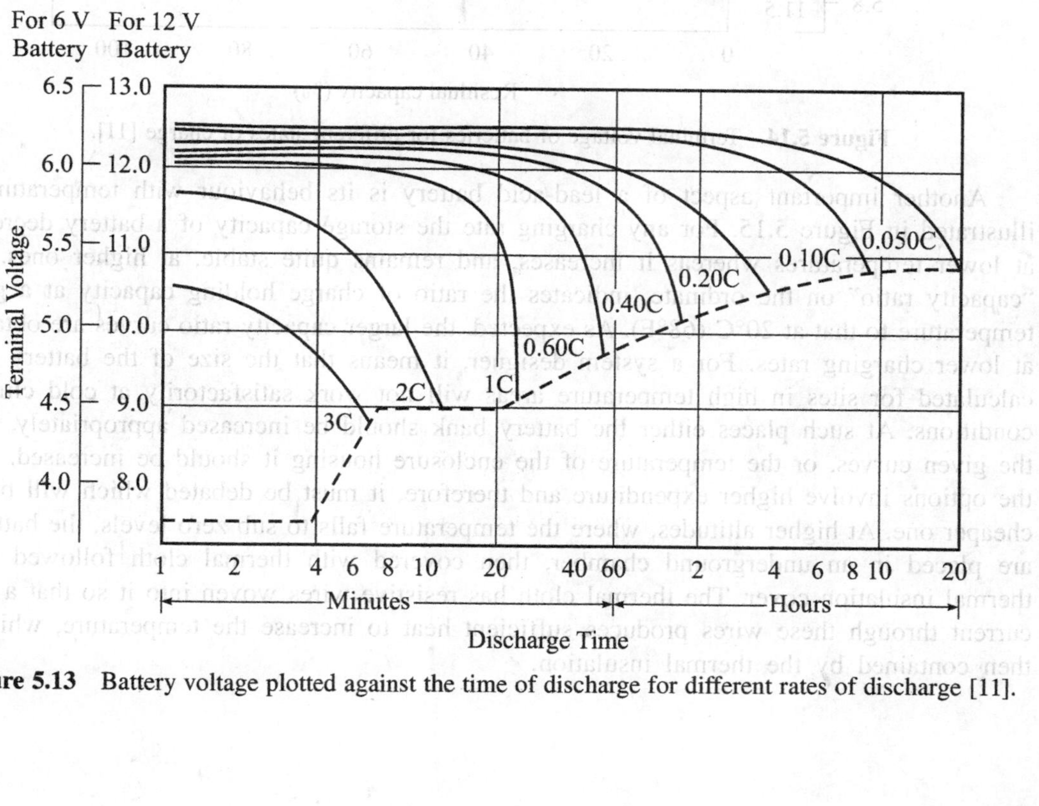

Figure 5.13 Battery voltage plotted against the time of discharge for different rates of discharge [11].

Figure 5.14 gives a plot of the battery voltage against the amount of charge left in it. When a battery is almost completely discharged, its terminal voltage is close to 11.5 V. This is a dangerous state, called deep discharge state, as the battery might not recover to store charge any further owing to sulphate formation on its plates. This means that the battery must not be allowed to discharge to such a level by a discharge controller. The battery should be charged at the earliest opportunity. On the other hand, a fully charged battery has a voltage of 2.23 V per cell and should be prevented from further charging to avoid gas formation.

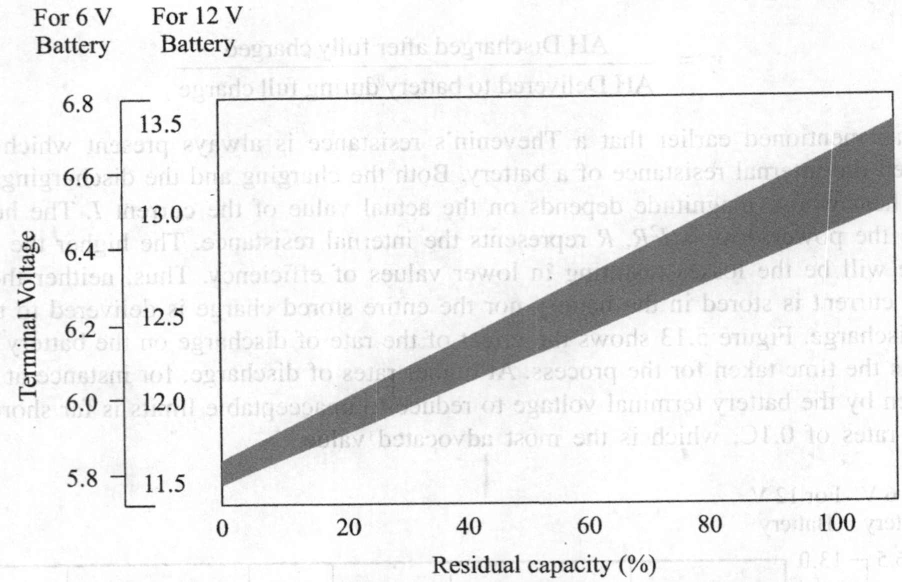

Figure 5.14 Terminal voltage of batteries for different states of charge [11].

Another important aspect of a lead-acid battery is its behaviour with temperature as illustrated in Figure 5.15. For any charging rate the storage capacity of a battery decreases at lower temperatures whereas it increases, and remains quite stable, at higher ones. The "capacity ratio" on the ordinate, indicates the ratio of charge holding capacity at a given temperature to that at 20°C (68°F). As expected, the larger capacity ratio curves are obtained at lower charging rates. For a system designer, it means that the size of the battery bank calculated for sites in high temperature areas will not work satisfactorily at cold climatic conditions. At such places either the battery bank should be increased appropriately, from the given curves, or the temperature of the enclosure housing it should be increased. Both the options involve higher expenditure and therefore, it must be debated which will be the cheaper one. At higher altitudes, where the temperature falls to sub zero levels, the batteries are placed in an underground chamber, then covered with thermal cloth followed by a thermal insulation cover. The thermal cloth has resistive wires woven into it so that a little current through these wires produces sufficient heat to increase the temperature, which is then contained by the thermal insulation.

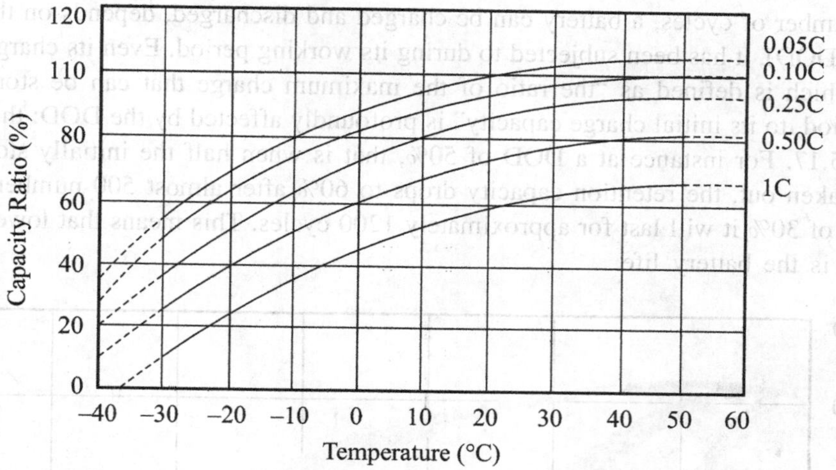

Figure 5.15 Curves of the capacity ratio of a battery plotted against temperature for different charging rates [11].

The lead-acid cell is quite sensitive to temperature in many respects such as the capacity retention as described above. Its shelf life is also affected as shown in Figure 5.16.

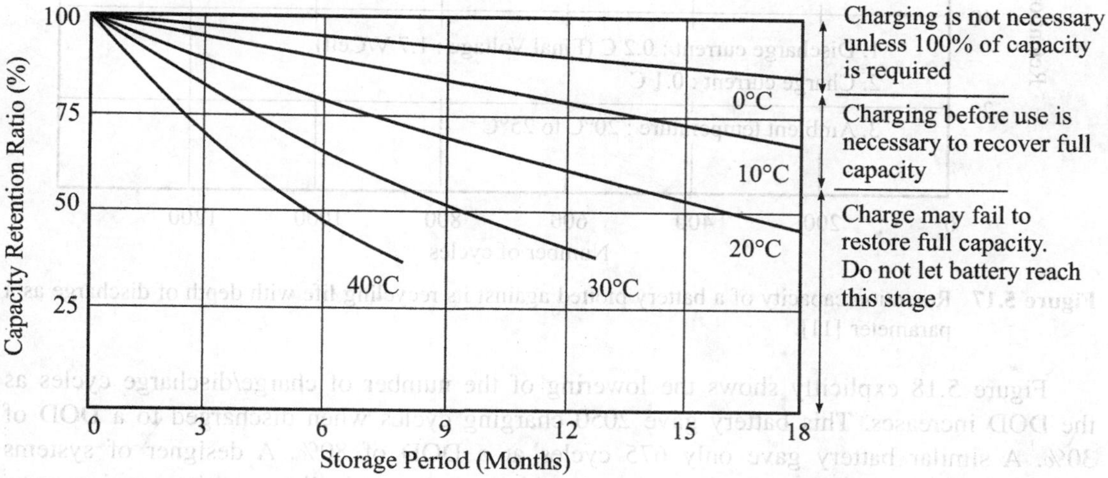

Figure 5.16 Charge capacity of a battery reduces at higher temperatures [11].

If a charged battery is left unused for a long time, it loses charge through its internal chemical mechanism which can be treated as its internal resistance. The decay is faster at higher temperatures. For example, at 40°C a battery is likely to discharge by more than 50% over a period of six months whereas it loses only 13% of its charge at 0°C even after eighteen months, which contradicts the process of charging. This shows that while it is preferable to store a fully charged battery at 0°C to enhance its shelf life, it is always desirable to charge or discharge it at a higher temperature of around 20°C.

The number of cycles, a battery can be charged and discharged, depends on the depth of discharge (DOD), it has been subjected to during its working period. Even its charge retention capacity which is defined as "the ratio of the maximum charge that can be stored, after a certain period, to its initial charge capacity" is profoundly affected by the DOD; this is shown in Figure 5.17. For instance at a DOD of 50%, that is when half the initially stored charge has been taken out, the retention capacity drops to 60% after almost 500 number of cycles. At a DOD of 30% it will last for approximately 1200 cycles. This means that lower the DOD the longer is the battery life.

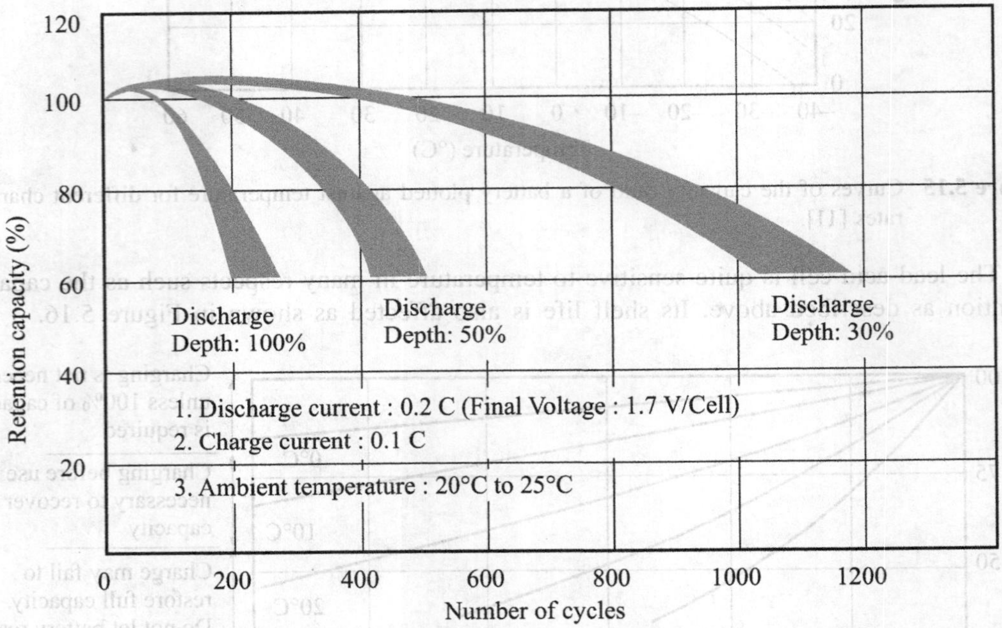

Figure 5.17 Retention capacity of a battery plotted against its recycling life with depth of discharge as a parameter [11].

Figure 5.18 explicitly shows the lowering of the number of charge/discharge cycles as the DOD increases. This battery gave 2050 charging cycles when discharged to a DOD of 30%. A similar battery gave only 675 cycles at a DOD of 80%. A designer of systems involving storage batteries has to contend with two diametrically opposite requirements. A PV system needs to discharge a battery to a depth of 80% but then the life reduces to only 375 cycles according to Figure 5.17. If the battery is charged every day to be used at the same night, it will last for only 1 year and 10 days which is very little indeed. On the other hand, if the battery is operated to a DOD of only 30%, to prolong its life, three of these will be needed in parallel thus enhancing the cost of storage. Fortunately, deep discharge batteries are available to comply with both the conditions.

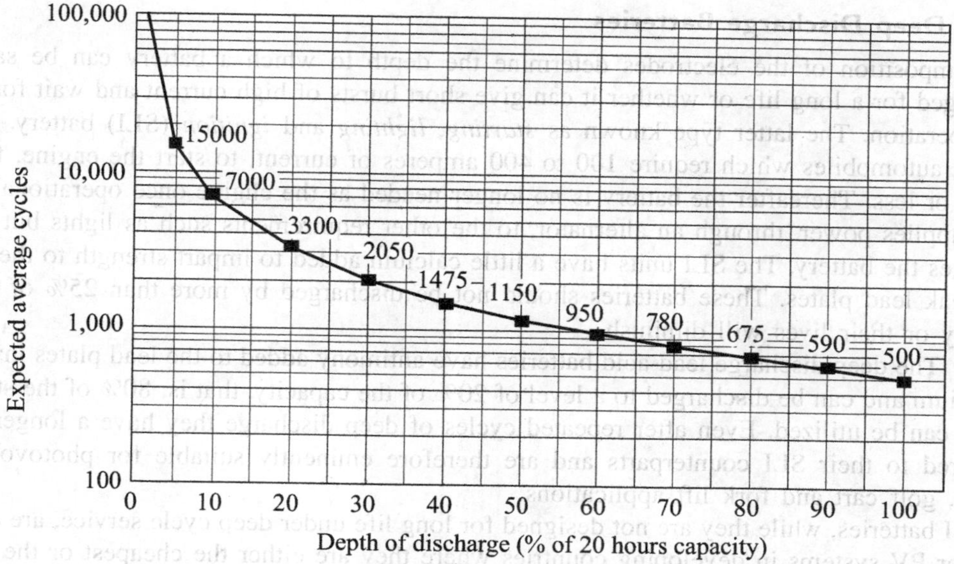

Courtesy: Electropaedia, battery and energy technologies [12]

Figure 5.18 Expected average charging cycles plotted against the DOD [12].

When the charge storage capacity of a battery reduces to 60% of its rated capacity when put into service as new, the life of the battery is considered to have ended. This is shown in Figure 5.19.

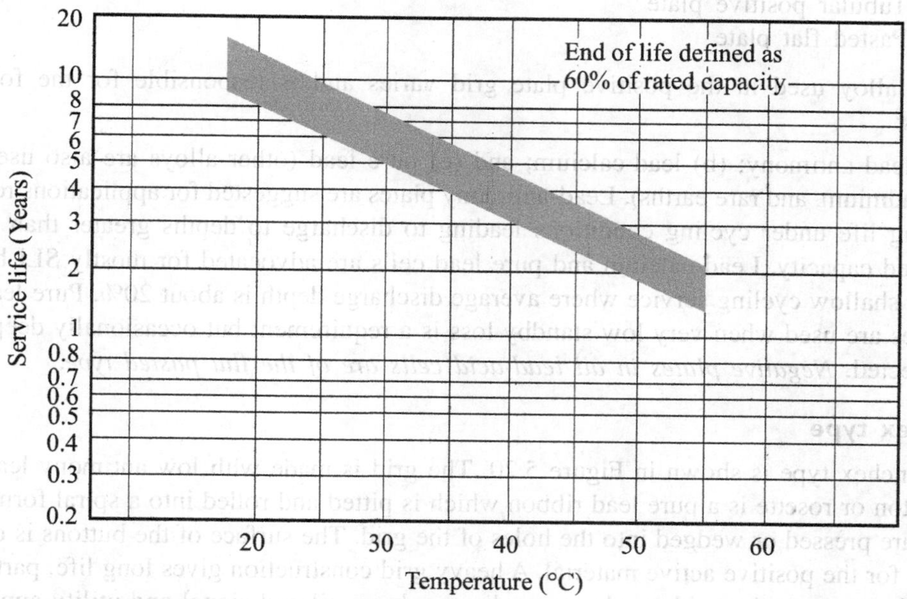

Figure 5.19 The service life of a battery plotted against temperature [11].

5.3.5 Deep Discharge Batteries

The composition of the electrodes determine the depth to which a battery can be safely discharged for a long life or whether it can give short bursts of high current and wait for the next operation. The latter type known as *starting*, *lighting* and *ignition* (SLI) battery. It is used in automobiles which require 100 to 400 amperes of current, to start the engine, for a second or less. Thereafter the battery is no longer needed as the engine once operational not only supplies power, through an alternator, to the other requirements such as lights but also recharges the battery. The SLI units have a little calcium added to impart strength to the soft and weak lead plates. These batteries should not be discharged by more than 25% of their capacity or their lives will diminish.

The deep discharge lead-acid batteries have antimony added to the lead plates instead of calcium and can be discharged to a level of 20% of the capacity, that is, 80% of the stored charge can be utilized. Even after repeated cycles of deep discharge they have a longer life compared to their SLI counterparts and are therefore eminently suitable for photovoltaic, marine, golf cart and fork lift applications.

SLI batteries, while they are not designed for long life under deep cycle service, are often used for PV systems in developing countries where they are either the cheapest or the only type available. These batteries may provide up to two years of useful service in small stand-alone PV systems where the average daily depth of discharge is limited to 10–20%, and the maximum allowable depth of discharge is limited to 40–60%, although they are not designed for such use.

Commercially available lead-acid battery types are classified by type of positive plate:

- Manchex
- Tubular positive plate
- Pasted flat plate

The alloy used in the positive plate grid varies and is responsible for the following subtypes:

(a) lead-antimony; (b) lead-calcium; and (c) pure lead (other alloys are also used, such as tin, cadmium, and rare earths). Lead-antimony plates are suggested for applications requiring very long life under cycling conditions leading to discharge to depths greater than 20% of their rated capacity. Lead-calcium and pure lead cells are advocated for mostly SLI batteries used for shallow cycling service where average discharge depth is about 20%. Pure lead alloy cell types are used when very low standby loss is a requirement but occasionally deep cycles are expected. *Negative plates in all lead-acid cells are of the flat pasted type.*

Manchex type

The Manchex type is shown in Figure 5.20. The grid is made with low antimony lead alloy. The button or rosette is a pure lead ribbon which is pitted and rolled into a spiral form. These in turn are pressed or wedged into the holes of the grid. The surface of the buttons is oxidized to PbO_2 for the positive active material. A heavy grid construction gives long life, particularly in standby type service with moderate cycling such as railroad signal and utility application. A life of 25 years is normal for utility operation.

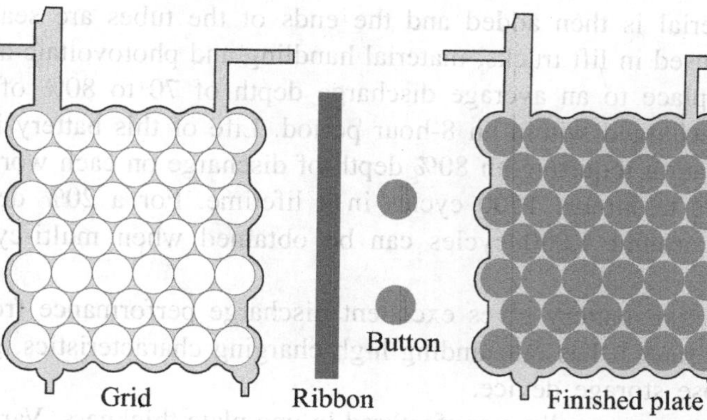

Courtesy: Arizona solar centre.

Figure 5.20 Manchex type positive plates [14].

Tubular positive plate

This construction of such a plate is shown in Figure 5.21. The grid, which carries the current, is made of a series of low antimony lead spines. Each spine is placed centrally inside a tubing of woven or porous plastic or glass material.

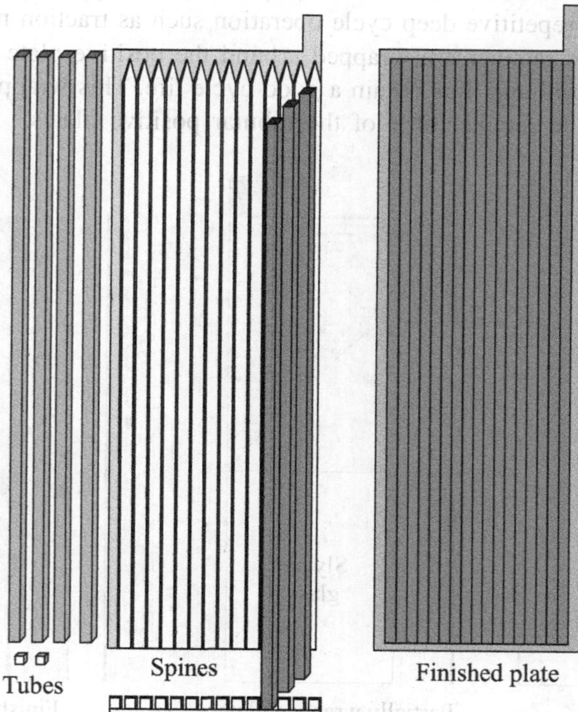

Courtesy: Arizona solar centre.

Figure 5.21 Construction of tubular anode lead-acid cell [14].

The active material is then added and the ends of the tubes are sealed. This type of battery is generally used in lift trucks, material handling and photovoltaic applications where deep cycling takes place to an average discharge depth of 70 to 80% of the 6-hour rated capacity and then recharged within an 8-hour period. Life of this battery is about six years in a lift truck application requiring an 80% depth of discharge on each working for 250 days a year; this amounts to almost 1500 cycles in a lifetime. For a 20% depth of discharge, shallow cycling of around 4,000 cycles can be obtained when multi-cycles per day are performed.

The tubular positive battery gives excellent discharge performance from diesel starting rates to the 24-hour rate. It has outstanding high charging characteristics, good standby life, and is a multipurpose storage device.

Tubular batteries are normally manufactured in one plate thickness. Variations in capacity are obtained by either increasing the number of tubes per plate and/or by varying the tube or plate height.

Flat pasted plate type

A typical pasted plate construction is shown in Figure 5.22. Depending on the size of the plate and the application the lattice grid is cast with pure lead, lead-calcium or lead-antimony. Active material is applied in the form of a wet paste and the plate is then cured, dried and formed. When used in repetitive deep cycle operation such as traction machinery, glass mats and a perforated plastic retainer are wrapped around the positive plate to minimize the loss of positive active material and thus obtain a good cycle life. This wrapping process performs the same function as the retainer tube of the tubular positive plates.

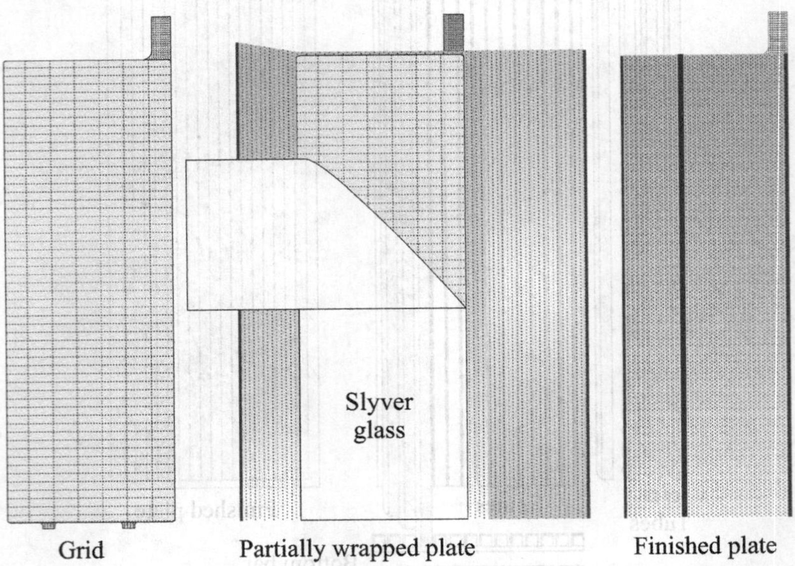

Courtesy: Arizona solar centre.

Figure 5.22 Flat pasted plate type construction for a lead-acid cell [14].

The real advantage to cells with lead-calcium, and pure lead positive grids, is in standby operations when long periods of float charging are expected between power failures and discharge cycles. In general, pure lead designs of the plates offer better cycle life than lead-calcium, but have weaker mechanical strength. During these periods cells are maintained at a constant charge voltage, and the charge current required to maintain lead-calcium and pure lead positive grid cells at voltage is much lower than the lead-antimony grid. A reduction in float charge current reduces water loss and the maintenance cost of adding water to battery cells.

Figure 5.23 gives the band of battery operation curves for different types of plates. It shows the relationship between the numbers of cycles in a lifetime versus the discharge depth. One cycle per day is more common in photovoltaic systems, however, exaggerated values of battery life may be arrived at from the figure if this value is taken. A closer match with the actual life can be obtained if the number of cycles is increased to an average of three cycles in two days. Battery life must therefore be verified experimentally at site.

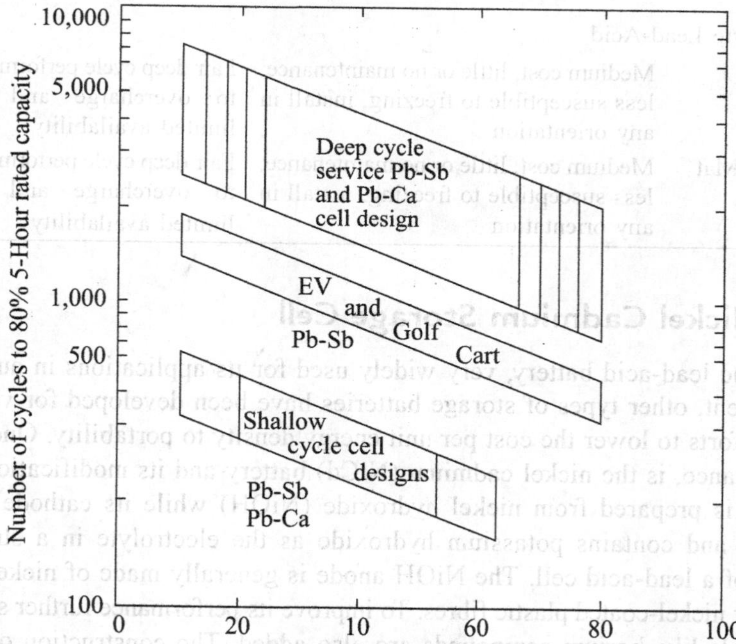

Courtesy: Arizona solar centre.

Figure 5.23 Estimated life cycle of lead-acid cells for different cell designs and depth of discharge at 25°C [14].

Table 5.2 gives a comparative chart of different lead acid batteries. From this chart batteries can be selected for different applications. It is clear that for photovoltaic applications the antimony-lead alloy plates are the best choice although they are more maintenance intensive as water loss is a problem. Details of many battery chemistries are given in [15].

Table 5.2 Characteristics of lead-acid batteries

Battery Type	*Advantages*	*Disadvantages*
Flooded Lead-Acid		
Lead-Antimony	Low cost, wide availability good deep cycle and high temperature performance, can replenish electrolyte	High water loss and maintenance
Lead-Calcium Open Vent	Low cost, wide availability, low water loss can replenish electrolyte	Poor deep cycle performance, intolerant to high temperatures and overcharge
Lead-Calcium Sealed Vent	Low cost, wide availability, low water loss	Poor deep cycle performance, intolerant to high temperature and overcharge, cannot replenish electrolyte.
Lead Antimony/ Calcium Hybrid	Medium cost, low water loss	Limited availability, potential for stratification
Captive Electrolyte Lead-Acid		
Gelled	Medium cost, little or no maintenance, less susceptible to freezing, install in any orientation	Fair deep cycle performance, intolerant to overcharge and temperatures, limited availability
Absorbed Glass Mat	Medium cost, little or no maintenance, less susceptible to freezing, install in any orientation	Fair deep cycle performance, intolerant to overcharge and temperatures, limited availability

5.4 The Nickel Cadmium Storage Cell

In addition to the lead-acid battery, very widely used for its applications in automobiles and traction equipment, other types of storage batteries have been developed for various reasons ranging from efforts to lower the cost per unit energy density to portability. One of these, that has wide acceptance, is the nickel cadmium (NiCd) battery and its modifications. The anode of a NiCd cell is prepared from nickel hydroxide (NiOH) while its cathode is made from cadmium oxide and contains potassium hydroxide as the electrolyte in a structure almost similar to that of a lead-acid cell. The NiOH anode is generally made of nickel fibres mixed with graphite- or nickel-coated plastic fibres. To improve its performance further small quantities of other materials like barium compounds are also added. The construction of a cylindrical cell is shown in Figure 5.24.

5.4.1 Chemistry of the Nickel Cadmium Storage Cell

NiCd batteries use nickel hydroxide for the anode plates and cadmium oxide for the cathode plates, in a structure similar to that of the lead-acid system. The electrolyte in the NiCd system is potassium hydroxide. The NiOH anode is generally made of nickel fibres mixed with graphite- or nickel-coated plastic fibres. Small quantities of other materials such as barium and cobalt compounds are also added to improve performance. Plastics coated with cadmium are used for cathode. A fibre structure is used for both the electrodes in an effort

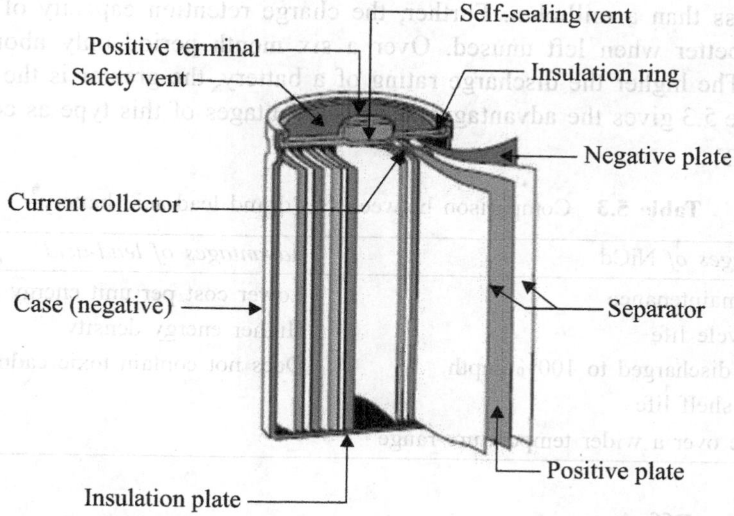

Figure 5.24 Cylindrical cell construction of a NiCd cell [16].

to increase the surface area while limiting the cost by reducing the amount of expensive nickel and cadmium. The reversible reaction during the processes of discharge and charge is given by

$$2NiOOH + 2H_2O + Cd \underset{\text{charge}}{\overset{\text{discharge}}{\rightleftharpoons}} 2Ni(OH)_2 + Cd(OH)_2 \qquad (5.9)$$

In the NiCd system the specific gravity of the electrolyte changes very little during battery operation in contrast to the lead-acid cell where the changes are drastic. The voltage of the fully charged cell is 1.29 V. In some cells, to increase the cycle life and also to improve higher temperature operation, LiOH is also added to the electrolyte.

5.4.2 Properties of the Nickel Cadmium Cell

NiCd cells are more temperature resilient than their lead-acid counterparts as they can endure freezing to high temperatures. They can also be fully discharged without any damage and withstand overcharging to some extent. Such qualities make them attractive for places where maintenance is difficult. The high initial cost of NiCd cells can be justified by low expenditure on maintenance.

The capacities of NiCd cells range from a few mAhr to more than 1200 Ahr and are widely accepted for military, space and industrial applications, where cost is the secondary consideration but the robust nature of the device is the primary concern. Here the thickness of the electrodes decides the discharge rates, as in the case of the lead-acid cell. The energy densities that are available for this type vary from 20 Whr/kg for the smaller cells to 50 Whr/kg for the fibrous plates. One advantage is the quick discharging rate of *C* that can feed almost 90% of the available charge to the load whereas the lead-acid cell gives a much lower quantity. Primarily this excess transfer of charge is due to a very low internal resistance of less than a milliohm. Further, the charge retention capacity of these charged cells is much better when left unused. Over a six month period only about 20% of the charge is lost. The higher the discharge rating of a battery, the greater is the loss of charge over time. Table 5.3 gives the advantages and disadvantages of this type as compared to the lead-acid battery.

Table 5.3 Comparison between NiCd and lead-acid battery

Advantages of NiCd	*Advantages of lead-acid*
Lower maintenance	Lower cost per unit energy storage
More cycle life	Higher energy density
Can be discharged to 100% depth	Does not contain toxic cadmium
Longer shelf life	
Reliable over a wider temperature range	

5.4.3 Memory Effect

The NiCd cells had the peculiar property that when discharged to a certain depth they tend to discharge to only this depth after further charging, that is, if discharged to say 30% and then recharged again, the cell will only discharge to 30% and not more. This is a nuisance particularly to a user who tops up the charge after every little use believing that a fully charged cell will work for a longer time. Some manufacturers had, in fact, introduced resistive strips to discharge the cell to 100% level before recharging. What happens in fact is that repeated shallow charges cause the crystalline structure of the electrodes to change and this causes the internal impedance of the cell to increase and its capacity to be reduced. Long slow charges such as trickle charging tend to promote this undesirable crystal growth also helped by high temperatures; and so these conditions should be avoided. However plastic bonded, fibre and pocket plate cells do not show this inclination. Nickel metal hydride cells suffer from the same problem but to a lesser extent.

5.4.4 Charging Characteristics

NiCd cells can be charged either by constant current or constant voltage or a combination thereof which is quite satisfactory for photovoltaic usage. Care must be taken to limit the current within acceptable limits during constant voltage charging to keep away from damage to the cell. Overcharging is not too damaging but should be avoided, if possible, as this leads to increased gassing and decomposition of water. Charging can be done from very low temperature of –50°C to quite high ones around 45°C but the charging efficiencies at these limits will reduce. Normally, charging of a fully discharged cell to full charge should be achieved in 5 to 7 hours, with a charging voltage between 1.50 and 1.65 volts per cell. The charging efficiency of the most efficient NiCd cell comes close to 85%. Other nickel based cells have been developed to replace the cadmium because its toxicity, when the cell is disposed off after use, has raised environmental issues.

The charging and discharging curves of NiCd cells [16] are given in Figure 5.25.

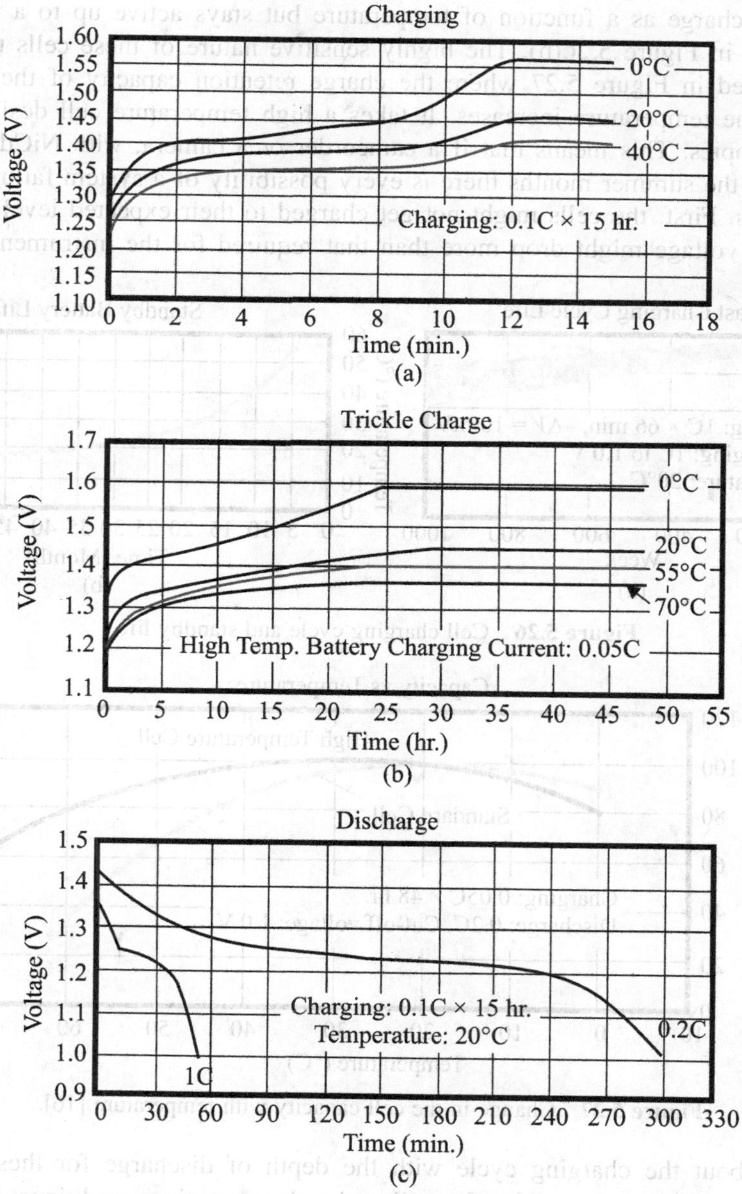

Figure 5.25 Charging, trickle charging and discharging curves of NiCd cells [16].

In Figure 5.25(a) the effect of temperature on the final charged voltage is considerable. A difference of 0.15 V or 9.7% is clearly seen when the cell is charged at the rate of *C*/10 over 0 to 40ºC. For a trickle charging rate of *C*/20, the end voltage at 0ºC rises to 1.6 V (Figure 5.25b). A slower discharge, in Figure 5.25(c), gives a steadier voltage over a longer time than that at a faster rate.

A life cycle of 1000 cycles can be expected when the cell is charged and discharged both at the rate of C as given in Figure 5.26(a). When a cell is left alone after charging, it certainly loses charge as a function of temperature but stays active up to a period of five years as shown in Figure 5.26(b). The highly sensitive nature of these cells to temperature is again reflected in Figure 5.27 where the charge retention capacity of the cells reduces drastically as the temperature increases. It takes a high temperature cell design to perform better in the tropics. This means that if a camcorder or a camera, with NiCd cells, is used in India during the summer months there is every possibility of a system failure. The reason is quite obvious. First, the cells might not get charged to their expected level and secondly, their operating voltage might drop more than that required for the instruments' operation.

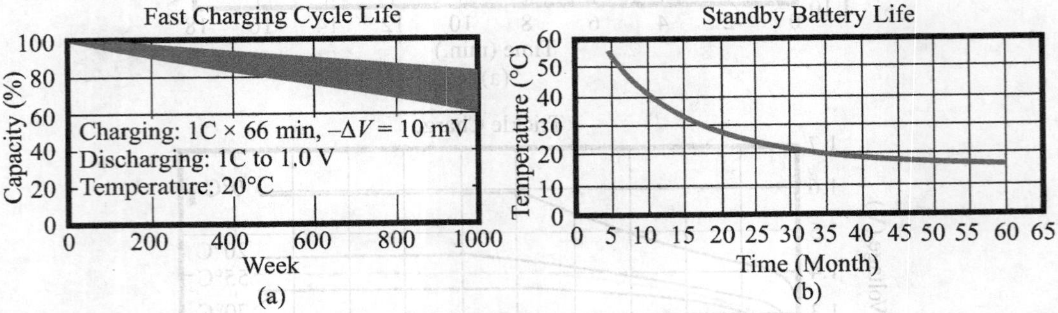

Figure 5.26 Cell charging cycle and standby life.

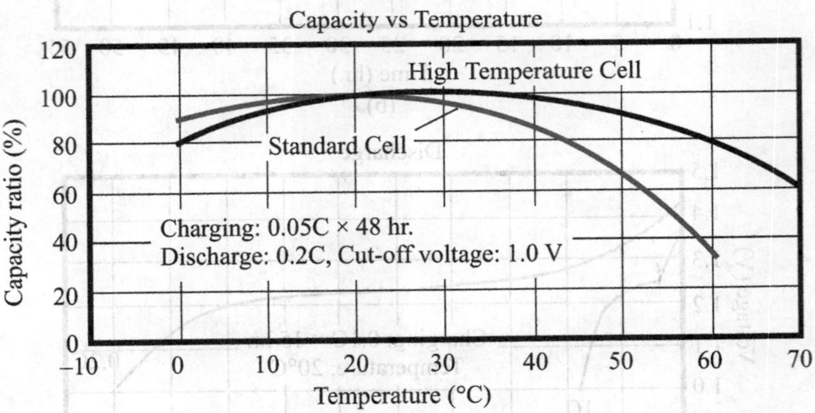

Figure 5.27 Change in the cell capacity with temperature [16].

An idea about the charging cycle with the depth of discharge for these cells is also necessary to ascertain the real life of a cell under the operating conditions. Reference [14] gives some values which are good indicators. Life measurements are not available to compare cycle life at various depths of discharge and operating temperatures. Life must be obtained from actual tests on the regime and under the situation of the application. Cycle life decreases with increasing discharge depth. Typical life data for a pocket plate NiCd cell are given in Table 5.4.

Table 5.4 Life cycle with depth of discharge for a pocket plate NiCd cell

Depth of discharge at 25°C % of 5 hr *rated capacity*	*Cycles to failure (Capacity below 80%* 5 hr *rated* Ahr)
20	2500
40	1700
60	1100
80	750
100	500

Courtesy: Arizona Solar Center [14]

5.5 Nickel-Metal Hydride (NiMH) Cells

Some other battery systems are presently in use, while others seem to show promise for use in future. The nickel-zinc battery, a combination of the NiCd and the CuZn systems, provides some attractive characteristics, including long life and a capacity advantage. This system has double the specific energy of the NiCd structure. The reversible chemical reaction is given by

$$2NiOOH + Zn + 2H_2O \rightleftharpoons 2Ni(OH)_2 + Zn(OH)_2 \tag{5.10}$$

If the battery is overcharged, it disintegrates the water in the KOH electrolyte, resulting in gasing and the release of hydrogen and oxygen. Therefore, charging must be managed to steer clear of overcharging. A cell when fully charged has a voltage of around 1.73 V and can be made in sizes in excess of 200 Ahr. At normal operating temperatures of 20°C and at a discharge rate of $C/3$, the cell voltage remains quite stable at about 1.6 V over most of the discharge range. Another technology that emerged recently is popular particularly in smaller applications such as computers, cellular phones and other portable consumer electronics applications in the nickel-metal hydride (NiMH) cell. This cell replaces the cadmium cathode with an environmental-friendly metal hydride cathode, resulting in higher energy density at the cathode and a longer lifetime or higher capacity, depending on the design. The anode is the same as in the NiCd cell and KOH is used as the electrolyte. The overall discharge reaction here is

$$MH + NiOOH \rightleftharpoons M + Ni(OH)_2 \tag{5.11}$$

The negative active material in the charged state is hydrogen in the form of metal hydride. This metal alloy can undergo reversible hydrogen absorption and desorption reaction when the cell is charged or discharged. On discharge most of the operating properties of these two systems are similar, as can be seen from literature on these cells [16]. Because of their relative high cost the NiCd and NiMH cells are used in low power photovoltaic instruments like rechargeable flash lights. A number of integrated circuits are available for charging this variety of cell chemistry. Among them is the LTC 4011 from Linear Technology Corporation [17]. Its design note, DN 380, is quite helpful for designing a four-cell charger.

5.5.1 Differences between Nickel Cadmium (NiCd) and Nickel-Metal Hydride (NiMH) Cells

The following points give some of the notable differences between the two systems:

- *Nickel Cadmium:* These cells, are an older form of rechargeable battery. While they tend to be less expensive than newer designs, the toxic cadmium used in their construction makes them harmful to the environment. They have lower self-discharge levels than NiMH cells.
- *Nickel-Metal Hydride:* These cells are more expensive to produce than NiCd cells, but offer a few advantages over the older design.
- *Charge Levels:* NiMH cells are capable of storing and holding a greater charge than their predecessors. For instance, in high drain devices like cameras they can offer up to twice the battery life.
- *Memory:* NiCd cells can develop a "memory" when not fully discharged. Crystals can develop inside the cell which interferes with charge levels thereby reducing the effective lifespan of the cell between charges and preventing a full recharge.
- *Cold Temperatures:* NiCd cells can be discharged over a wider temperature range (–4°F to 149°F) or (–20°C to 65°C) than NiMH cells (14°F to 122°F) or (0°C to 50°C).
- *Service Life:* While NiCd cells must be charged carefully and discharged to maintain their usable lifespan, they can remain in service considerably longer than NiMH cells. NiMH cells wear out much faster.

There are many more rechargeable cells discussed in [1] but all of them may not be suitable for photovoltaic applications primarily from the cost considerations. Also the specific energy of the cells plays a major role in their selection.

5.6 Lithium Cells

With an energy density of 150 Whr/kg and output voltage of up to 4 V the lithium technologies appear quite attractive. These also have a very long charge retention and shelf life. The arguments against them are: low cycle life, relatively poor low-temperature performance, relatively poor high discharge rate performance, capacity fading and potential safety problems. In the recent past, there were some reports of the lithium based batteries catching fire while in operation in laptops resulting in their recall by some manufacturers. This problem is potentially dangerous when laptops are used on board aircraft. The origin of this hazard is associated with metallic lithium which is highly reactive and is quite dangerous if it comes in contact with water or humans. The lithium metal negative electrode has since been discarded in favour of those made of lithiated carbon. Researchers at NASA have found a different version of lithium cell useful for flying unmanned aerial vehicles (UAV) [18]. The safety record of these cells coupled with their large energy density has made such a mission possible. Manufacturers are also targeting the electric automobile sector, that one day might include the photovoltaic technology, with lithium chemistry based batteries. These cells have recently achieved widespread acceptance in camcorders and other gadgets which require longer usage after each recharging.

Lithium ion is an emerging technology that has the possibility to bring high energy, long life power sources, having a minimal detrimental impact on the environment, to the consumer electronics industry. This new technology, however, still requires improvement in several areas before its use becomes more widespread. These areas include improving the capacity of the anode while reducing its passivation, and increasing the capacity, cycle life and rate capability of the cathode.

The advantage of a lithium ion cell its negative electrode uses a lithiated carbon material instead of lithium metal as stated earlier. For the positive plate a lithiated transition metal compound is used while the electrolyte is generally an organic solution. The reaction at the positive electrode is:

$$LiMO_2 \longleftrightarrow Li_{1-x}MO_2 + xLi^+ + xe \tag{5.12}$$

and at the negative electrode

$$C + xLi^+ + xe \longleftrightarrow Li_xC \tag{5.13}$$

the overall equation for this type of chemistry is

$$LiMO_2 + C \longleftrightarrow Li_xC + Li_{1-x}MO_2 \tag{5.14}$$

The right-hand side of Equation (5.14) appears during charging whereas the left-hand side represents the discharging states. $LiMO_2$ is the lithiated metal oxide intercalation compound.

The best gravimetric energy of 0.518 kWhr/kg is obtained when the metal used in the cathode is cobalt. The cathode material in that case is $LiCoO_2$ [19].

5.6.1 Charging Level

The cycle life of lithium batteries can be increased by reducing the charging cut-off voltage. This essentially gives the battery a partial charge instead of fully charging it, which is similar to working at a lower DOD. The graph in Figure 5.28 shows the typical cycle life improvements possible with decreased full charge voltage.

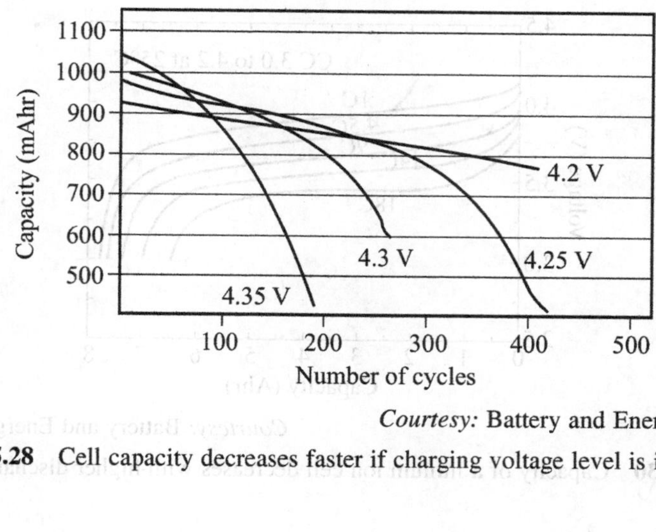

Courtesy: Battery and Energy Technologies [12]

Figure 5.28 Cell capacity decreases faster if charging voltage level is increased.

The graph in Figure 5.29 shows typical discharge curves for cells of different cell chemistries discharged at the rate of 0.2C. Each cell of different chemistry has its own characteristic nominal voltage and discharge curve. The lithium ion system has a fairly flat discharge curve while others such as lead acid have a definite slope.

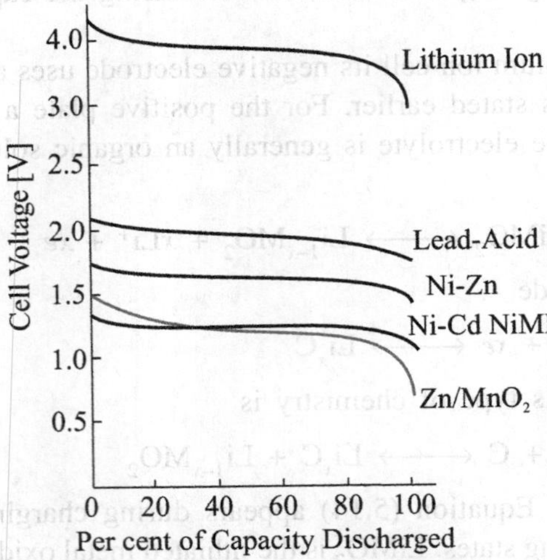

Courtesy: Battery and Energy Technologies [12]

Figure 5.29 Discharge characteristics of cells with different chemistries.

5.6.2 Discharge and Charge Rates

The discharge curves for a lithium ion cell, in Figure 5.30, show that the effective capacity of the cell is reduced if the cell is discharged at very high rates (or conversely increased with low discharge rates). This is called the capacity offset and the effect is common to most cell chemistries.

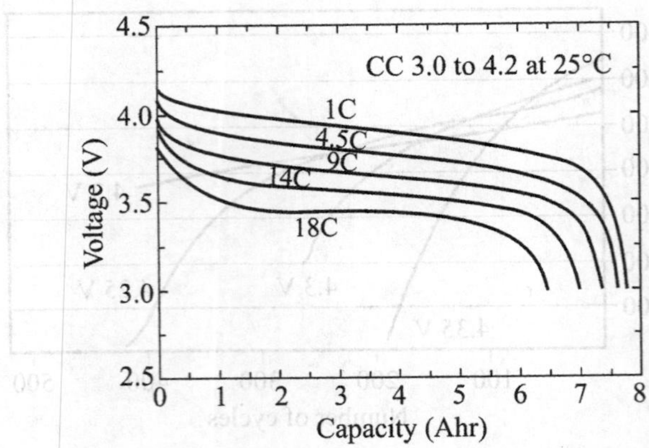

Courtesy: Battery and Energy Technologies [12]

Figure 5.30 Capacity of a lithium ion cell decreases with higher discharge rates.

Special precautions are needed to ensure the cell is fully charged while at the same time overcharging should be avoided. To safeguard the cell, it is recommended that the charging method switches to constant voltage, slightly below the maximum voltage, before the cell voltage reaches its upper limit.

As the charge approaches completion, the current decreases to a trickle charge. Normally the current is monitored so that when the charging current reaches a predetermined trickle value the charging source changes to a constant voltage one. The current and the voltage variation with time are shown in Figure 5.31. This method is used for lithium and SLA batteries.

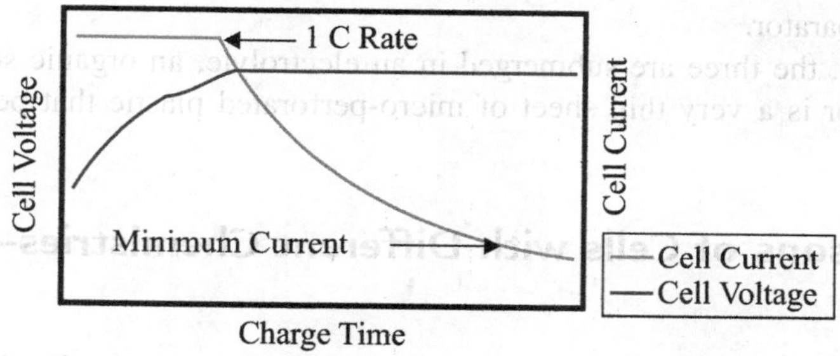

Courtesy: Battery and Energy Technologies [12]

Figure 5.31 Charging characteristics of lithium ion cell.

The LM 3622 lithium-ion battery charge controller from National Semiconductor, U.S.A. [20] can charge one or two of these cells from a DC voltage source. This takes into account the constant current or voltage necessary to prevent overcharging, and therefore damage to the cells. Another circuit, MAX8903A lithium-ion battery charge controller, is also used for this purpose.

The cut-away view of a cylindrical lithium ion cell shown in Figure 5.32 clearly identifies the different parts used in its construction.

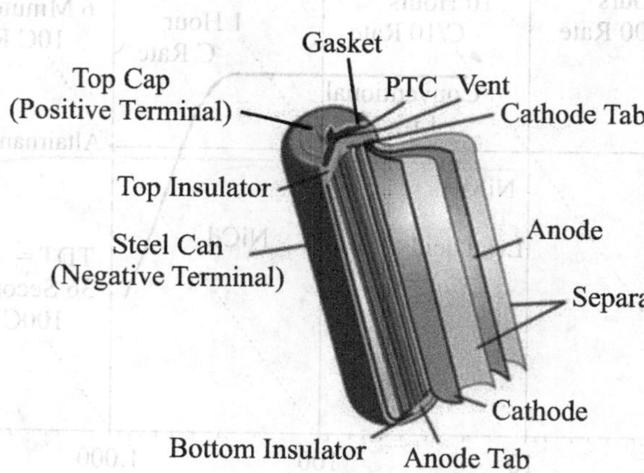

Courtesy of HowStuffWorks.com

Figure 5.32 Cut-away view of lithium ion cell [21].

Like most batteries, there is an outer case made of metal. Since the battery is pressurised a steel case is needed for strength against possible rupture. This metal case has some sort of pressure-sensitive release hole acting as a safety valve. In an unlikely situation of the cell coming to a state of explosion due to excessive internal pressure resulting from overheating, this vent will release the extra pressure. Of course, the cell might be useless afterwards but the vent will save any personnel nearby from coming to harm; so this is something to avoid. For the same reason the positive temperature coefficient (PTC) switch is kept to stop the battery from overheating.

Three electrodes, in the form of sheets, are rolled into a cylinder and placed inside the metal case. These are namely, a positive electrode, a negative electrode and these two are separated by a separator.

Inside the case, the three are submerged in an electrolyte, an organic solvent, commonly ether. The separator is a very thin sheet of micro-perforated plastic that permits the passage of the ions.

5.7 Comparisons of Cells with Different Chemistries—The Ragone Plots

The Ragone plot is useful for characterizing the trade-off between effective energy storage capacity (Whr) and the ability to deliver power. These comparisons are plotted on logarithmic scales. The graph in Figure 5.33 shows the superior specific energy density of lithium ion cells. Note also that lithium ion cells with lithium titanate anodes (Altairnano) deliver a very high power density but at a lower energy density.

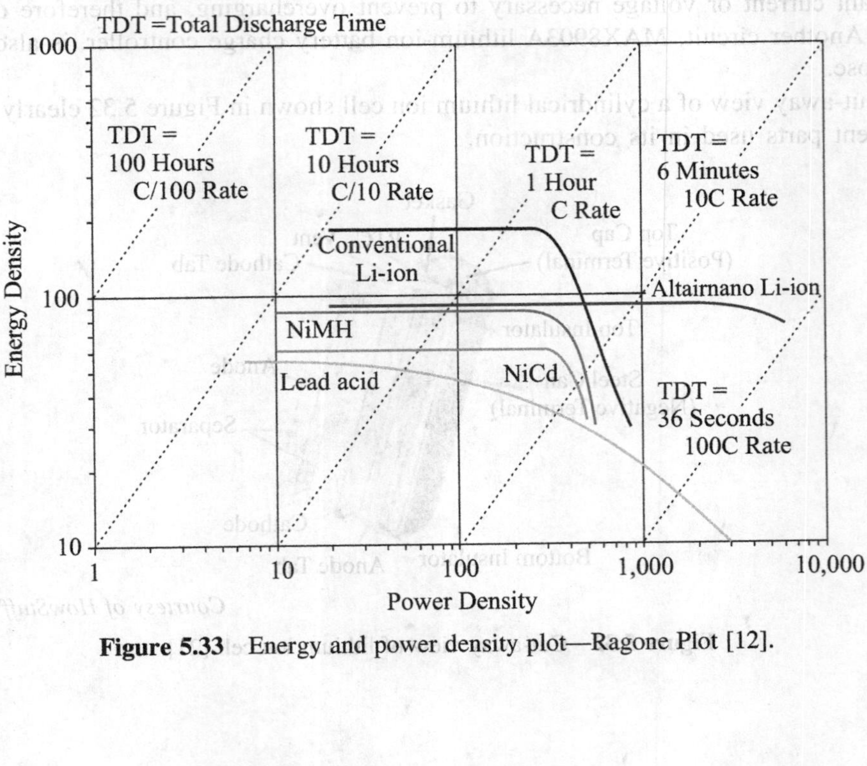

Figure 5.33 Energy and power density plot—Ragone Plot [12].

The Ragone plot shown in Figure 5.34 compares the performance of a variety of electrochemical devices. It shows that ultra capacitors (super capacitors) can deliver very high power but the storage capacity is very limited. On the other hand, fuel cells can store large amounts of energy but have a relatively low power output. Because of this ability, the ultra capacitor, with values in several farads, is used to start trucks and other vehicles in winters when the lead-acid batteries cannot deliver the high starting current to crank the engine. In such cases the ultra capacitor is placed in parallel with the battery so that charge can accumulate slowly in the capacitor. When the ignition switch is turned on, the ultra capacitor discharges quickly after giving several hundred amperes of current sufficient to start the engine. The starting current can be anywhere in between 200 and 500 amperes. This property also makes them ideal for capturing regenerative braking energy in electric vehicle (EV) applications.

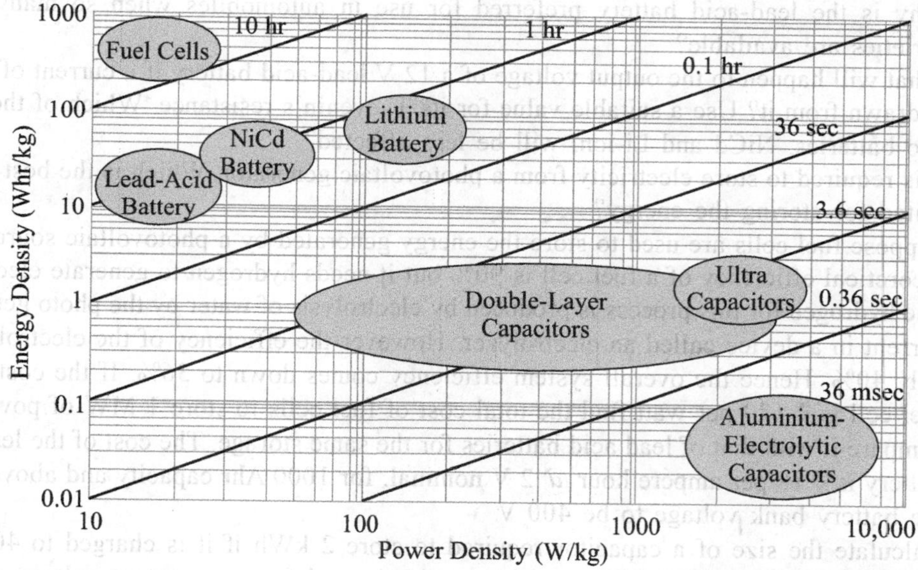

Source: US Defence Logistics Agency

Figure 5.34 Ragone Plot of Electrochemical Devices [12].

The sloping lines on the Ragone plots indicate the relative time to get the charge in or out of the device. The fuel cells are at the other extreme of performance where larger amount of energy can be produced over long periods but are not very effective for delivering short-term high power. The lithium cells are somewhere in between and provide a reasonable compromise between the two.

Looking at the present scenario of storage batteries, it appears that although the lead-acid devices continue to dominate the market in the automotive and heavy traction sector, mainly because of their low energy/cost, it will not be long when they have to give way to the lithium based ones. Lithium ion batteries, in the range of 3.6 V per cell, are increasingly being used in laptop computers even replacing their NiMH/NiCd counterparts and other portable equipment like the camcorders. These have also been used in UAVs and cars where they seem to be the future powering sources for replacing petrol and diesel. The lithium titanate based batteries show great promise even in military use such as submarines, where high power densities are required, combined with fairly high energy densities. Effort is now directed towards developing a battery which will be charged at 50°C and will also be capable of discharging at almost the same rate. Such fast storage capability will change the concept of replacing batteries in cars plying over long distances. These batteries can then simply be charged at stations much like those used for filling petrol. The charging stations can use photovoltaic power.

Questions

1. What is the size of a capacitor in farads for furnishing a current of 400 A_{DC}, from a 12 V source, for 1 second to start a car engine?
2. Which is the best of the three capacitors: NiCd, NiMH or the Li-ion? Give reasons.
3. Why is the lead-acid battery preferred for use in automobiles when so many other varieties are available?
4. What will happen to the output voltage of a 12 V lead-acid battery if a current of 400 A is drawn from it? Use a suitable value for its Thevenin's resistance. Which of the other two batteries, NiCd and Li-ion, will be less affected?
5. It is required to store electricity from a photovoltaic generator. Which is the best device suited for storing the energy?
6. Suppose fuel cells are used to store the energy generated by a photovoltaic source. The theoretical efficiency of a fuel cell is 90% but it needs hydrogen to generate electricity. The hydrogen for this process is produced by electrolysis of water by the photo generated current in a device called an electrolyzer. However the efficiency of the electrolyzer is only 40%. Hence the overall system efficiency comes down to 36%. If the cost of the fuel cell is ₹ 180 per watt find the total cost of fuel cells to store 1 MW of power and compare it with that of lead acid batteries for the same storage. The cost of the lead acid battery is ₹ 18 per ampere hour @ 2 V nominal, for 1000 Ahr capacity and above. Take the battery bank voltage to be 400 V.
7. Calculate the size of a capacitor required to store 2 kWh if it is charged to 400 V.
8. What happens if a 50% discharged battery is charged from a constant voltage source? Will the battery be damaged? What kind of damage will occur?
9. If a constant voltage source is available to charge a lead-acid battery what precautions should be taken to avoid damage to the battery?
10. Find out any integrated circuit other than the one given in this chapter for charging a lead-acid battery.

References

[1] David Linden, editor in chief, *Handbook of Batteries*, 2nd ed., McGraw–Hill, Inc., New York, 1995.

[2] Data Sheet of UC3906, http://focus.ti.com/docs/prod/folders/print/uc3906.html.

[3] Application Note, U-104: Improved Charging Methods for Lead-Acid Batteries Using the UC 3906, http://focus.ti.com/lit/an/slua115/slua115.pdf.

[4] Data sheet of LM 317. http://www.national.com/ds/LM/LM117.pdf.

[5] A.K. Mukerjee and Vinay Gupta, Communication System using Photovoltaic Power, *Energy Conversion and Management*, vol. **41**, 2000, pp. 551–557.

[6] Solar Photovoltaic Power Source for Single Channel VHF and Similar Systems, Specification No. P 520S93, Power Plant Issue II, September 1993. Telecommunication Engineering Centre, Khurshid Lal Bhawan, Janpath, New Delhi.

[7] D. Berndt, In: *Maintenance-free batteries*. Taunton, Somerset, England: Research Studies Press, 1993, pp. 165–204.

[8] David F. Stout, Milton Kaufman, Editors, *Operational Amplifiers Circuit Design*, New York: McGraw-Hill, 1976, pp. 5–13.

[9] A.K. Mukerjee and K.D. Chaudhuri, Design of Transistor Regulated Power Supplies, *I.E.T.E. Students Journal*, vol. **15**(2), 1974.

[10] R.D. Middlebrook, Design of Transistor Regulated Power Supplies, *Proceedings of I.R.E.*, vol. **45**, 1957, p. 1502.

[11] "Technical Manual", Power-Sonic Corporation, San Diego, California, www.power-sonic.com/.

[12] Electropaedia, Battery and Energy Technologies, http://www.mpoweruk.com/performance.htm.

[13] Maintenance and Operation of Stand-Alone Photovoltaic Systems, Sandia National Laboratories, Albuquerque, New Mexico, U.S.A., www.sandia.gov/pv/docs/DOC/Order%20Form.doc.

[14] Contractor Report: SAND81-7135, Unlimited Release, *Handbook of Secondary Storage Batteries and Charge Regulators in Photovoltaic Systems*, Final Report Originally Printed August 1981. Prepared by: Exide Management and Technology Company, 19 West College Avenue, P.O. Box 336, Yardley, Pennsylvania 19067, U.S.A.
Work Performed for: The U.S. Department of Energy Sandia National Laboratories, Albuquerque, New Mexico 87185 Under Contract No. 13–2202.

[15] Batteries and Charge Control in Stand-Alone Photovoltaic Systems, *Fundamentals and Application*, January 15, 1997. *Prepared for*: Sandia National Laboratories, Photovoltaic Systems Applications Department, PO Box 5800, Albuquerque, NM 87185–0752, U.S.A. *Prepared by*: James P. Dunlop, P.E. Florida Solar Energy Center, 1679 Clearlake Road Cocoa, FL 32922–5703, U.S.A.

[16] Nickel Cadmium and Nickel-Metal Hydride Rechargeable Batteries and Chargers, Power-Sonic Corporation, San Diego, California, U.S.A. www.power-sonic.com/.

[17] Linear Technology Corporation, LTC4011 data sheet. Also Design Note 380. http://cds.linear.com/docs/Datasheet/4011fb.pdf.

[18] Performance Characterization of a Lithium-Ion Gel Polymer Battery Power Supply System for an Unmanned Aerial Vehicle, National Aeronautics and Space Administration, Glenn Research Center, November 2004. Available from: NASA Center for Aerospace Information 7121 Standard Drive, Hanover, MD 21076. Available electronically at http://ntrs.nasa.gov/archive/nasa/casi.ntrs.nasa.gov/20050019219_2005000107.pdf.

[19] Pleck Batteries, Lithium Ion Batteries, Technology Manual. http://pleckbatteries.com/PLECK%20Batteries.files/Techinfo/PLECK%20LI-ION%20TECH%20MANUAL.pdf.

[20] Data sheet of LM 3622 lithium-ion battery charge controller. //www.national.com/analog.

[21] http://electronics.howstuffworks.com/lithium-ion-battery1.htm.

CHAPTER 6

BOS Inverters for AC Loads

6.1 Introduction

Photovoltaic power consists of DC current and voltage but to use it with AC gadgets it needs to be converted into AC; an inverter does so. Some gadgets can be operated undoubtedly from a DC source but the majority of them is available for AC operation and is therefore comparatively less expensive. It is a viable idea to install DC modules at almost inaccessible locations, where it is costlier to send grid power. Devices such as the PV operated lanterns; home lighting systems, with a small fan, and solar street lights for external illumination are commonly found at such places but are expensive initially. To use the mass produced items the DC power has to be converted to AC, particularly in such places where utility power is present but is not available for long periods because of 'load shedding', a phrase used to indicate that the available power is inadequate for all the loads present in the area and therefore is distributed in turn. This multiplexing is routine in the cities where consumers use either inverters or diesel generators to augment their needs. The latter generates green house gases apart from noise pollution while the former first stores the power in batteries and later converts the stored power to AC during outages. The cycle efficiency of an inverter is roughly 60% which means that 40% of the utility energy is lost so that the purpose of load shedding is defeated. If PV energy is used instead, during the outages, this problem can be mitigated partially. In case, there is a mixed load of DC and AC appliances, the inverter can play a useful role. Whether an inverter is used with PV or as a standby replacement for the utility during outages, it must have a high efficiency so that during the process of conversion excessive energy is not wasted. In the following sections the process of inversion is discussed in its fundamental form since extensive treatments are available in books on the subject [1] and [2].

6.2 Conversion and Inversion

The process of changing one waveform to another is broadly termed *conversion*. In multi converter power electronic systems, different converters such as DC to DC (choppers), DC to AC (inverters), and AC to DC (rectifiers) are used in source, load and distribution subsystems to provide power at different voltage levels and forms. Thus the MPPT, in its primary form, is a chopper while the act of converting the voltage level from the line voltage of 220 V (rms) to lower levels of DC, to drive different circuits, is called *rectification*. The process of rectification is usually followed by filtration of the ensuing waveform, to reduce the amplitudes of the primary frequency and its generated harmonics, for providing a cleaner DC voltage. The resulting 'clean' DC voltage is further purified, and its voltage stabilized against the voltage excursions of the feeding source or load variations, by introducing electronic regulators. The third process called *inversion* means a conversion of DC power to AC power at the same or a different voltage level. The topic of the current discussion is *inversion* and the instrument used to achieve this is called an *inverter*. The inverter is an essential part of a photovoltaic system because the source here supplies power in DC form while the load usually operates on 220 V (rms). The DC loads are few and far between. A grid interactive inverter on the other hand must supply *PV* power to the grid at the voltage, frequency and phase of the utility.

6.3 Sine and Square-wave Inverters

All the instruments, domestic, commercial or industrial, run on 50 or 60 Hz AC power. Although the lower powered ones use single phase it is convenient to operate the high powered ones on three phase electricity. If square wave electricity is used to drive inductive loads, the harmonics will cause problems. For instance the Fourier series for a square wave is shown in Figure 6.1.

$$f(t) = \sum_{n=1,3,5...} \left\{ \frac{4}{n\pi} \sin (n2\pi ft) \right\} \tag{6.1}$$

or, $$f(t) = \frac{4}{\pi}\left\{ \sin (\omega t) - \frac{1}{3}\sin (3\omega t) + \frac{1}{5}\sin (5\omega t) - \frac{1}{7}\sin (7\omega t) + \frac{1}{9}\sin (9\omega t)...... \right\} \tag{6.1a}$$

where $4/\pi$ is the amplitude of the fundamental frequency $\sin (\omega t)$, $\omega = 2\pi f$, and the others are its harmonics. It is to be noticed that:

1. The even harmonics are absent
2. The amplitudes of the harmonics reduce as their order increases
3. The harmonics are alternately positive and negative.

If the amplitude of the square wave is *A* instead of 1, then the right-hand side of Equation (6.1a) should be multiplied by *A*. This means that the peak amplitude of the fundamental wave

$$V_{o\,(\text{peak})_1} = \frac{4A}{\pi} \tag{6.1b}$$

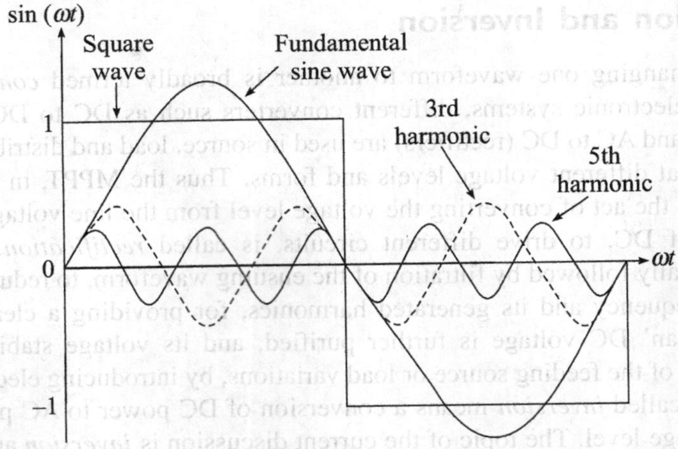

Figure 6.1 A square wave broken up to show its components.

Equation (6.1a) is plotted in Figure 6.1 to show the frequency and amplitude of the harmonics. When any instrument is operated on a square wave power, the harmonics interfere with its operation. For an inductive load, the higher harmonics receive higher reactance. A third harmonic opposes the fundamental current flow and consequently tends to reduce the speed of an electric motor besides causing audible core vibrations. The sound of the core vibration is particularly felt in tube lights with inductive ballasts. However, for capacitive loads a large harmonic current flows but having lower amplitudes this is contained within limits in some cases although the power factor is affected. Still, inverters with square wave output were made and sold and the reason for this is not inconceivable. To understand this, it is necessary to know the efficiencies of class *A*, class *B* and class *D* amplifiers. Consider a class *B* amplifier. This amplifier conducts over half a cycle.

The transistor in Figure 6.2 does not conduct when there is no signal present at its input. If the upper half of a sine wave is now given at its base, this emitter follower configuration,

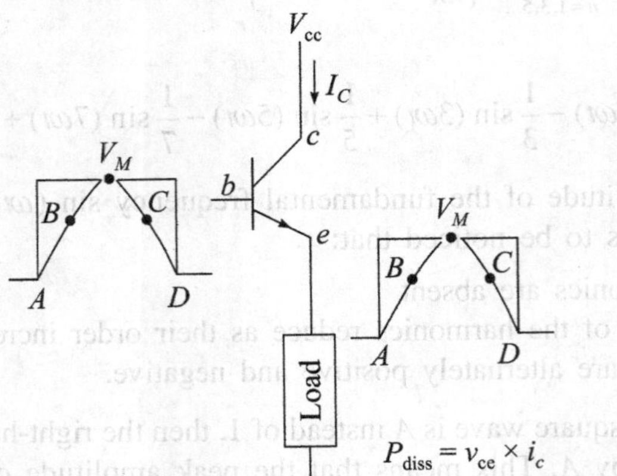

Figure 6.2 Efficiencies with sine and square waves.

with a gain of almost unity, will give the same output at its emitter. Consequently, the current starts rising from zero to a maximum of I_M, giving rise to a voltage from point A to V_M at the output. When at A, the current input is 0 A so that the power dissipation of the transistor is 0 W. At point B, the input current is finite, I_B amperes, while the voltage has risen to V_B V. The dissipation here is $I_B \times (V_{cc} - V_B)$ W. Similarly, at point V_M, the dissipation of the transistor is $I_M \times (V_{cc} - V_M)$ W. This shows that along the sinusoidal path from A to V_M there are finite currents and voltages which when summed up will contribute to a lot of power dissipation in the transistor which will manifest as heat. Such wastage of power will naturally reduce the efficiency of the amplifier. Photovoltaics being an expensive power source, a designer will obviously try to reduce or even eliminate such losses.

Now, consider the square wave which has been shown on top of the sine wave. At point A, both the voltage and the current are zero and hence their product is also zero meaning thereby that no power is wasted. The next instant the current rises to I_M, but the voltage across the transistor falls to $(V_{cc} - V_M)$. $(V_{cc} - V_M)$ being very small, normally it is the transistor saturation voltage, the dissipation is also small and continues to be so, till the top of point D, where it abruptly falls to zero. Thus, the power wastage here is far lower than in the case of a sine wave. The efficiency of a class B circuit is typically $\pi/4$ or 78.5%, whereas that in the case of a square wave exceeds 90%. A switching amplifier that conducts over 180° is called a class D amplifier. This is the reason for the survival of the square wave inverter. The second reason is that the earlier switching devices such as the silicon controlled rectifiers had a limitation by way of uncontrolled switching. These could be switched ON but switching them OFF was extremely difficult. The situation took a turn for the better with gate controlled devices such as the Gate Turn off Thyristors, power MOSFETs and IGBTs. The latter two can also switch reasonably high currents at high speeds. Thus, all half and full bridge switching circuits can be considered as forms of class D amplifiers.

6.4 Pulse Width Modulation

The next problem is how to produce a high efficiency sine wave inverter, without creating harmonics, with a switching amplifier. The concept of pulse width modulation becomes crucial for achieving this goal. Normally a train of pulses, with each pulse having the same width, is produced by conventional pulse generators as shown in Figure 6.3. Each pulse has the same t_{on}, its top end. The t_{on} can be different from t_{off}, the bottom end, but if these two quantities are equal a square wave results as shown below the train. Now, consider the circuit of Figure 6.4.

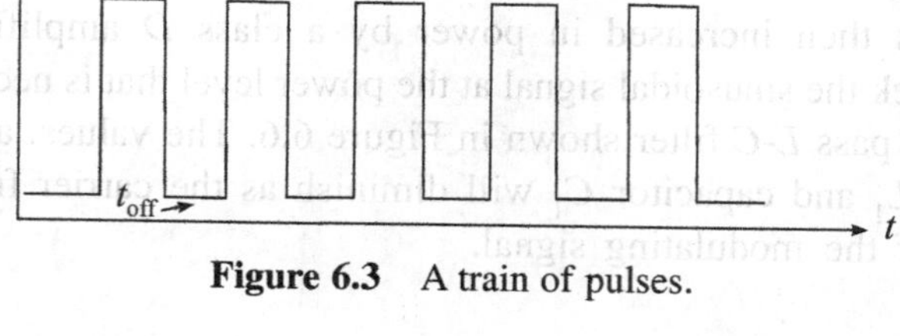

Figure 6.3 A train of pulses.

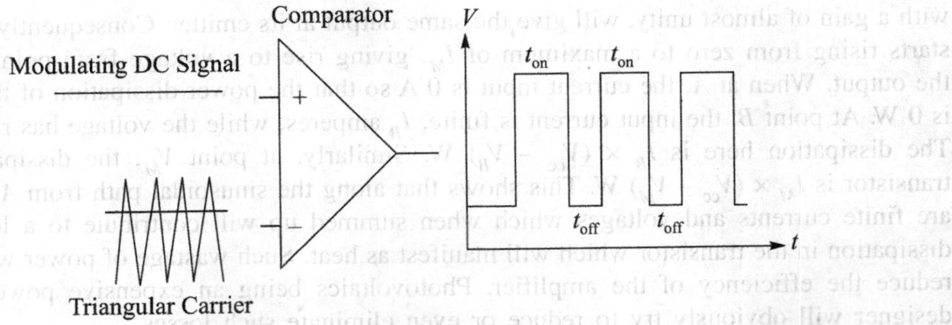

Figure 6.4 A DC modulated pulse train.

With a triangular carrier wave at one input of a comparator and a DC voltage at the other, called the *modulating signal*, the pulse train that appears at the output has a larger t_{on} and a smaller t_{off}. The line running through the triangle represents the DC voltage. By lowering or raising the DC voltage the points of intersection between the two will give different pulse widths at the output. If now a sinusoidal signal replaces the DC as the modulating signal, the voltage variation too will be of the same nature and the output signal will have a modulation accordingly, as shown in Figure 6.5. It is important to remember that the frequency of the triangular wave is more than that of the modulating signal by a factor of several hundreds or even thousands. So that the PWM that emerges after modulation contains a large number of pulses per cycle of the sine wave.

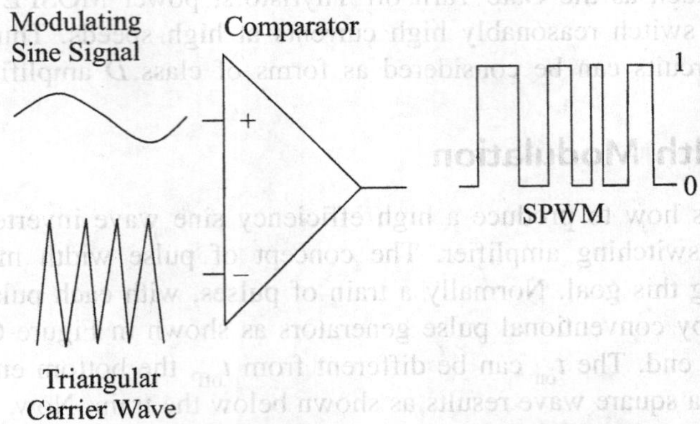

Figure 6.5 Production of sine pulse width modulated signal.

6.5 Sine Pulse Width Modulation (SPWM)

The SPWM signal is then increased in power by a class *D* amplifier and subsequently demodulated to get back the sinusoidal signal at the power level that is necessary. Demodulation here is done by a low pass *L-C* filter shown in Figure 6.6. The values, and therefore the size, of both the inductor L_1 and capacitor C_1 will diminish as the carrier frequency increases in comparison to that of the modulating signal.

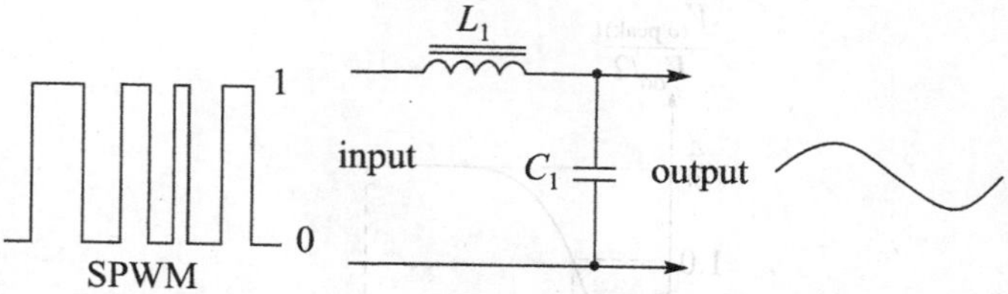

Figure 6.6 Sine waveform restored after filtration.

All modulated signals, whether amplitude or frequency, have sidebands and so does this one. If the comparator operates from a power source of $\pm V_1/2$ V, then the resulting SPWM will fluctuate roughly between these two values; this is shown in Figure 6.7. The SPWM so produced is used to operate the switches of the half bridge inverter of Figure 6.9. The inverter gives an output pulse train similar to the input switching signal but at a higher power level. This is then demodulated to retrieve the sine wave.

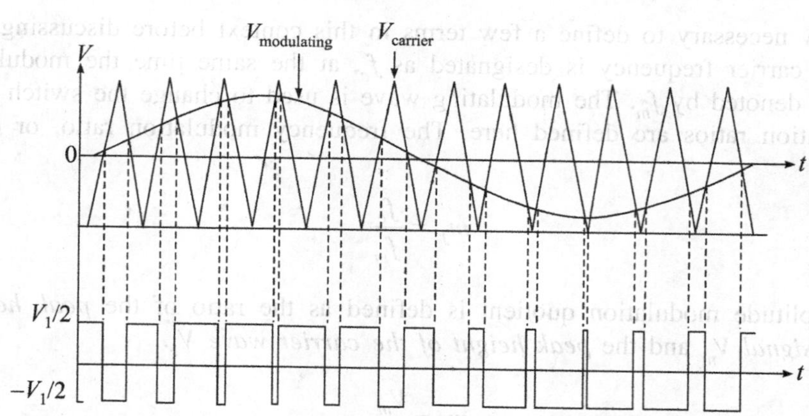

Figure 6.7 The SPWM formation in the comparator of Figure 6.5.

Several features of this type of modulation are as follows:

- The amplitude of the triangular carrier wave is kept constant while that of the modulating wave can vary.
- The frequency of the carrier is kept constant for any particular application. However, it can reduce for very high voltage and power inversion. Similarly, it can increase for low voltage functions.
- For linear modulation, the amplitude of the modulating signal should always be lower than that of the carrier.
- If the amplitudes of the two signals are reversed, then it results in an unwanted situation called *over modulation*.

This can be seen in Figure 6.8 where a curve between the ratio of the peak amplitude of the fundamental and $V_{dd}/2$ is plotted against m_a. The output remains linear as long as m_a is confined within 1 but starts to bend as m_a is further increased showing that the output no longer maintains linear proportionality. This region is to be avoided as it introduces distortion. Beyond about m_a equal to 3.24, the output gives a square wave modulated waveform.

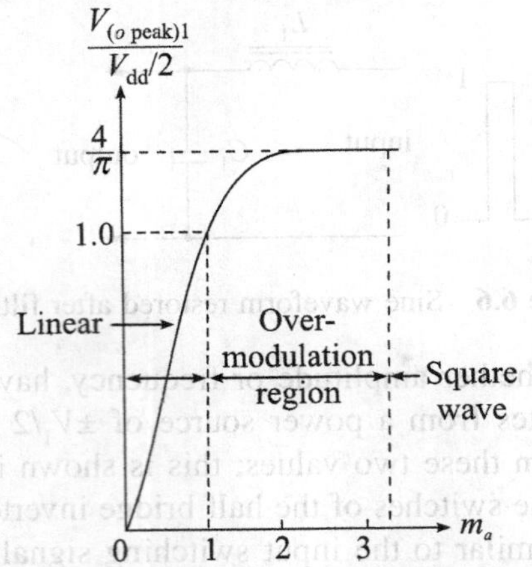

Figure 6.8 Different regions of modulation.

It is now necessary to define a few terms in this context before discussing PWM any further. The carrier frequency is designated as f_c, at the same time the modulating wave frequency is denoted by f_m. The modulating wave is used to change the switch duty cycle. Two modulation ratios are defined here. The frequency modulation ratio, or quotient, is defined as:

$$m_f = \frac{f_c}{f_m} \tag{6.2}$$

The amplitude modulation quotient is defined as the ratio of the *peak height of the modulation signal* V_m and the *peak height of the carrier wave* V_c.

$$m_a = \frac{V_m}{V_c} \tag{6.3}$$

6.6 Single Phase Half Bridge Inverter

Coming back to the inverter of Figure 6.9 its operation will be described in some detail. To start with it is assumed that the point O, situated at the junction of capacitors C_2 and C_3, is at a voltage of $V_{dd}/2$ where V_{dd} is the stable input DC voltage. The two switches SW_1 and SW_2 work alternately. When a pulse turns on SW_1, its inverted form turns off SW_2 for the same time duration. As a result the current i_o flows into the load. The voltage at point P rises to $V_{dd}/2$. At the end of the pulse SW_2 is turned on and SW_1 is switched off. Now the point P is grounded through SW_2. Throughout the process the other end of the load is connected to O at a DC voltage of $V_{dd}/2$, whereas P goes from V_{dd} to almost zero voltage. If both the capacitors C_2 and C_3 are large enough to offer very low reactance to the modulating signal, the voltage at O will remain constant at $V_{dd}/2$. For filtering out the harmonics, it is necessary to know how far the first one lies from the fundamental or modulating signal and how large is its amplitude.

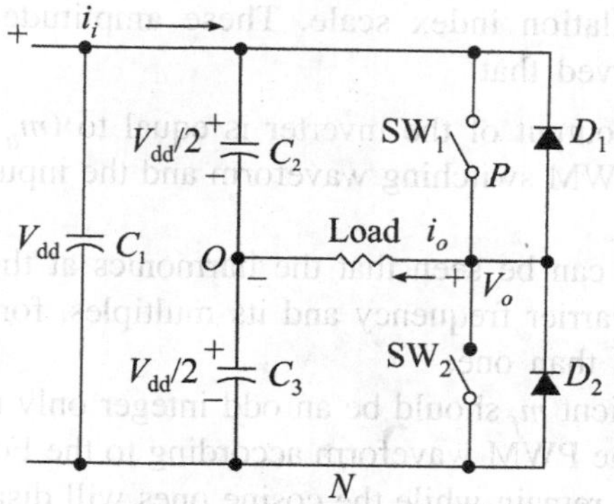

Figure 6.9 A half bridge inverter.

Example 6.1: In the circuit of Figure 6.9, V_{dd} = 200 V, m_a = 0.6 and m_f = 15. The modulating frequency is 50 Hz. Find the values of the peak amplitudes of the fundamental frequency voltage and those of some of the harmonics of the PWM output waveform.

Solution: A chart for the variation of amplitudes of harmonics with m_a is given in [1]. For different values of m_f the amplitudes of the harmonics are given below.

$$\text{Amplitude of the fundamental } V_{o\,(\text{peak})1} = 0.6 \times \frac{V_{dd}}{2} \tag{6.3a}$$

or $V_{o\,(\text{peak})1} = 0.6 \times 100 \text{ V} = 60 \text{ V}$ lies at 50 Hz

The first harmonic's amplitude, called $V_{o\,(\text{peak})15}$ because m_f equals to 15, is

$$V_{o\,(\text{peak})15} = 1.006 \times \frac{V_{dd}}{2} = 1.006 \times 100 = 100.6 \text{ V lies at a frequency of 750 Hz}$$

The m_fth harmonic lies at a frequency of $m_f \times 50$ Hz. Similarly, the 3 m_fth one lies at 2250 Hz. Also

$V_{o\,(\text{peak})17} = \mathit{0.131} \times 100 = 13.1$ V at 850 Hz

$V_{o\,(\text{peak})13} = \mathit{0.131} \times 100 = 13.1$ V at 650 Hz

$V_{o\,(\text{peak})30} = \mathit{0} \times 100 = 0$ volt because the second harmonic $2m_f$ is absent at 1500 Hz.

$V_{o\,(\text{peak})29,\,31} = \mathit{0.37} \times 100 = 37$ V at 1450 and 1550 Hz

$V_{o\,(\text{peak})27,\,33} = \mathit{0.071} \times 100 = 7.1$ V at 1350 and 1650 Hz

$V_{o\,(\text{peak})45} = \mathit{0.083} \times 100 = 8.3$ V at 2250 Hz

$V_{o\,(\text{peak})43,\,47} = \mathit{0.203} \times 100 = 20.3$ V at 2150 and 2350 Hz

$V_{o\,(\text{peak})41,\,49} = \mathit{0.047} \times 100 = 4.7$ V at 2050 and 2450 Hz

Figure 6.10 shows the relative peak amplitudes, and their location, of the fundamental and the harmonics plotted against m_f for m_a equals to 0.6 and when the carrier frequency is greater than the modulating signal by a factor of 15. These figures are printed in italics in the above example. As expected the amplitude of the fundamental frequency, for example, 50 or 60 Hz, is 0.6 times the output PWM waveform peak. It must be noticed that first harmonic amplitude is 1, which is more than that of the modulating signal although located

15 times away on a modulation index scale. These amplitudes of the harmonics change with m_a. It should be observed that:

1. The peak sine wave output of the inverter is equal to ($m_a \times V_{dd}/2$) since the output is similar to the input PWM switching waveform and the input itself follows the definition of Equation (6.3).
2. From Figure 6.10, it can be seen that the harmonics at the inverter output materialize as sidebands of the carrier frequency and its multiples, for example, m_f, $2m_f$, $3m_f$, etc. so long as m_a is less than one.
3. The modulation quotient m_f should be an odd integer only then the even harmonics will be eliminated from the PWM waveform according to the Fourier series. Further only the sine components will remain while the cosine ones will disappear. The carrier frequency f_c may well be above the audible limit of 20 kHz if possible to avoid discomfort if the inverter were to be placed in a working place. However, f_c cannot be arbitrarily large as the switching losses will increase proportionally particularly at high values of V_{dd}. Thus, for practical purposes the starting value for m_f is 21.
4. A desirable feature of the modulation process is the use of synchronous f_c and f_m waveforms which, in turn, requires m_f to be an integer. In case, this is not so then sub-harmonics of the fundamental will be formed. If the f_m is 50 Hz, then the f_c must be 50×21 = 1050 Hz. The amplitudes of sub-harmonics will be very low for m_f more than 21 and can be tolerated for the process of inversion if a transformer is used to isolate the load such as the power grid or an AC motor; otherwise heavy currents at frequencies close to zero will flow. For this reason synchronous PWM is preferred and triggered carrier waves are generated for use with grid interactive inverters. In PCUs for grid applications the m_f can vary because it is dependent on the grid, therefore in such cases the f_c cannot be an integral multiple of the f_m but the two always remain synchronized.
5. Over modulation occurs when the amplitude of the f_m exceeds that of f_c, that is, m_a more than one. This is a condition that may well be avoided.

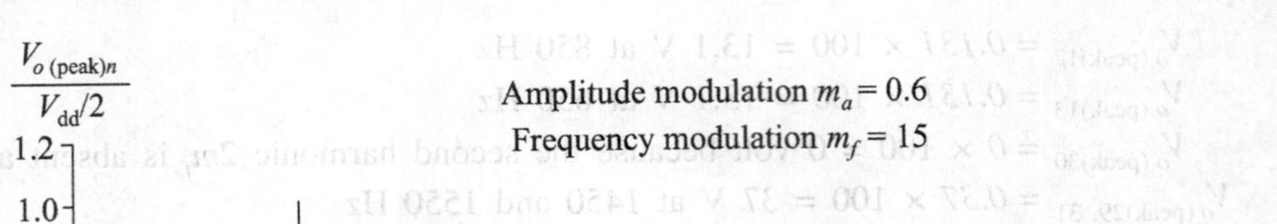

Figure 6.10 Location of harmonics for an SPWM wave from a half bridge inverter.

The above discussion indicates that farther the harmonics are situated from the fundamental the better will be the filtration, that is, m_f should be as large as practicable. Although the amplitudes of the harmonics change with m_a the filter will contain these within acceptable limits. Another benefit of a large separation between the two signals is the reduced size of the filter components which will diminish their cost in addition to the weight.

6.7 Single Phase Full Bridge Inverter

In a full bridge inverter, shown in Figure 6.11, there are four states of operation of the four switches and their accompanying anti-parallel diodes as listed in Table 6.1.

Table 6.1 Status of switches for the inverter of Figure 6.10.

Switch status ON	*Condition if*	v_{BO}	v_{AO}	$v_o = v_{BO} - v_{AO}$
SW_1 and SW_4 D_1 and D_4	$I_o > 0$ $I_o < 0$	$V_{dd}/2$	$-V_{dd}/2$	V_{dd}
SW_2 and SW_3 D_2 and D_3	$I_o < 0$ $I_o > 0$	$-V_{dd}/2$	$V_{dd}/2$	$-V_{dd}$
SW_1 and D_3 D_1 and SW_3	$I_o > 0$ $I_o < 0$	$V_{dd}/2$	$V_{dd}/2$	0
D_2 and SW_4 D_4 and SW_2	$I_o > 0$ $I_o < 0$	$-V_{dd}/2$	$-V_{dd}/2$	0

For bipolar switching, the first two operations are used to generate the PWM but in the case of the unipolar one, all the four states are used. Thus, three voltage states of $V_{dd}/2$, $-V_{dd}/2$ and 0 are obtained in the latter mode of inversion instead of only two, $V_{dd}/2$ and $-V_{dd}/2$, as in the bipolar case.

6.7.1 Bipolar Voltage Switching for SPWM

The single phase full bridge inverter circuit is given in Figure 6.11, where four switches are employed instead of two as used in the half bridge circuit. It can be proved that the output

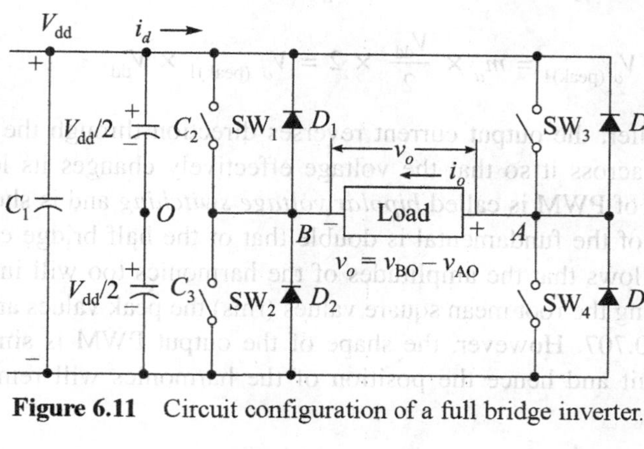

Figure 6.11 Circuit configuration of a full bridge inverter.

v_o is twice that of the previous case. These are two half bridge inverters with the load connected between their two outputs.

An equivalent circuit of the full bridge inverter arrangement is shown in Figure 6.12. The switches work alternately in pairs. When SW_1 and SW_2 conduct SW_3 and SW_4 remain closed. The process is repeated with the two pairs reversing their roles. When SW_1 and SW_2 are in the ON state current I_1 from the high tension supply V_{dd} passes through the load indicated by the arrow. When SW_3 and SW_4 are ON, the current I_2 passes through the load in the reverse direction. The two directions of the current flowing through the load produce a voltage across the load that is effectively twice the V_{dd}. This fact is an argument in favour of the full bridge over the half bridge configuration although it uses two more switches. The generation and sequential application of sine wave pulse width modulation (SPWM) is discussed in this context.

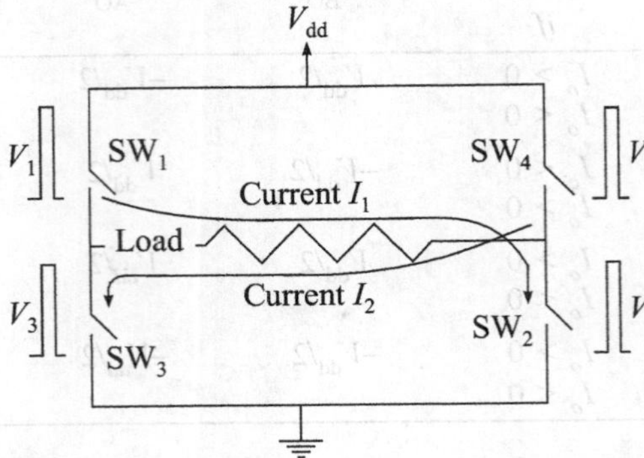

Figure 6.12 Switches arranged as a full bridge inverter.

From Figure 6.11

$$v_{BO}(t) = -v_{AO}(t) \tag{6.4}$$

Therefore,

$$v_o(t) = v_{BO}(t) - v_{AO}(t) = 2v_{BO}(t) \tag{6.5}$$

The peak of the fundamental component is therefore given by Equations (6.4), (6.5) and the expression used in Equation (6.3a), for m_a less than one, as

$$V_{o\,(peak)1} = m_a \times \frac{V_{dd}}{2} \times 2 = V_{o\,(peak)1} \times V_{dd} \tag{6.6}$$

As observed earlier, the output current reverses direction through the load and therefore doubles the voltage across it so that the voltage effectively changes its levels from $+V_{dd}$ to $-V_{dd}$. Thus, this type of PWM is called *bipolar voltage switching* and is shown in Figure 6.13. Since the amplitude of the fundamental is double that of the half bridge circuit, according to Equation (6.6), it follows that the amplitudes of the harmonics too will increase by the same amount. For calculating the root mean square values (rms) the peak values are simply multiplied by $1/\sqrt{2}$ equal to 0.707. However, the shape of the output PWM is similar to the one for the half bridge circuit and hence the position of the harmonics will remain the same.

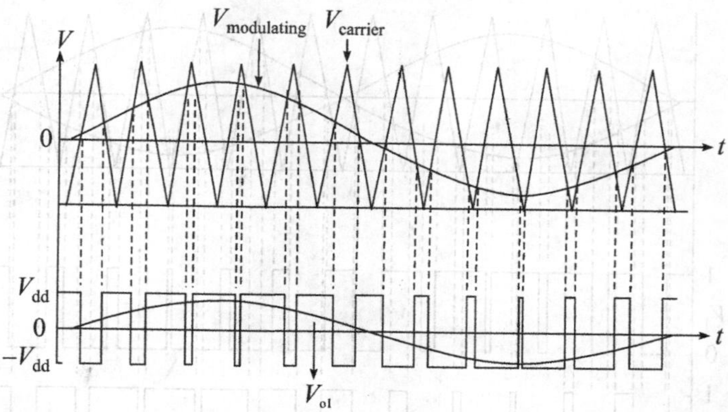

Figure 6.13 Waveforms of the full wave inverter with bipolar switching.

6.7.2 Unipolar Voltage Switching for SPWM

In unipolar voltage type of switching the diagonal switches are not turned ON or OFF for the same duration as in the earlier case but are independently operated by signals produced by comparing two modulating signals, of the same frequency but of opposite phase, with the same carrier wave. A detailed sequence of SPWM pulses, so produced, is shown in Figure 6.14 where not only the generation is clearly shown but also the operation of the switches is given to raise power of the input modulating sine wave. There are four switches to be operated hence four signals have to be generated. When SW_1 is ON, SW_3 must be OFF and vice versa otherwise, if both are simultaneously ON, V_{cc} will short circuit to ground damaging both SW_1 and SW_3 in the process. The same argument holds for the other two switches SW_2 and SW_4.

V_1, V_2, V_3 and V_4 are the voltages used to drive the four switches. The pulses in the train V_1 are created when the sine wave intersects the triangular carrier. This is promptly inverted, by an integrated circuit CD 4069, to form the pulses shown in V_3. Thus, when SW_1 is ON SW_3 is automatically OFF for the same duration and vice versa. To control the other two switches either the carrier polarity has to be reversed or the modulating signal should be inverted. In the diagram recourse to the latter has been taken to produce V_2 and V_4. This action produces a string of PWM pulses, as shown in V_4, which are shifted in phase from the ones shown as V_1. The pulses in V_2 are produced by the phase inversion of those in V_4. It is noteworthy that the diagonal switches do not conduct for the same time duration when they are ON; in fact, they may not perform simultaneously at all. The current flows through the diagonal switches only when both of them are ON. Therefore, an AND multiplication is necessary to determine the current flow. The $V_1 \cdot V_2$ and $V_3 \cdot V_4$ operations are performed and presented in Figure 6.14. It is clear that the two AND products make pulse trains that represent half the modulating sine wave. Since $V_1 \cdot V_2$ and $V_3 \cdot V_4$ are pulses that close the diagonal switches to make current flow in the reverse directions, the voltage that is impressed across the load can be shown as $V_{effective}$ in the Figure 6.14. $V_1 \cdot V_2$ and $V_3 \cdot V_4$ give voltages across the load, therefore the voltage variations in the figure are shown between 0 and V_{dd}. The pulses shown in the $V_{effective}$ train denote a voltage change between either 0 and V_{dd} or 0 and $-V_{dd}$. For this reason, this type of PWM switching is called *unipolar* as against the bipolar scheme where the voltage changes from V_{dd} to $-V_{dd}$.

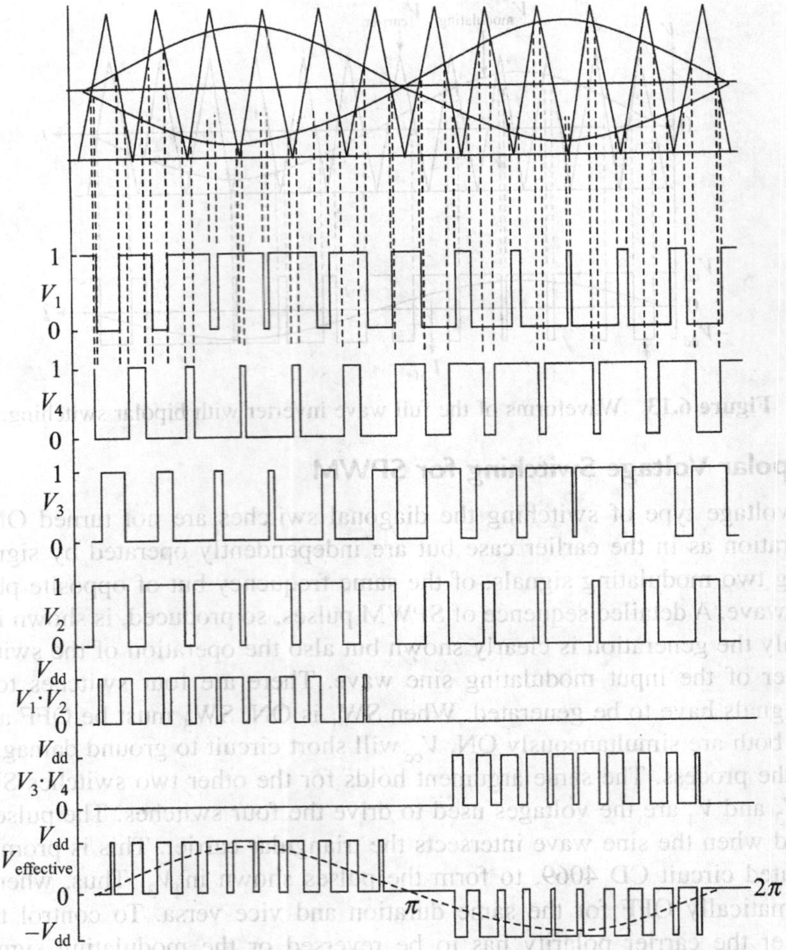

Figure 6.14 SPWM generation and timing sequence of the switches are shown.

Although the peak to peak voltage excursion in both the schemes is $2V_{dd}$, the unipolar one has the distinct advantage of doubling the harmonic frequencies, that is, the lowest harmonic occurs at twice the carrier frequency and all the other harmonics appear as its sidebands. In a single phase inverter with unipolar switching the carrier frequency is an even multiple of that of the modulating signal unlike in the bipolar switching where it is an odd multiple. Further it can be shown that if the above assumption of even m_f remains then:

1. The amplitudes of all the harmonics m_f, $2m_f$, $3m_f$, $4m_f$, etc. are reduced to zero.
2. The sidebands of the odd harmonics such as m_f, $3m_f$ and so on, also cancel out.
3. However, the sidebands of the even harmonics remain.

Example 6.2: Calculate the position and peak amplitude of the fundamental and the first four important harmonics of a unipolar switched full bridge inverter with the specifications listed as follows:

(i) Modulating frequency = 50 Hz (ii) Amplitude modulation index m_a = 0.8
(iii) Carrier frequency = 20 kHz (iv) V_{dd} = 300 V_{DC}.

Solution: The frequency modulation index m_f = 20,000/50 = 400. Therefore, the location of the 1st, 2nd, 3rd and the 4th harmonics will be at

1. m_f = 400 × 50 Hz
2. $2m_f$ = 800 × 50 Hz
3. $3m_f$ = 1200 × 50 Hz, and
4. $4m_f$ = 1600 × 50 Hz
5. The fundamental of course will be situated at 50 Hz.

The amplitudes of all the four harmonics will be zero in a voltage scale as discussed above, however, the even numbered ones will have side bands of finite values.

1. The amplitude of the fundamental = 0.8 × 300 = 240 V
2. Harmonic # ($2m_f$ +1) situated at (801 × 50) Hz = 0.314 × 300 = 94.2 V
3. Similarly, $2m_f$ – 1 at (799 × 50) Hz = 0.314 × 300 = 94.2 V
4. $2m_f$ + 3 at (803 × 50) Hz = 0.139 × 300 = 41.7 V
5. $2m_f$ – 3 at (797 × 50) Hz = 0.139 × 300 = 41.7 V
6. $4m_f$ + 1 at (1601 × 50) Hz = 0.105 × 300 = 31.5 V
7. $4m_f$ – 1 at (1599 × 50) Hz = 0.105 × 300 = 31.5 V
8. $4m_f$ + 3 at (1603 × 50) Hz = 0.115 × 300 = 34.5 V
9. $4m_f$ – 3 at (1597 × 50) Hz = 0.115 × 300 = 34.5 V

The harmonics and their separation from the fundamental are shown in Figure 6.15 as calculated in Example 6.2. Here the first harmonic is 400 times away from the fundamental on a frequency scale but its amplitude is theoretically zero. The others are farther away. This vast separation helps to design a low pass filter, to extract the fundamental, which is relatively small and inexpensive. If these frequencies were retained for a bipolar switching the first harmonic would be about 200 times away, with an amplitude comparable to that of the fundamental, although the frequency would be slightly different—an odd integer.

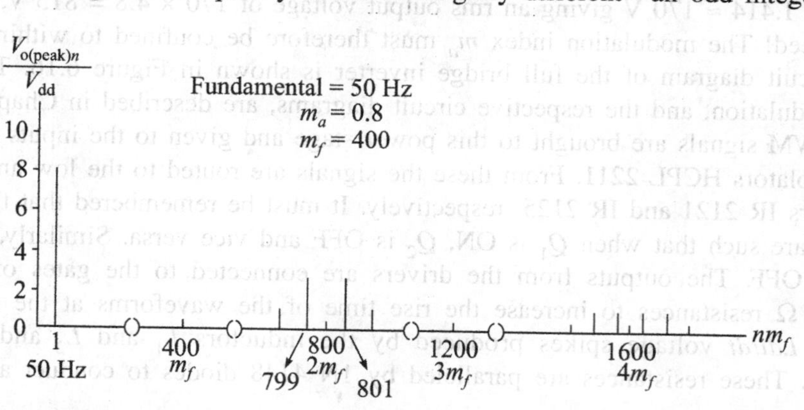

Figure 6.15 Position and amplitudes of unipolar voltage switching of PWM full bridge inverter.

6.8 Design of an SPWM Inverter

It is desired to design and fabricate a full bridge inverter using unipolar switching to produce a 230 V (rms), 50 Hz output for up-loading power to the grid of 220 V (rms). The other specifications are given below:

1. Input voltage = 300 V_{DC}
2. Switching frequency = 20 kHz
3. Output power (rms) = 600 W
4. Amplitude modulation index 0.8 maximum

To proceed with the design work, first the frequency modulation index must be determined. The frequency modulation index, m_f = 20,000/50 = 400. The amplitudes and the position of the harmonics are found next. The required values are calculated in Example 6.2 and are shown in the Figure 6.15. This will help in the determining the components of the LC filter later on.

The topology evidently is that of a high speed switching inverter with a low pass filter to give out a 50 Hz signal. This signal at a low voltage is then passively amplified by a transformer to the required level of 230 V (rms). The primary objective of the transformer is to provide isolation from the grid so that DC and low frequency sub-harmonics generated accidentally during the pulse width modulation process, do not get coupled to the grid. Also for the sake of safety, it is advisable not to have any direct contact between the two, although some manufacturers make directly connected systems. This configuration of the inverter was chosen because a high power and high frequency transformer is very expensive. A description of the high frequency inverter is given in [3]. The details of the available transformer are given below:

1. Input voltage = 48 V
2. Output voltage = 220 V
3. Power handling capacity = 2 kVA

This transformer was selected, like the other magnetic parts of the circuit, to upgrade the inverter to a 2 kW system merely by changing the IGBT switches. However, this device has a much higher turns ratio of 4.8 so that if m_a = 0.8 the rms input voltage will become 0.8 × 300 ÷ 1.414 ≈ 170 V giving an rms output voltage of 170 × 4.8 = 815 V; a very high voltage indeed! The modulation index m_a must therefore be confined to within 0.3.

The circuit diagram of the full bridge inverter is shown in Figure 6.16. The different steps of modulation, and the respective circuit diagrams, are described in Chapter 10. Four trains of PWM signals are brought to this power stage and given to the inputs of four very fast opto isolators HCPL 2211. From these the signals are routed to the low and high level IGBT drivers IR 2121 and IR 2125, respectively. It must be remembered that the phases of the signals are such that when Q_1 is ON, Q_2 is OFF and vice versa. Similarly, when Q_3 is ON, Q_4 is OFF. The outputs from the drivers are connected to the gates of the IGBTs through 15 Ω resistances to increase the rise time of the waveforms at the collectors to reduce the *Ldi*/*dt* voltage spikes produced by the inductors L_1 and L_2 and other stray inductances. These resistances are paralleled by 1N 4148 diodes to conduct away charges

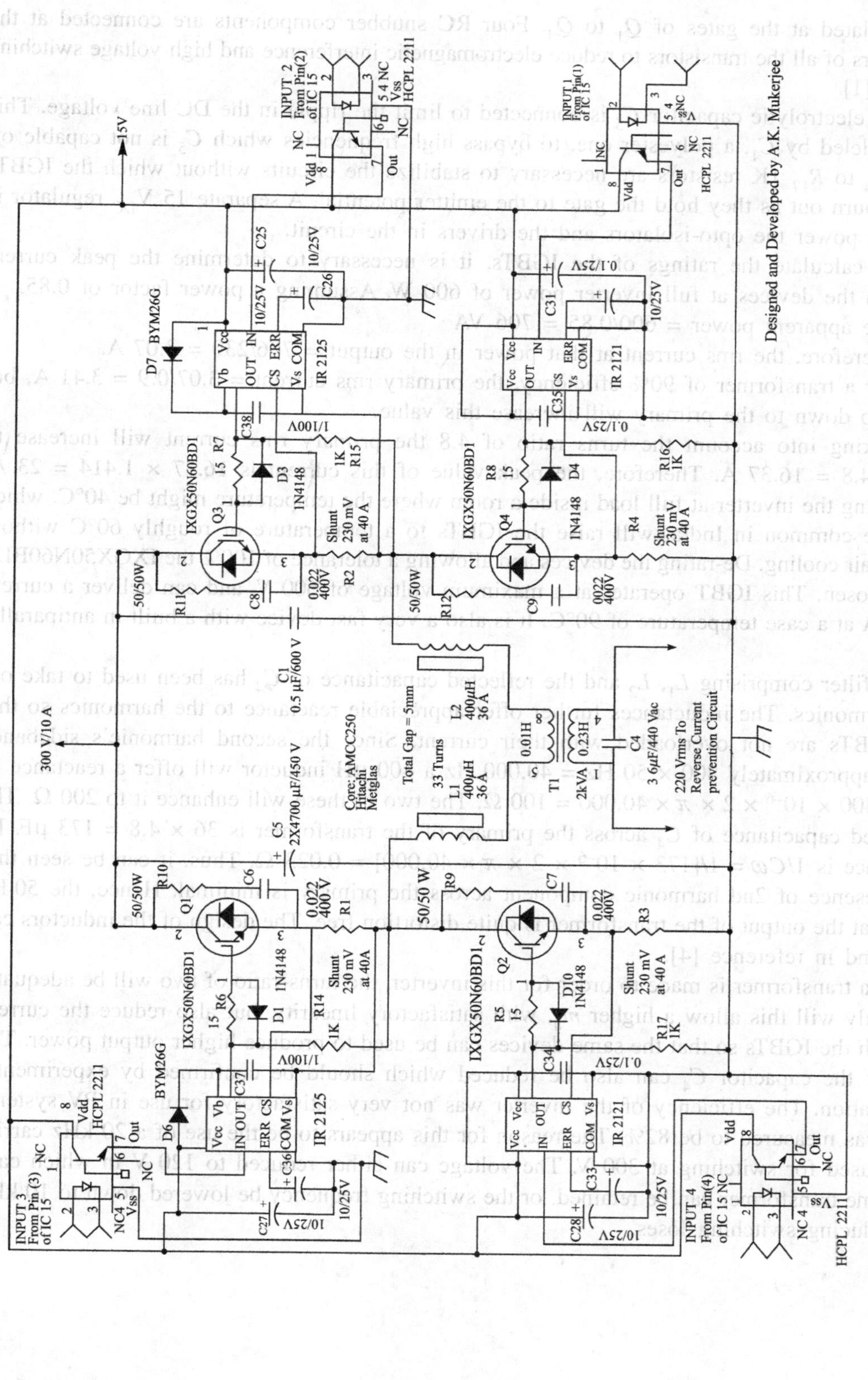

Figure 6.16 Circuit diagram of a 600 W full bridge inverter.

accumulated at the gates of Q_1 to Q_4. Four RC snubber components are connected at the collectors of all the transistors to reduce electromagnetic interference and high voltage switching spikes [1].

An electrolytic capacitor C_5 is connected to limit the ripple in the DC line voltage. This is paralleled by C_1, a polyester one, to bypass high frequencies which C_5 is not capable of. The R_{14} to R_{17} 1K resistors are necessary to stabilize the circuits without which the IGBTs might burn out as they hold the gate to the emitter potential. A separate 15 V_{DC} regulator is used to power the opto-isolators and the drivers in the circuit.

To calculate the ratings of the IGBTs, it is necessary to determine the peak current through the devices at full inverter power of 600 W. Assuming a power factor of 0.85,

The apparent power = 600/0.85 = 706 VA

Therefore, the rms current at that power in the output = 706/230 = 3.07 A.

For a transformer of 90% efficiency, the primary rms current = 3.07/0.9 = 3.41 A, but the step down to the primary will increase this value.

Taking into account the turns ratio of 4.8 the primary rms current will increase to 3.41× 4.8 = 16.37 A. Therefore, the peak value of this current is 16.37 × 1.414 = 23 A. Operating the inverter at full load inside a room where the temperature might be 40°C, which is quite common in India, will raise the IGBTs to a temperature of roughly 60°C without forced air cooling. De-rating the devices and allowing a tolerance of 100% the IXGX50N60BD1 was chosen. This IGBT operates at a maximum voltage of 600 V, and can deliver a current of 50 A at a case temperature of 90°C. It is also a very fast device with a built in antiparallel diode.

A filter comprising L_1, L_2 and the reflected capacitance of C_4 has been used to take out the harmonics. The inductances further offer appreciable reactance to the harmonics so that the IGBTs are not overloaded with their current. Since the second harmonic's sidebands occur approximately 800 × 50 Hz = 40,000 Hz a 400 μH inductor will offer a reactance of $L\omega = 400 \times 10^{-6} \times 2 \times \pi \times 40{,}000 = 100\ \Omega$. The two of these will enhance it to 200 Ω. The reflected capacitance of C_4 across the primary of the transformer is 36 × 4.8 = 173 μF. Its reactance is $1/C\omega = 1/[173 \times 10^{-6} \times 2 \times \pi \times 40{,}000] = 0.023\ \Omega$. Thus, it can be seen that the presence of 2nd harmonic component across the primary is minimal. Hence, the 50 Hz signal at the output of the transformer is quite distortion free. The design of the inductors can be found in reference [4].

If a transformer is made to order for this inverter, the turns-ratio of two will be adequate. Not only will this allow a higher m_a, with satisfactory linearity, but also reduce the current through the IGBTs so that the same devices can be used to produce higher output power. The size of the capacitor C_4 can also be reduced which should be confirmed by experimental verification. The efficiency of the inverter was not very satisfactory for use in PV systems as it was measured to be 82%. The reason for this appears to be the use of a 20 kHz carrier wave used for switching at 300 V. The voltage can either reduced to 120 V in which case the same transformer can be retained, or the switching frequency be lowered down to 10 kHz for reducing switching loses.

Questions

1. For a modulation index $m_a = 3$, find the position and amplitudes of the fundamental and first four harmonics of a bipolar switched full bridge inverter working at 50 Hz. Take $m_f = 100$, $f_c = 10$ kHz and $V_{dd} = 100$ V_{DC}.
2. Repeat Question 1 for a half bridge inverter.
3. A full bridge bipolar inverter made to deliver 1 kW at 220 V (rms) into a resistive load requires suitable switches. Find the voltage and current ratings of the switches. The other variables remain the same as Question 1.
4. Calculate the values of the LC filter for the above inverter so that a 40 dB voltage reduction of the 1st harmonic is obtained as compared to the fundamental.
5. Calculate the values of an LC filter to extract a 50 Hz sine wave from a 50 Hz square wave so that a 40 dB voltage attenuation of the 1st harmonic, with respect to the fundamental, is obtained.
6. Find the wire area of the inductor and the minimum working voltage of the capacitor required in Question 5 for a 1 kW, 220 V (rms) output circuit. Will an electrolytic capacitor work in the filter?
7. Calculate the values of the RC snubber for the above.
8. What other alternatives are available for the IR 2121 and IR 2125 IGBT drivers?
9. What should be the rise time requirement of any optocoupler associated with such an inverter as mentioned in section 6.8?
10. What should be the value of V_{dd} if a transformer-less operation of the inverter is decided to supply 220 V across any load? Take $m_a = 0.8$ maximum.

References

[1] Ned Mohan, Tore M. Undeland and William P. Robbins, *Power Electronics: Converters, Applications and Design*, John Wiley & Sons, New York, 1989, p. 109.

[2] Muhammad H. Rashid, *Power Electronics: Circuits, Devices and Applications*, 2nd ed., Prentice-Hall of India, New Delhi, 2000, pp. 356–412.

[3] P.C. Sen, *Modern Power Electronics*, Wheeler Publishing, New Delhi, 1998, pp. 891–896.

[4] http://www.metglas.com/design/.

CHAPTER 7

BOS Maximum Power Point Trackers

7.1 Maximum Power Point Trackers

Maximum power point trackers (MPPTs) are electronic circuits, used to draw the maximum power produced by a photovoltaic generator. In Chapter 4 a photovoltaic system was described in some length. To recapitulate, it consists of various components that constitute the balance of system (BOS). These include converters, inverters, batteries, battery charge controllers and other components like MPPT. Since the existing solar cell technology does not allow appreciably high conversion efficiencies, it is always endeavoured to have the BOS designed to minimize power losses. Also, several mechanical and electronic systems are deployed for enabling the PV source to deliver the maximum possible power to the load. Figure 7.1 shows the block diagram of a hybrid system with PV as one of its components. The switches can turn to other sources of power in absence of the sun. Generally, a diesel generator produces AC that has to be stepped down and converted to DC if storage is required.

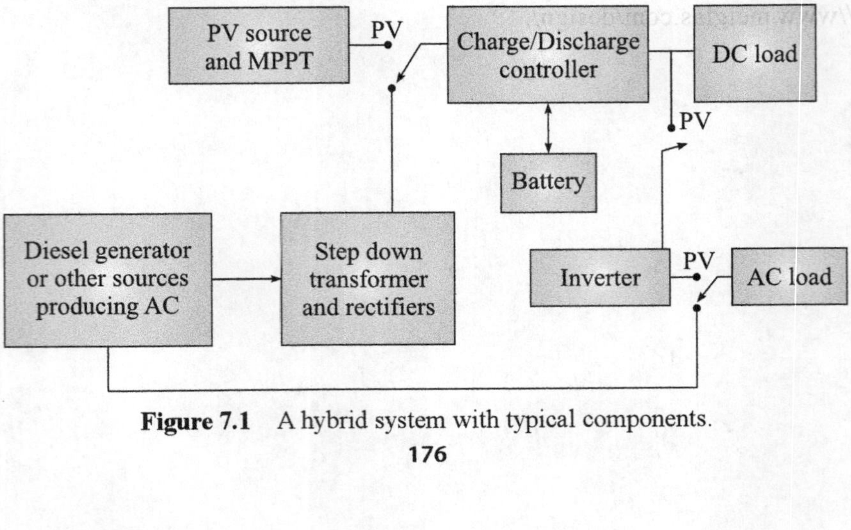

Figure 7.1 A hybrid system with typical components.

To achieve this aim, apart from using electro-mechanical fixtures such as fixed, single or double-axis trackers that track the direction of the sun, certain electronic circuits are also used to ensure operation of the PV source at the maximum power point during different environmental conditions. Such electronic instruments are essentially DC to DC converters called *Maximum Power Point Trackers* (MPPTs). MPPTs ensure that the PV array provides the correct amount of current for operation at the PV MPP so that the load is always supplied with the maximum possible power generated under the given atmospheric conditions. A relatively high cost MPPT is a viable option in high power systems where the cost of the gain in power is higher as compared to the price of the MPPT unit. In effect, the MPPTs reduce the cost of the PV array to a minimum for the required power output. Unlike the sun-tracking systems, the electronic maximum power point trackers do not involve any mechanical tracking but use only power electronic circuitry with a dedicated logic of operation. This chapter is devoted to explaining the concept, problems and possible solutions related to MPP tracking and its advantages in PV systems.

7.1.1 The Concept of MPPT

The power available from a PV source peaks at a particular operating point as explained in Sections 3.3 and 4.6a. This peak power and the corresponding operating voltage (and PV current) also varies with atmospheric conditions such as insolation, temperature etc. Hence, the changing maximum (or peak) power needs to be continuously tracked for optimum operation. In other words, the array should be continually forced to operate at the voltage corresponding to the MPP. Figure 7.2 illustrates the load mismatch that results in operation at a sub-optimal point. The maximum power hyperbola intersects the *I-V* curve (at a particular insolation level), at MP point *M* and the actual load curve intersects the source curve at point *I* and the hyperbola at I_{max}, where it is supposed to be utilizing the entire available power. Tracking algorithms thus have the task of virtually varying the slope of the load line to intersect the *I-V* curve at *M*. An ideal load would always intersect the point of intersection of the *I-V* curves and the corresponding MPP hyperbola. The ideal load curve is also shown in Figure 7.2.

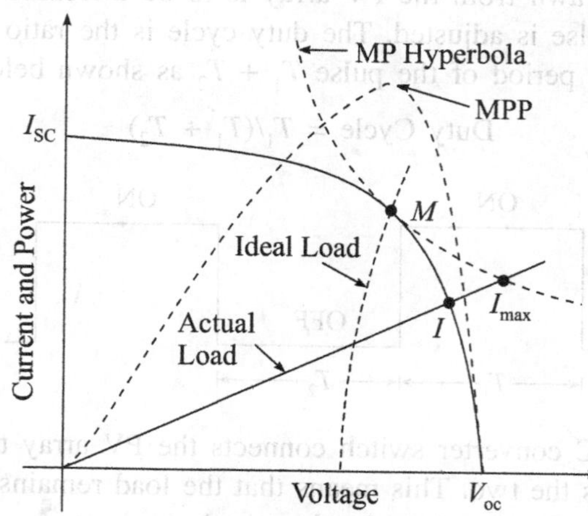

Figure 7.2 The concept of load-mismatch and MPP tracking.

7.1.2 Topology of an MPPT

Basically, an MPPT feeding any DC load like a resistor or a battery is a DC-DC converter with the control circuit ensuring correct switching of the converter for drawing the optimum current from the PV source. The duty cycle of the converter switching signal is manipulated and controlled by a suitable control algorithm that vies for the optimum operating voltage and current required for operation at the maximum power point. With the continuous change in insolation and temperature, this operating point also keeps changing thus requiring the controller to find newer optimum operating points continuously. Figure 7.3 shows an overall block diagram of a typical MPPT.

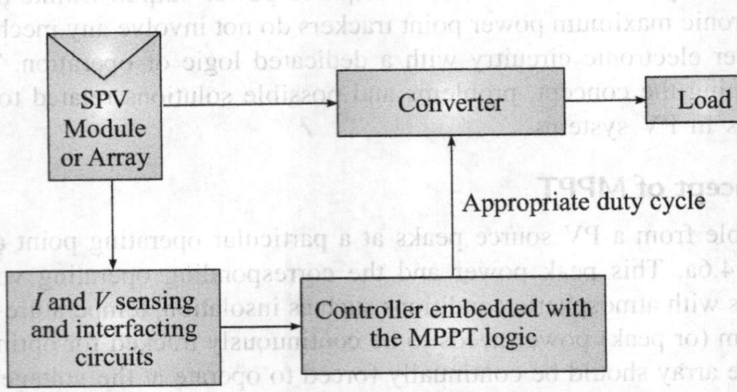

Figure 7.3 Block diagram of a typical MPPT based SPV system.

The controller circuit is either based on analog components or digital ones such as a microcontroller or a digital signal processor (DSP). Depending upon the required accuracy, flexibility and affordability, digital or analog based circuitry can be chosen. The current and voltage from the PV array entering the DC-DC converter are measured by Hall Effect based sensors and fed to the microcontroller in which an appropriate algorithm is embedded. Depending on whether the current drawn from the PV array is to be increased or decreased, the duty cycle of the switching pulse is adjusted. The duty cycle is the ratio of the ON time of the pulse T_1 to the total time period of the pulse $T_1 + T_2$ as shown below, that is,

$$\text{Duty Cycle} = T_1/(T_1 + T_2)$$

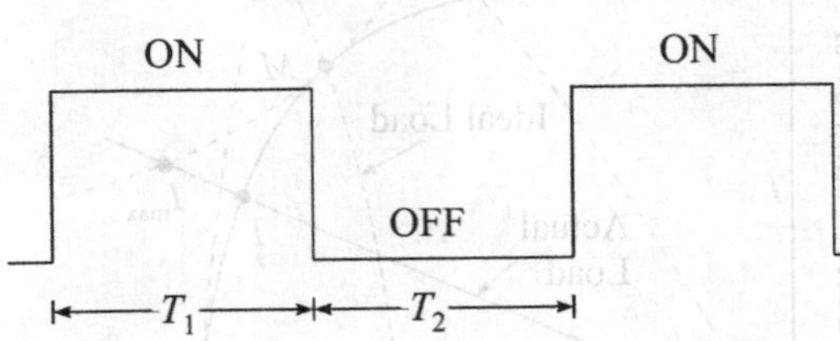

During T_1 the DC-DC converter switch connects the PV array to the load while in the OFF period it disconnects the two. This means that the load remains connected for a longer period when T_1 increases, that is when the duty cycle is more.

Figure 7.4 shows the elaborate block diagram of a typical PV system involving an MPPT shown within the dashed lines. Appropriate voltage and current sensors are involved in sensing the PV array voltage and current which are then sampled at the sensing inputs of the controller that is programmed with particular logic algorithms. With every sampling cycle, the output of the algorithm decides on the magnitude of the duty cycle of the switching signal of the converter. This signal is routed to the converter switch gate by an output port of the controller. Algorithms designed to locate and continually track the peak power point are discussed in the subsequent sections of this chapter.

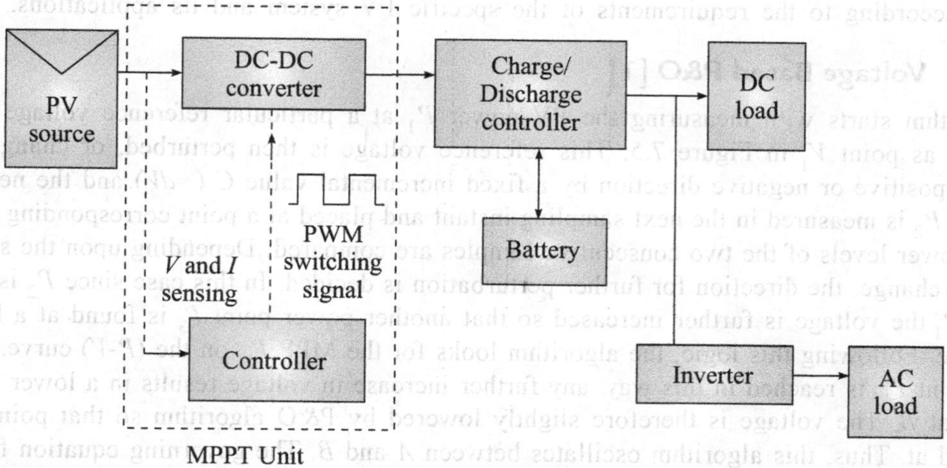

Figure 7.4 Block diagram of a PV system with the components of an MPPT.

There are various factors affecting the power output of PV sources. Partial shading is a key factor affecting the output of a solar cell. Some customized MPPT techniques exist that address such situations and provide solutions for optimal operation and maximum utilization of the PV source in such situations. Such mechanisms and other special techniques of MPPT are discussed in Chapter 8.

7.1.3 Overview of Control Algorithms

A number of algorithms exist that, apart from tracking the MPP, also address specific issues related to tracking and the related electronics. The fundamental mechanism of tracking is that the duty cycle of the converter is started at either a very low or very high or an arbitrary value and the power obtained at that operating point is first observed. The value of the operating point voltage or current is then altered and the new power is observed. Depending upon the direction of increase on power, the duty cycle or the operating point is then altered to approach the optimal point. This is the underlying principle of the Perturb and Observe (P&O) algorithm, the basic mechanism of MPP tracking. Several other algorithms were subsequently developed that address specific aspects of tracking meant to improve the overall tracking efficiency. The most important and widely-used algorithms with their specific mechanisms and approach are discussed in the following sections.

7.2 The Perturb and Observe (P&O)

The primary and classical algorithm implemented in the MPPTs is the P&O method which involves perturbing the point (voltage) of operation and measuring the power delivered by the PV array. This process is followed by comparing the power values at two consecutive instants that leads to the decision on the direction of perturbation in the subsequent sampling cycle. Such a course of action leads to climbing of the hill-like *P-V* curve to ultimately reach the MPP and oscillate around it until the *P-V* curve changes due to the next change in atmospheric conditions. Several variants of the classical P&O exist and can be judiciously used according to the requirements of the specific PV system and its applications.

7.2.1 Voltage Based P&O [1]

Algorithm starts with measuring the PV power P_1 at a particular reference voltage (V_{ref}) shown as point V_1 in Figure 7.5. This reference voltage is then perturbed, or changed, in either positive or negative direction by a fixed incremental value C ($=dV$) and the new PV power P_2 is measured in the next sampling instant and placed at a point corresponding to V_2. The power levels of the two consecutive samples are compared. Depending upon the sign of power change, the direction for further perturbation is decided. In this case since P_2 is more than P_1 the voltage is further increased so that another power point P_3 is found at a higher voltage. Following this logic, the algorithm looks for the MPP P_M on the (*P-V*) curve. Once the point P_M is reached in this way, any further increase in voltage results in a lower power point at *B*. The voltage is therefore slightly lowered by P&O algorithm so that point *A* is arrived at. Thus, this algorithm oscillates between *A* and *B*. The governing equation for the algorithm is

$$V_{ref}(k) = V_{ref}(k-1) \pm C \tag{7.1}$$

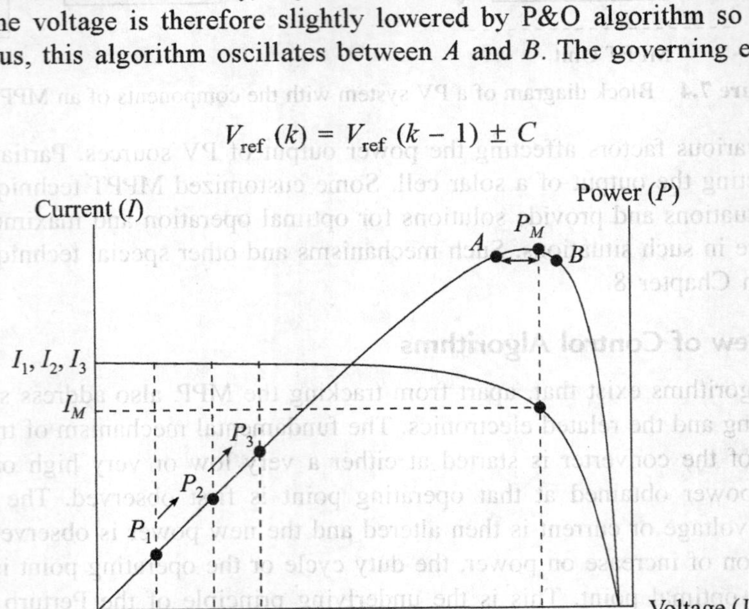

Figure 7.5 Mechanism of P&O alogrithm.

Flowchart of P&O algorithm is given in Figure 7.6, where k is the instant of measurement.

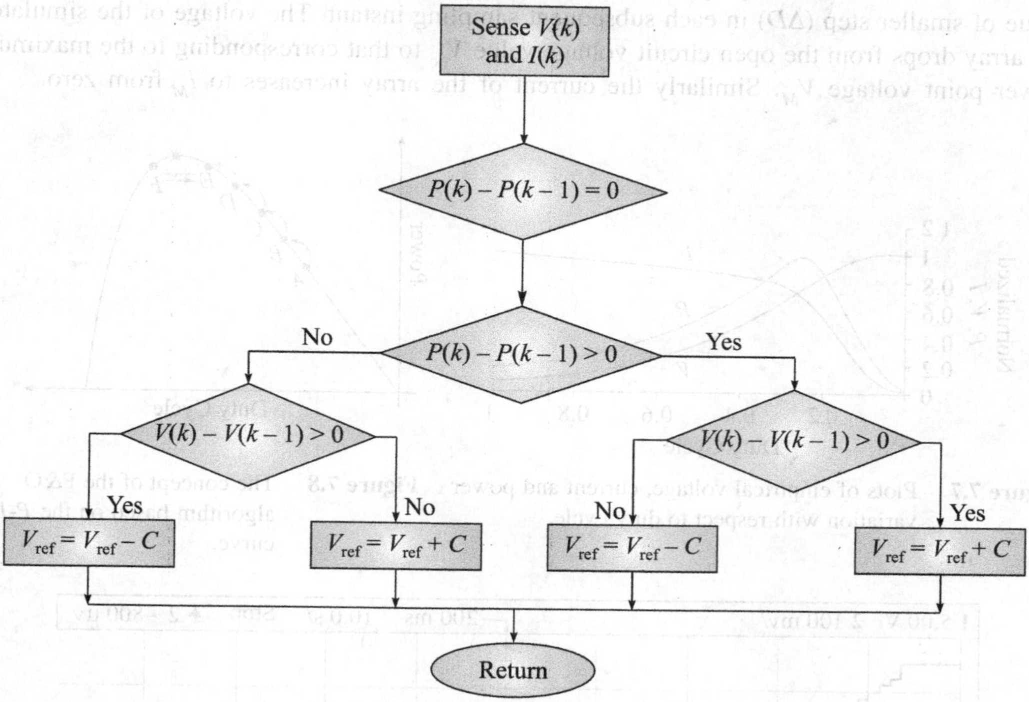

Figure 7.6 Flowchart of the P&O algorithm.

7.2.2 Duty Cycle Based P&O Algorithm [2]

The above explained voltage based algorithm requires two control loops viz one for voltage and the other for power. This makes the sampling loop and the algorithm coding long. Further research came up with a variant of P&O where only one loop is involved. This is based on tracking the power versus duty cycle (*P-D*) curve. Figure 7.7 shows some empirical curves of voltage, current and power of a PV module with respect to duty cycle D. As is observed, similar to the *P-V* curve, the *P-D* curve also maximizes at the maximum power point of the PV source. Thus it is much simpler to directly perturb the duty cycle and observe the values of the corresponding PV power and then follow the same mechanism of the classical P&O algorithm. Figure 7.8 shows the mechanism of the duty cycle based P&O. Whenever the operating point reaches the MPP, the logic forces it to oscillate about the peak with constant amplitude. This is decided by the step size of perturbation to which the duty cycle variable is subjected. The duty cycle based P&O method has the clear advantage of being simple to implement as programming-codes in the digital control devices such as microcontrollers. The governing mechanism is given by the following control equation

$$D(k) = D(k-1) \pm C \tag{7.2}$$

Figure 7.9 shows oscilloscope traces of this algorithm implemented on a buck converter based MPPT where the duty cycle was large in the initial stages and then reduced to a fixed value of smaller step (ΔD) in each subsequent sampling instant. The voltage of the simulated PV array drops from the open circuit voltage value V_{oc} to that corresponding to the maximum power point voltage V_M. Similarly the current of the array increases to I_M from zero.

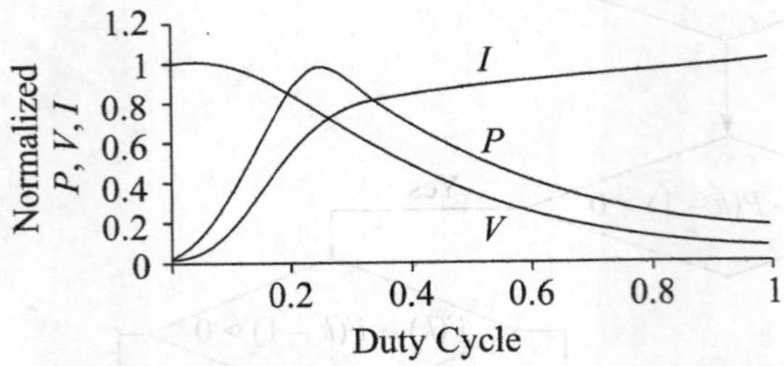

Figure 7.7 Plots of empirical voltage, current and power variation with respect to duty cycle.

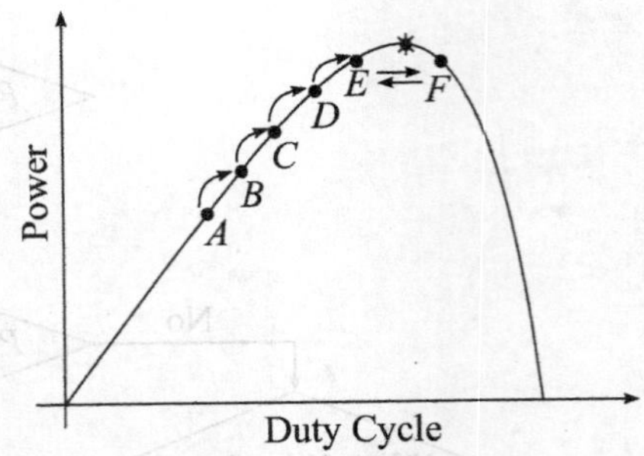

Figure 7.8 The concept of the P&O algorithm based on the *P-D* curve.

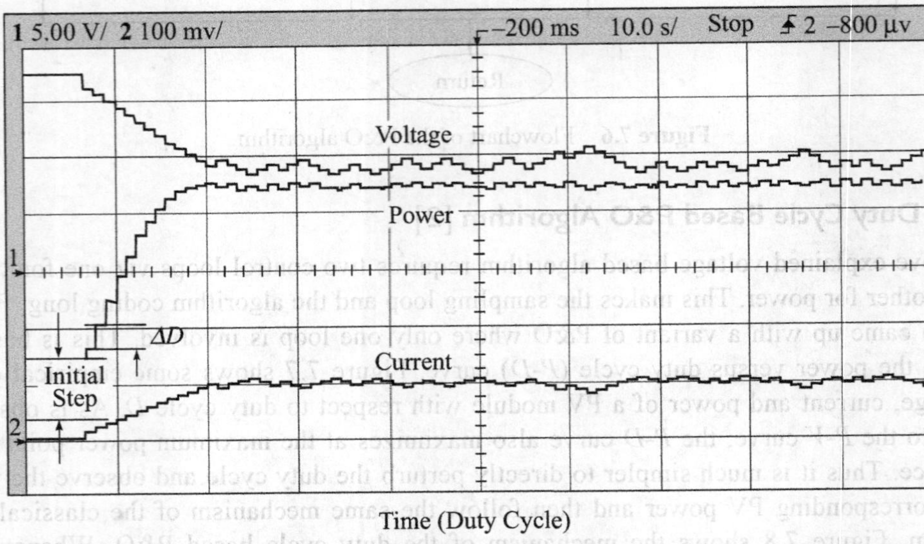

Figure 7.9 Empirical curves showing MPPT working with duty cycle *D* based P&O algorithm.

As expected, the power curve rises comparatively fast to the MPP P_M because of large changes of the duty cycle initially but then stabilizes almost there. The ripple or the variations there exist because of algorithm shifting the power point from *E* to *F* of Figure 7.8. The ripple is small here due to a reduction of the duty cycle step size but would have been large if the initial large perturbation in the duty cycle had been retained.

7.2.3 Variants of the P&O Algorithm

A fixed step size in perturbation makes a major compromise on the dynamic and steady state performance in terms of power loss and speed of response. On one hand, a large step size results in larger oscillations during steady state operation with the consequent loss of power although the MPP is speedily reached. On the other hand, a shorter step size impedes the startup and dynamic response to changing atmospheric conditions by slowing the process of acquiring the optimal operating point. This problem was addressed to a large extent by introducing variable step size methods that eliminate the necessity of this trade-off [3]–[6]. The general equation governing the variable step size P&O methods is the following:

$$\text{Control parameter } (k) = \text{control parameter } (k - 1) + \text{step size parameter} \times \text{slope} \quad (7.3)$$

where,

the parameter slope represents (–1 or +1) that decides the direction of perturbation, and step size parameter is the suitable derivative that decides the amplitude of the subsequent perturbation. The slope of the *P-V* curve and that of the *P-D* curve are eminently used as amplitude parameters while adjusting V_{ref} and *D*. These are apparent from the following equations.

$$V_{ref}(k+1) = V_{ref}(k) + \frac{|P(k) - P(k-1)|}{|V(k) - V(k-1)|} \times \text{slope} \quad (7.4)$$

$$D(k+1) = D(k) + \frac{|P(k) - P(k-1)|}{|D(k) - D(k-1)|} \times \text{slope} \quad (7.5)$$

$$D(k+1) = D(k) + \frac{|P(k) - P(k-1)|}{|V(k) - V(k-1)|} \times \text{slope} \quad (7.6)$$

The underlying principles of the other available variants of this concept are as follows:

- Making step size proportional to *dP* and the slope sign (direction) of perturbation dependent on *dP*/*dV*
- Step size made proportional to *dD* and slope sign (direction) of perturbation dependent on *dP*.

Several such variants of the variable step size hill-climbing method are available that lessen the problem of trade-off between steady state oscillations and the quickness of response to changing conditions. These modifications ensure that the perturbation size is larger when the operating point is away from the MPP but steadily decreases as the MPP is approached, thus significantly reducing the amplitude of oscillations near the peak.

7.2.4 Three Point Search Method

Another novel method of approaching the problem of oscillations and power-loss at the peak point is to observe the power at three consecutive operation points in each sampling instant. As shown in Figures 7.10 and 7.11, power at a sampling instant *k* and one on either direction $(k - 1)$ and $(k + 1)$ are observed. These points are searched at nominal search steps $(\pm\alpha)$.

If any of the conditions other than the following are observed, then perturbation of the operating point in the next sampling instant is skipped. This ensures that there are no unnecessary search perturbations at the MPP thus avoiding oscillations and consequent loss in power.

$$P(k-1) < P(k) \leq P(k+1) \tag{7.7}$$

$$P(k-1) \geq P(k) > P(k+1) \tag{7.8}$$

$$P(k-1) < P(k) > P(k+1) \tag{7.9}$$

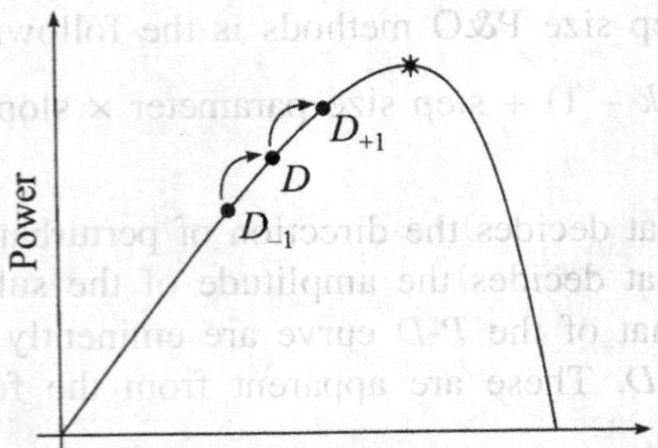

Figure 7.10 Power vs duty cycle curve.

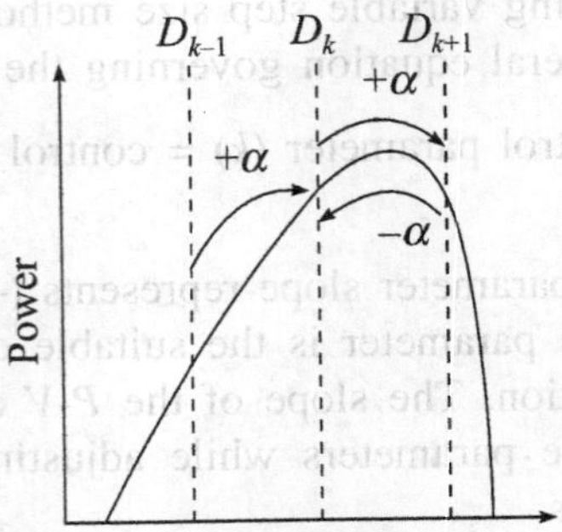

Figure 7.11 Test methodology of three-point algorithm.

7.2.5 History of Two Points to Avoid Oscillations

Another variant of multiple point search method is that of checking the history of search of the two previous operating points in each sampling instant. In this method, the previous two perturbations are checked for deciding whether to take or skip the next step. In case, it is observed that the previous two perturbations do not lead to climbing of the hill, the perturbation of the present sample is not done thus avoiding unnecessary oscillations.

7.2.6 The Concept of Drift

A major challenge faced by most tracking algorithms is that of the tendency of tracking in the wrong direction. This drift of operating point in the wrong direction takes place often during rapidly changing atmospheric conditions.

The drift phenomenon can be explained from Figure 7.12. Assume that initially the operating point is oscillating between points *A* and *B*. The incident solar radiation starts increasing at the instant the operating point shifts from *A* to *B*. The change in duty cycle takes the system's operating point to *C* instead of *B* because the power level has increased. Sensing increase in power, the duty cycle is perturbed in the same direction leading to the operating point being gradually drifted away from MPP (from point *C* to *F*).

Classical P&O algorithm and its variant suffer from the lack of a solution for addressing the situation of drift. Research has yielded potential solutions for this problem. Incremental Conductance (INCC) method is a well-known algorithm that can eminently address this issue. This algorithm is discussed in the next section. Research with the endeavour to overcome the inherent limitations of the classical P&O method has yielded several improved and robust

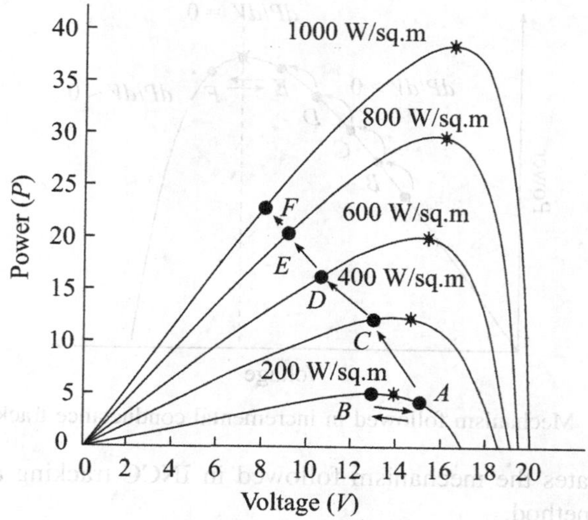

Figure 7.12 The mechanism of drift during MPP tracking.

algorithms. The advantage of this hill-climbing method is the involvement of simple algorithms those are easy to implement. To summarize, the limitations of the classical P&O technique are:

- Oscillation of the operating point around the MPP resulting in loss of power.
- Slow or impeded response during changing atmospheric conditions due to fixed search step size—alleviated by the introduction of variable search step methods.
- Tendency of the operating point of drifting towards the wrong direction resulting in the invention of INCC algorithm.

7.3 Incremental Conductance (INCC)

As is obvious from the *P-V* curve of solar modules, the slope is positive, negative and zero on the left, right and the peak point respectively. This observation is utilized while devising the INCC [6] algorithm. Following this logic, the operating point is said to reach the MPP when the following condition holds:

$$\frac{\Delta P}{\Delta V} = 0 \tag{7.10}$$

Minimizing the Equation (7.10), the following condition holds for the MPP:

$$\frac{\Delta I}{\Delta V} = -\frac{I}{V} \qquad \text{(at the MPP)} \tag{7.11}$$

and the following for the left and right sides of the curve:

$$\frac{\Delta I}{\Delta V} > -\frac{I}{V} \quad \text{and} \quad \frac{\Delta I}{\Delta V} < -\frac{I}{V} \tag{7.12}$$

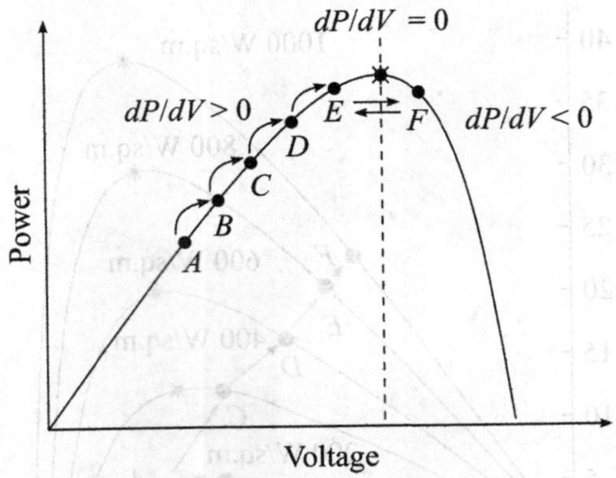

Figure 7.13 Mechanism followed in incremental conductance tracking method.

Figure 7.13 illustrates the mechanism followed in INCC tracking and Figure 7.14 gives the flowchart of the method.

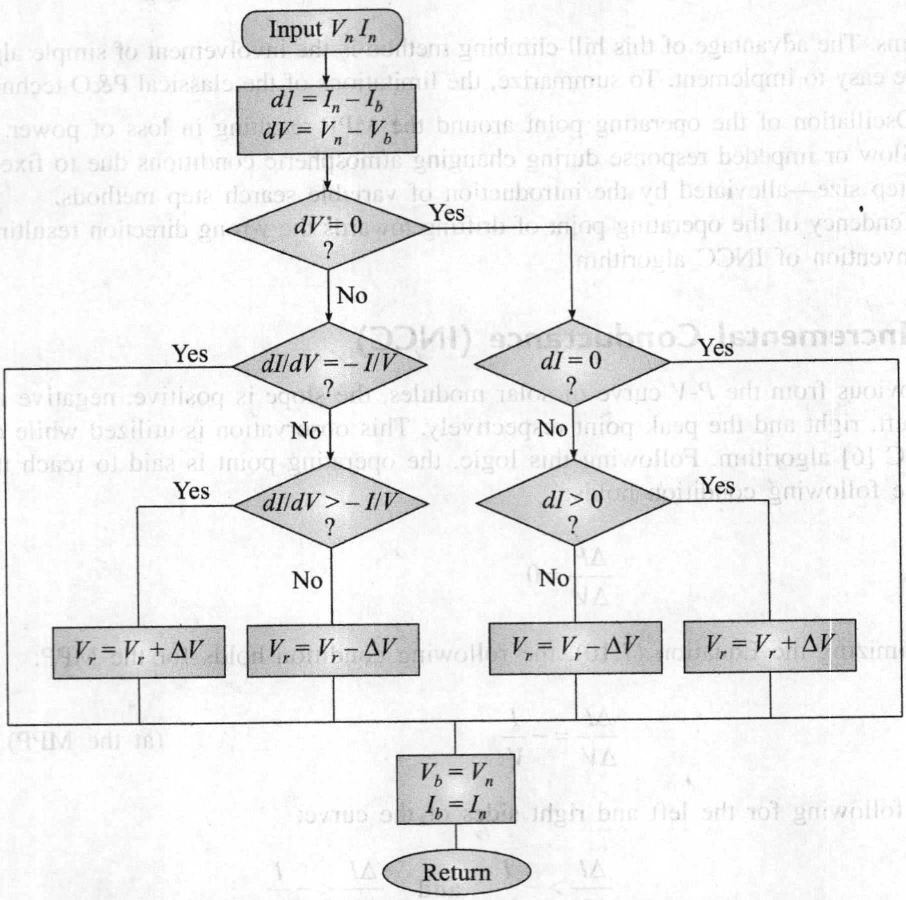

Figure 7.14 Flowchart of the incremental conductance method.

Observation of Equations (7.11) and (7.12) reveals that only the voltage and current values are used to carry out tracking, thereby avoiding the need to calculate power. Since this method observes which side of the peak point the system is working on, this avoids the hill-climbing mechanism. Therefore, in this case, the perturbation direction is not decided based on the comparison of the power value of the previous instant. This is therefore an eminent drift-free tracking technique. Proposed as a more advanced algorithm, this method too has some disadvantages. Involving division operations, its hardware and software implementation is relatively complex. To alleviate the drawback of involving complex calculations that add to the length of implementation codes, a handful of variations and improvements were introduced in the INCC algorithm.

7.3.1 Improvements in INCC

Several improvements to trim down the computational complexity and application requirements have been suggested. Reduction in the number of operations to be carried out in digital processor is desirable. For this purpose, the following simpler equations have been proposed in [7], [8] and [9] respectively while retaining all the advantages of the original INCC.

$$I + \frac{\Delta I}{\Delta V} V = 0 \tag{7.13}$$

and

$$V\Delta I = -I\Delta V \tag{7.14}$$

As can be observed, the above equations involve lesser computations and thus simplify the implementation code of the algorithm.

Several other improvements have been suggested to make INCC more robust while responding to changing atmospheric conditions. For example, in [10], during analog implementation, only the power slope has been used for tracking while the calculation of the voltage derivative is eliminated.

7.4 Hybrid/Complex Methods

For handling the disparate requirements at startup, steady state and changing conditions, hybrid algorithms have evolved. Using combinations of classical methods utilizing their individual advantages, more robust schemes are innovated.

7.4.1 Hybrid Algorithm I

As proposed by [11] a hybrid algorithm consisting of two loops can be used for tracking the MPP more efficiently. Figure 7.15 shows the flowchart for such an algorithm. Here, the measure of amplitude of the change in insolation is recorded with respect to a predetermined value ε. This is a measure of the incremental change in current that reflects the change in insolation. During changing atmospheric conditions, if the insolation changes rapidly, then the algorithm follows the 'insolation loop' for tracking. On the other hand, if the insolation changes slowly, then the temperature loop is followed. Check is conducted in each sampling cycle and the loop is chosen accordingly.

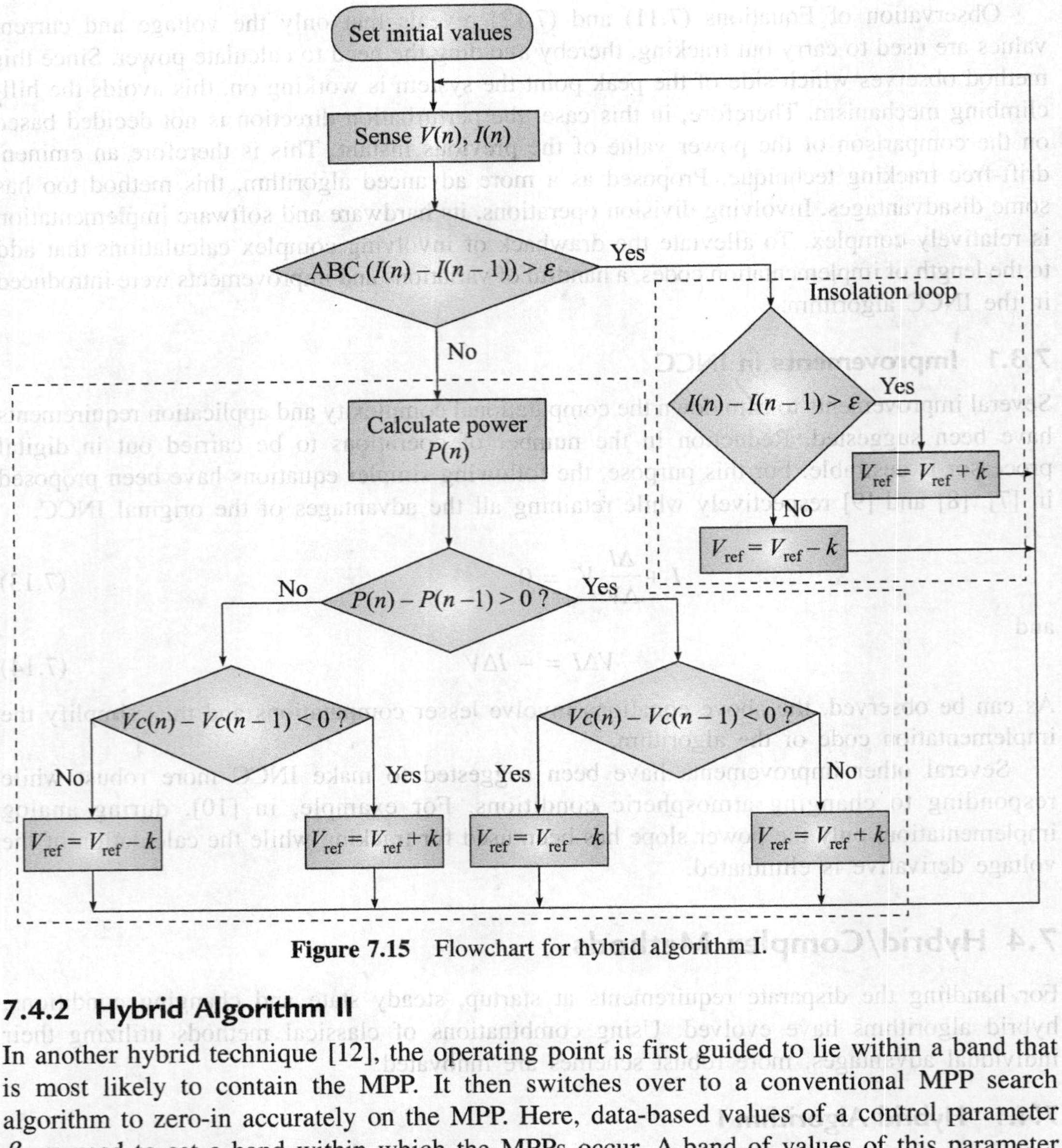

Figure 7.15 Flowchart for hybrid algorithm I.

7.4.2 Hybrid Algorithm II

In another hybrid technique [12], the operating point is first guided to lie within a band that is most likely to contain the MPP. It then switches over to a conventional MPP search algorithm to zero-in accurately on the MPP. Here, data-based values of a control parameter β are used to set a band within which the MPPs occur. A band of values of this parameter β, given by Equation (7.15) is collected over a wide range of temperature and insolation conditions.

$$\beta = \ln\left(\frac{i_{PV}}{v_{PV}}\right) - c \times v_{PV} \tag{7.15}$$

where c is a parameter dependent upon Boltzmann's constant, electronic charge, ambient temperature, etc. The MPPs are observed to be usually located within this band. β_{min} and β_{max}

values are determined by considering the entire range of possible insolation and temperature variations. Figure 7.16 shows the flowchart of the mechanism of this hybrid method.

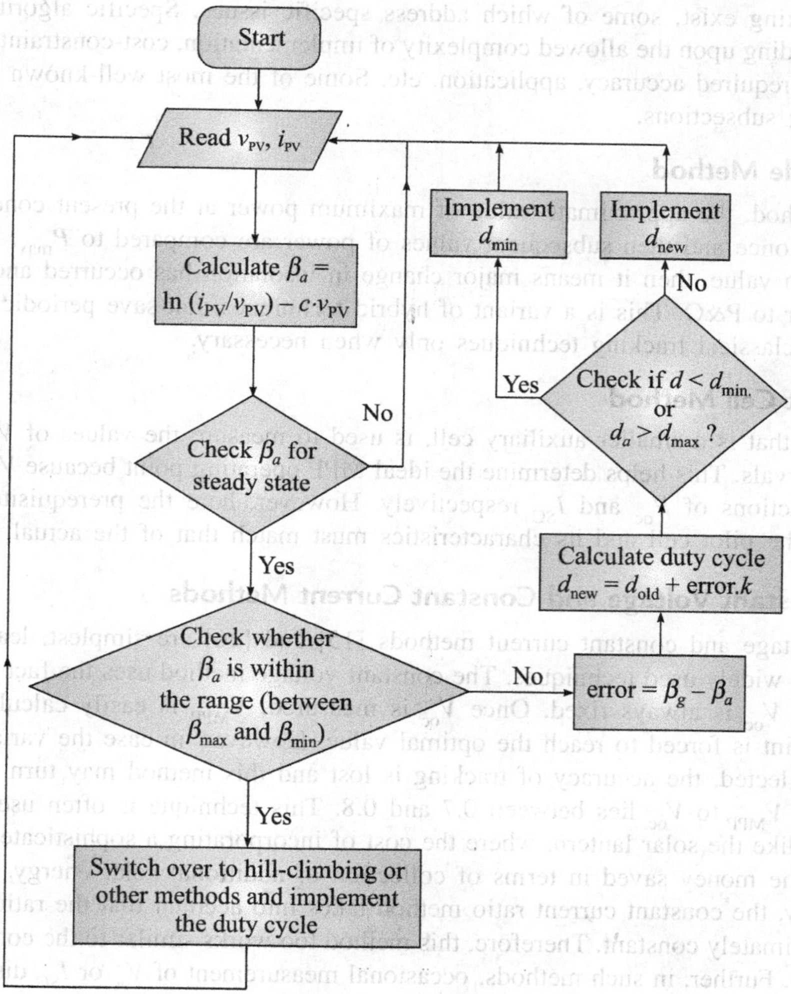

Figure 7.16 Flow chart for hybrid Algorithm II.

7.4.3 Other Hybrid Algorithms

Another adaptation of hybrid techniques [13] suggests a combination of the advantages of P&O and INCC algorithms. Here, normal INCC is periodically interrupted by P&O method.

Bozetepe [14] proposed, in yet another novel hybrid method, to start the converter logic at a voltage close to a pre-determined fraction of V_{oc} that lies close to the MPP. This way the MPPT locates the optimum operating point faster after the logic switches over to a classical algorithm. This method however requires a prior knowledge of V_{oc}.

7.5 Data Based Techniques and Other Approximate Methods

Apart from the various algorithms categorized in the previous sections, several other methods of MPP tracking exist, some of which address specific issues. Specific algorithms can be chosen depending upon the allowed complexity of implementation, cost-constraints, restriction in hardware, required accuracy, application, etc. Some of the most well-known are given in the following subsections.

7.5.1 Crude Method

In crude method, the approximate value of maximum power at the present condition (P_{max}) is calculated once and then subsequent values of power are compared to P_{max}. If it is more than a certain value, then it means major change in insolation has occurred and the system switches over to P&O. This is a variant of hybrid techniques that save periodic calculations and use the classical tracking techniques only when necessary.

7.5.2 Pilot Cell Method

A pilot cell, that is a smaller auxiliary cell, is used to measure the values of V_{oc} and I_{SC} at periodic intervals. This helps determine the ideal MPP operating point because V_{MPP} and I_{MPP} are fixed fractions of V_{oc} and I_{SC} respectively. However, here the prerequisite is that the material of the pilot cell and its characteristics must match that of the actual panel.

7.5.3 Constant Voltage and Constant Current Methods

Constant voltage and constant current methods [15] and [16] are simplest, least expensive and the most widely used techniques. The constant voltage method uses the fact that the ratio of V_{MPP} and V_{oc} is always fixed. Once V_{oc} is measured, V_{MPP} is easily calculated and the operating point is forced to reach the optimal value. However, in case the variations in this ratio are neglected, the accuracy of tracking is lost and this method may turn approximate. The ratio of V_{MPP} to V_{oc} lies between 0.7 and 0.8. This technique is often used in cheaper instruments like the solar lantern, where the cost of incorporating a sophisticated MPPT will far exceed the money saved in terms of collection of additional solar energy.

Similarly, the constant current ratio method takes into account that the ratio of I_{MPP} and I_{SC} is approximately constant. Therefore, this method too works similar to the constant voltage ratio method. Further, in such methods, occasional measurement of V_{oc} or I_{SC} disrupts normal tracking and causes momentary loss in power drawn.

7.5.4 Look-up Table Method

Here prior knowledge of the panel characteristics and data are required and this method turns out to be very source specific. Every time the PV source is changed, a new look-up table needs to be generated and stored.

Although simple but crude techniques such as those given above exist, the classical algorithms are widely preferred for deployment. Depending upon the required accuracy, allowed complexity of implementation, number of allowed sensors, application and cost constraints, particular MPP tracking algorithms are chosen by manufacturers. Further categories of tracking methodologies and comparisons are discussed in the next chapter.

7.6 Instrument Design

7.6.1 Buck Topology-Based System

The block diagram of a buck topology-based system is illustrated in Figure 7.17. The current of the solar array is sensed by a Hall Effect based sensor while the voltage is measured either by a similar device or simply a voltage divider to reduce the cost. These are then buffered by an operational amplifier before being fed to a microcontroller. Inside this device the two are multiplied to give power and then further processed according to the algorithm embedded into it. Appropriate power supplies energize the controller. These are universal in operation meaning thereby that their input voltage can vary over a range of 100 V to 400 V_{DC}, while keeping the output voltage constant, to tolerate the variations of the PV array. The output pulses from the microcontroller are then optically isolated by an HPCL 2211 isolator and fed to the input of the MOSFET driver IR 2121. This is required to provide electrical isolation from the MOSFET which operates at a high voltage.

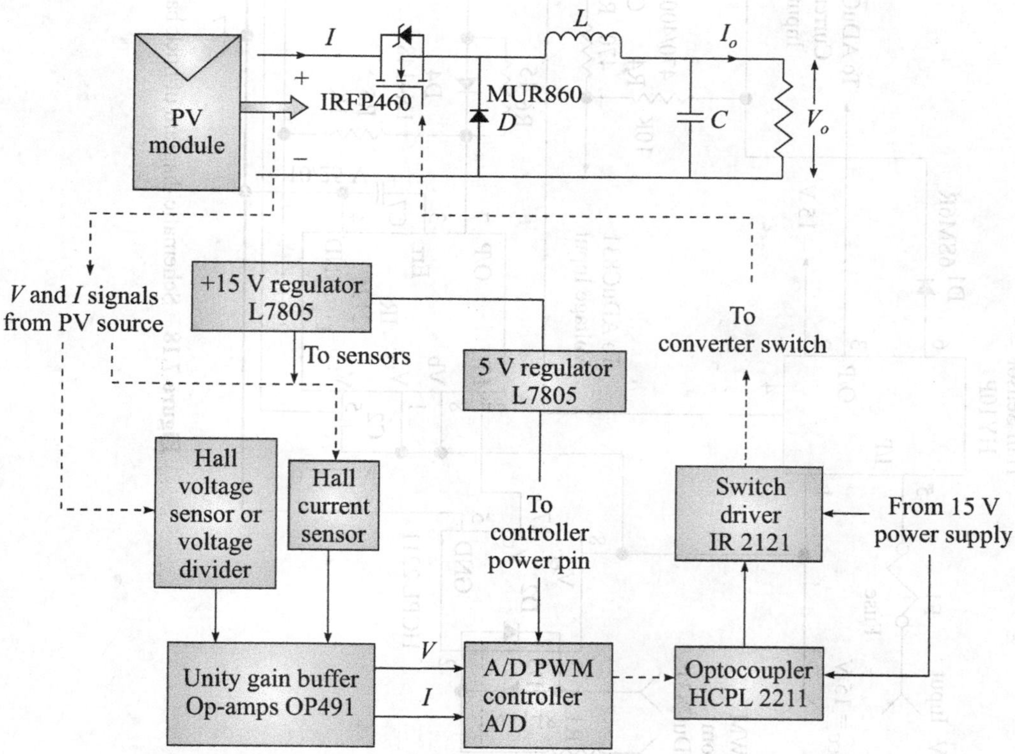

Figure 7.17 A buck converter based MPPT with supporting blocks.

The schematic diagram is given in Figure 7.18. Photograph of such a system is shown in Figure 7.19. The switch and diode are IRFP460 (Power MOSFET 500 V, 20 A) and MUR860 (Ultrafast Rectifier 8 A, 600 V) respectively. The Hall Effect sensors are useful for

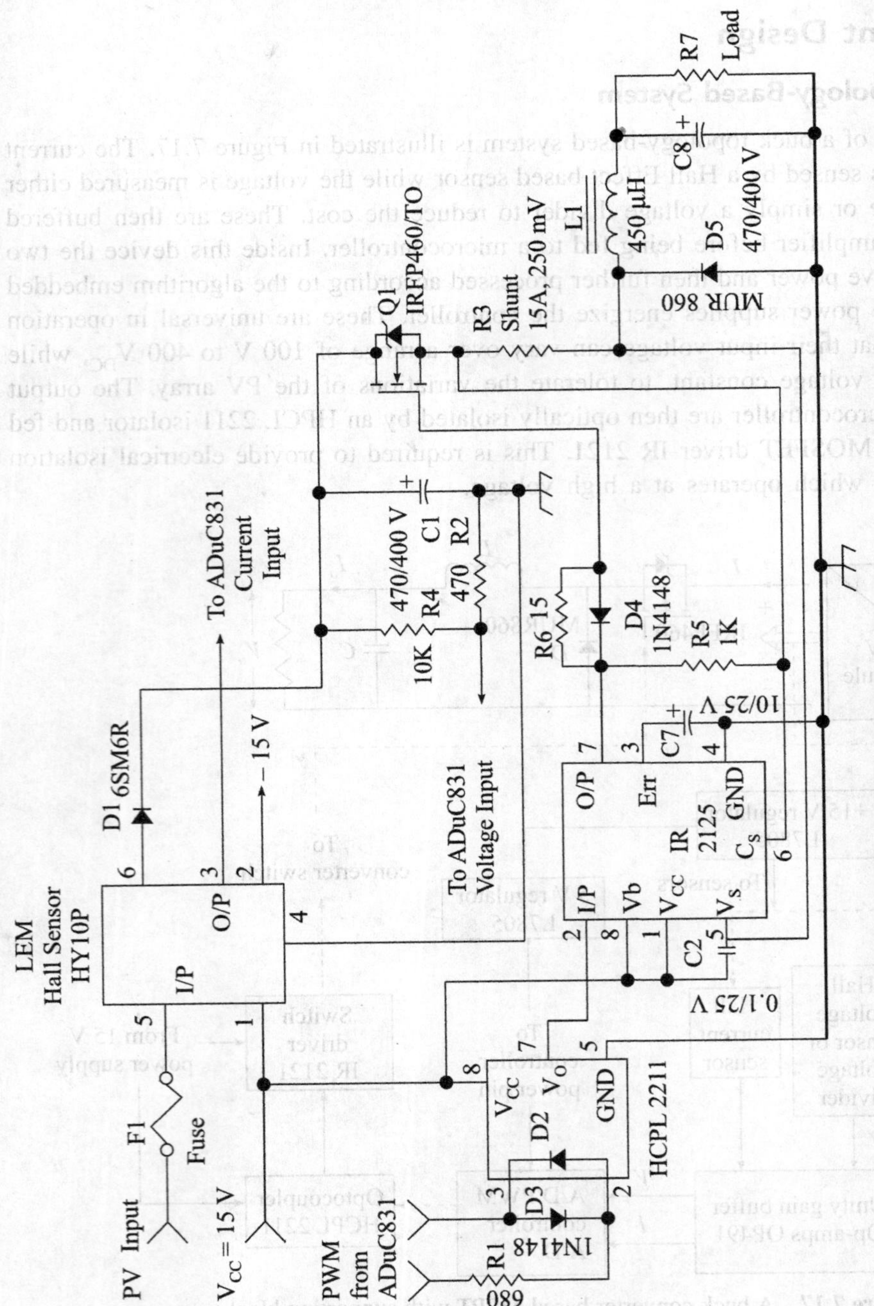

Figure 7.18 Schematic diagram of buck based MPPT system.

not only providing high voltage insulation but also for giving zero resistance to the current path so that there is no voltage drop. The diode D_2 provides protection to the circuit from accidental reverse polarity of the PV connection. The diode D_3 saves the LED inside the opto-coupler HPCL 2211 from reverse biasing while D_4 drains away any charge that might accumulate at the gate of the MOSFET IRFP 460; this makes the circuit safer. The shunt resistance placed at the source of the MOSFET generates a voltage to turn off the driver IR 2125 at an overload condition so that the instrument is protected even if its output short circuits.

Example 7.1: Calculate the values of the inductor L and the capacitor C for the buck based MPPT of Figure 7.19. The specifications are given below:

1. Input voltage range (V_i) = 7–50 V_{DC}
2. Switching frequency (f) = 10 kHz
3. Output voltage (V_o) = 6 V_{DC} (within range to charge a 6 V battery)
4. Output current (I_o) = 4 A
5. Minimum load current, I_{min} = 0.8 A
6. ($\Delta V_o/V_o$) = 0.005 is the fraction of the output ripple ΔV_o

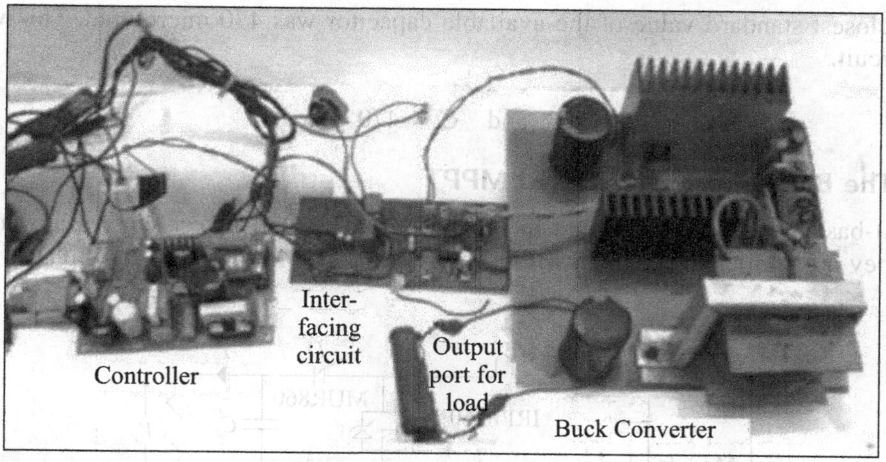

Figure 7.19 Photograph of the buck based MPPT.

The value of the inductor is calculated by the following formula for continuous conduction mode (CCM) for a DC–DC converter [17].

$$L \geq \frac{V_o(V_i - V_o)}{2V_i f I_{min}} \tag{7.16}$$

where

$V_o/V_i = D$ the duty cycle for such converters since V_o proportional to the duty cycle times V_i.

Substituting these values in Equation (7.16)

$$L \geq \frac{6(50-6)}{2\times 50\times 10\times 1000\times 0.8} = 330\ \mu\text{H}$$

The value of the inductor used was 450 μH to ensure that it always remained in the conduction mode.

The value of the capacitor was calculated from the following expression [18]:

$$C = \frac{(1-D)}{8L(\Delta V_o/V_o)f^2} \tag{7.17}$$

here for the worst case

$$D = D_{LOW} = 6/50 = 0.12$$

where D_{LOW} is its lowest value.

Therefore:

$$C = \frac{(1-0.12)}{8\times 450\times 10^{-6}\times 0.005(10\times 1000)^2} = 489\times 10^{-6}\ \text{F} = 489\ \mu\text{F}$$

The closest standard value of the available capacitor was 470 microfarad. This was used in the circuit.

Thus;

$$L = 450\ \mu\text{H} \quad \text{and} \quad C = 470\ \mu\text{F}.$$

7.6.2 The Boost Converter-Based MPPT

The boost-based system is illustrated in Figure 7.20. The Hall Effect sensors are not shown here as they are similar to those shown in Figure 7.17. The MOSFET is now placed between

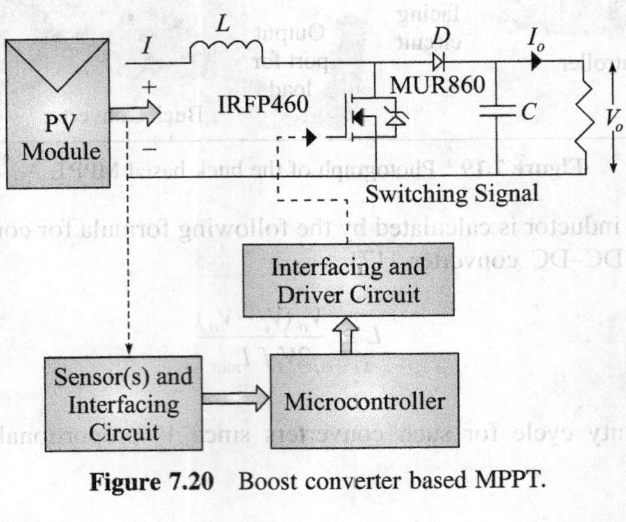

Figure 7.20 Boost converter based MPPT.

the high voltage and the ground so that when it is switched ON, high current i flows through the inductor. When the switch is thrown open the current is interrupted suddenly giving rise to high voltage pulses at its drain according the equation

$$V = L\,(di/dt) \tag{7.18}$$

where

L = value of the inductor

di/dt = the rate of change of the interrupted current.

These pulses charge up the capacitor C so that the voltage across itself exceeds the value at the input. Since this circuit increases the voltage level it is called a *boost converter.*

Example 7.2: Why is a boost converter based MPPT used? Calculate the values of components of such a circuit.

A buck converter is one that always gives an output voltage lower than that at its input. For example, if the lowest input voltage is 10 V, then the output voltage will never exceed this value. In fact considering the voltage drop across the switch, which might be a MOSFET or an IGBT, the output will always be lower than 10 V. The boost converter based circuit, on the other hand, gives an output higher than the one present at its input or simply boosts the input voltage. Consider the maximum voltage V_M versus time curve for one day given in Figure 7.21. The buck converter operates on the line AB and gives an output less than this voltage. Hence, the time of day covered here is less than that covered by CD where the boost converter operates. Thus, the boost converter not only functions for a longer time but it also works when the insolation is low and transfers more enegy than its buck counterpart. The two when incorporated in one unit is called a buck-boost converter based MPPT and has a definite advantage over either. However, its efficiency is lower than their individual ones.

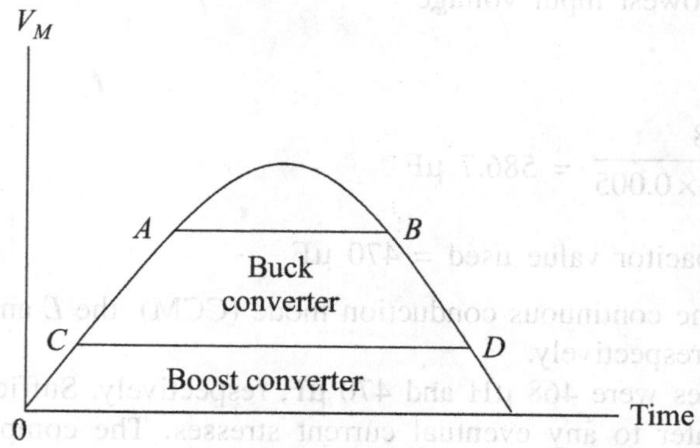

Figure 7.21 Maximum voltage V_M curve showing the areas covered by the two types of MPPTs.

Calculation of L and C values.

The step-up converter inductor value is governed by Equation (7.19)[18].

$$L \geq \frac{D(1-D)^2 R_{\max}}{2f} \tag{7.19}$$

Input voltage range (V): 7–50 V_{DC}

Switching Frequency (f): 10 kHz

Output voltage (V_o): 60 V

$$R_{max} = 60\ \Omega.$$

Output current I = 1 A

Allowed inductor ripple current $(\Delta V_o/V_o) = 0.005$

$$D = D_{LOW} = 1 - (V_{inp,\ high}/V_o) = 1 - (50/60) = 0.17$$

where $V_{inp,\ high}$ is the highest input voltage.

Therefore,

$$L \geq \frac{0.17(1-0.17)^2 \times 60}{2 \times 10 \times 10^3} = 351.33\ \mu H$$

An inductor of 468 μH was used.

Capacitor calculation is done according to Equation (7.20):

$$C = \frac{D_H T}{R_{min}(\Delta V_o/V_o)} \tag{7.20}$$

$$D_H = 1 - (V_{inp,\ low}/V_o) = 1 - (7/60) = 0.88$$

where $V_{inp,\ low}$ is the lowest input voltage

$$R_{min} = 30\ \Omega$$

$$C = \frac{0.88}{10000 \times 30 \times 0.005} = 586.7\ \mu F$$

Therefore, the capacitor value used = 470 μF

For operation in the continuous conduction mode (CCM), the L and C values used were 468 μH and 470 μF, respectively.

The L and C values were 468 μH and 470 μF, respectively. Sufficient margin was kept for the inductor to cater to any eventual current stresses. The complete schematic of the converter and the control circuitry is shown in Figure 7.22. Figure 7.23 shows the photograph of the boost converter.

Caution should be taken with a boost converter because its output continues to rise in absence of a voltage limiting circuit. The output voltage can rise to dangerous levels where the MOSFET and the capacitors might be damaged. To avoid any such situation a load capable of consuming at least 1 A must be always connected to the output. Such MPPTs are very useful for charging batteries.

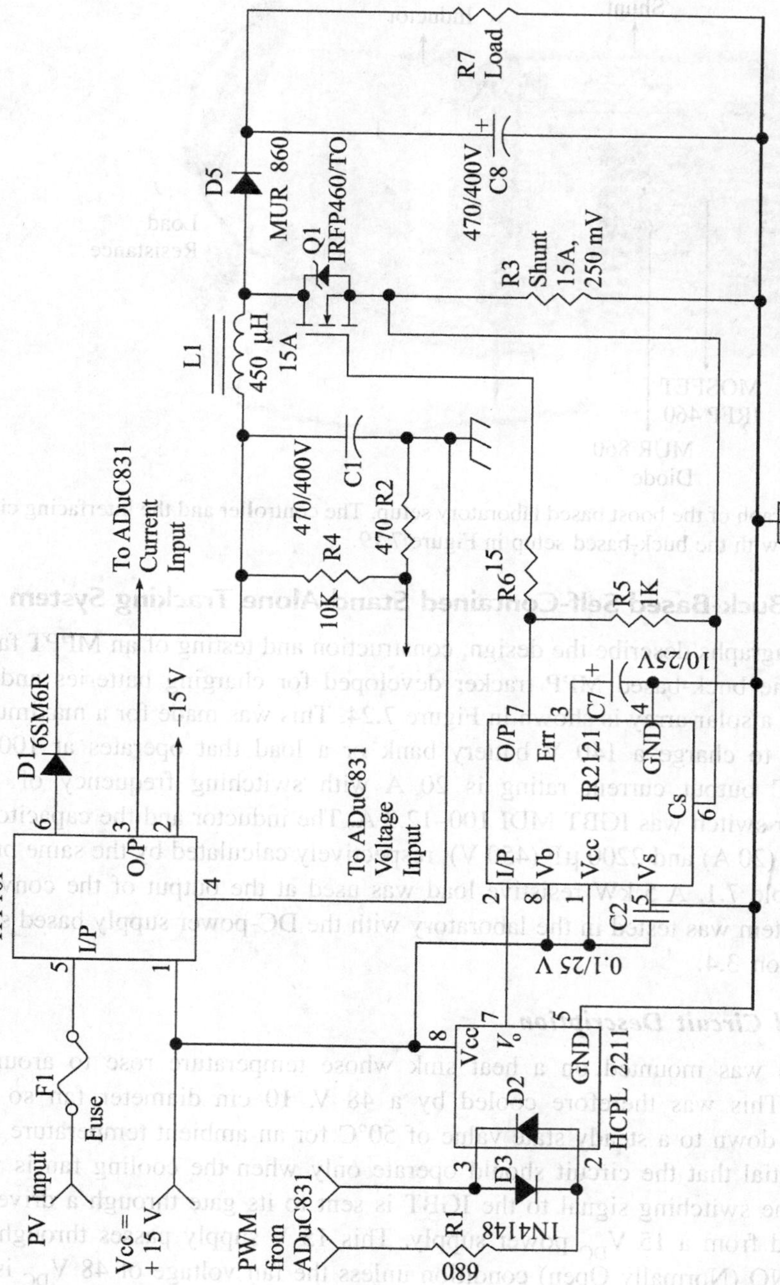

Figure 7.22 Schematic of the boost converter based MPPT.

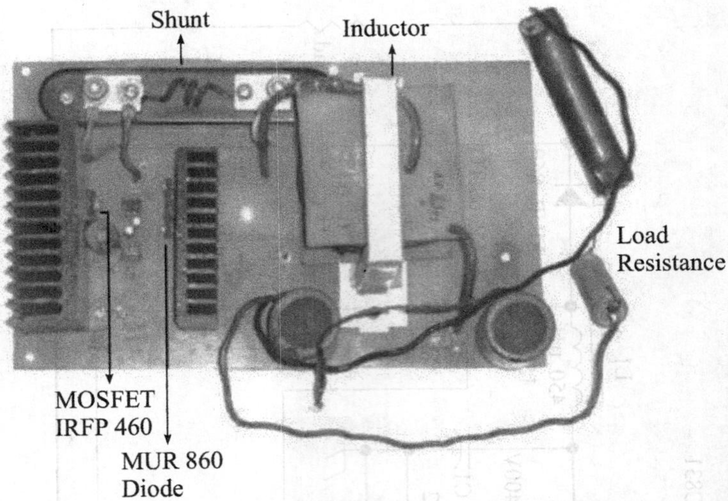

Figure 7.23 Photograph of the boost based laboratory setup. The controller and the interfacing circuits are shown with the buck-based setup in Figure 7.19.

7.6.3 2.5 kW Buck-Based Self-Contained Stand-Alone Tracking System

The following paragraphs describe the design, construction and testing of an MPPT fabricated by the authors. The buck-based MPP tracker developed for charging batteries and driving 2.5 kW loads from a solar array is shown in Figure 7.24. This was made for a maximum input voltage of 210 V to charge a 140 V battery bank or a load that operates at 100–140 V. The maximum DC output current rating is 20 A with switching frequency of 10 kHz. The semiconductor switch was IGBT MDI 100–12.3 A. The inductor and the capacitor values used were 400 μH (20 A) and 2200 μF (450 V), respectively calculated by the same procedure as given in Example 7.1. A 5 kW resistive load was used at the output of the converter for testing it. This system was tested in the laboratory with the DC-power supply based simulator described in Section 3.4.

Construction and Circuit Description

The IGBT switch was mounted on a heat sink whose temperature rose to around 70°C without cooling. This was therefore cooled by a 48 V, 10 cm diameter fan so that the temperature came down to a steady state value of 50°C for an ambient temperature of 35°C. Hence, it is essential that the circuit should operate only when the cooling fan is working. For this reason, the switching signal to the IGBT is sent to its gate through a driver, which in turn is operated from a 15 V_{DC} power supply. This 15 V supply passes through a relay, which is kept in NO (Normally Open) condition unless the fan voltage of 48 V_{DC} is applied across it. The relay then switches on and transmits the operating voltage to the driver and the optocoupler preceding it. If somehow the fan supply fails, the IGBT would stop functioning. Besides these two, there is one more power supply of ±15 V to drive Hall-Sensors used to measure the voltage and the current. All these power supply sub-assemblies work from an

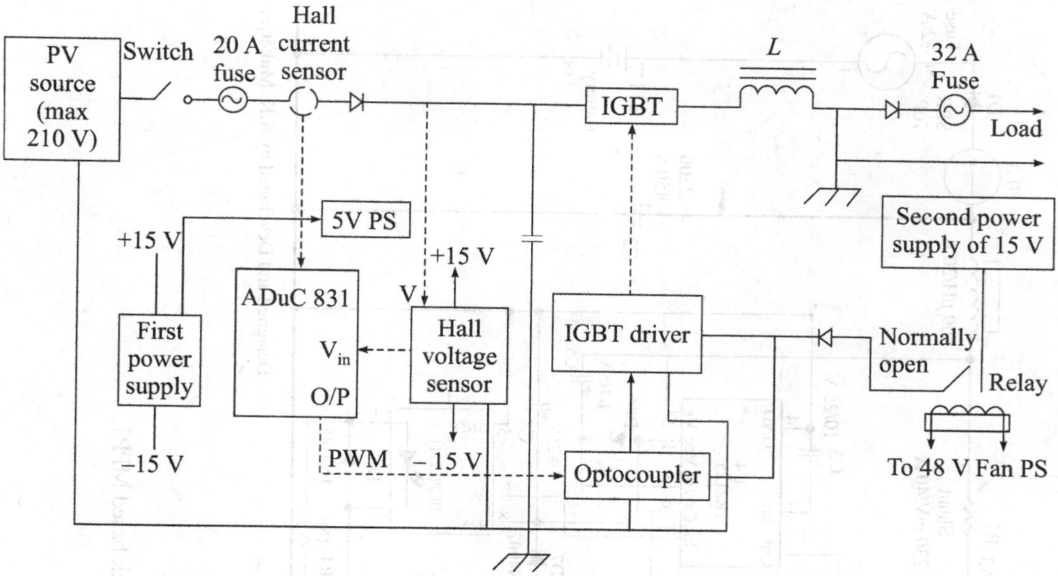

Figure 7.24 Block diagram of the 2.5 kW MPPT with the power supplies.

input of 100 V_{DC} to 350 V_{DC} which is adequate for the PV array voltage output from morning to evening. The photograph of the system, taken from top, is shown in Figure 7.25.

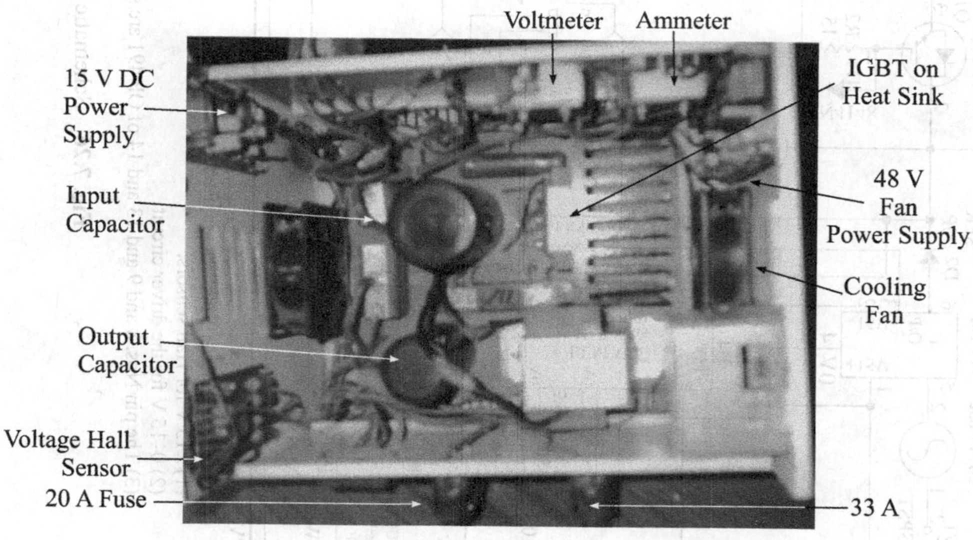

Figure 7.25 The 2.5 kW buck topology based MPPT.

The Control Unit and Interfacing Circuit

The 8051-core microcontroller (μC) that was employed for this work had the model number ADuC831 and manufactured by *Analog Devices* [19]. It contains an 8-channel, 12-bit ADC, 62 KB program memory, 4 KB data memory, 256 bytes of RAM and a dual-output 16-bit PWM. The controller is programmed to read the analogue voltage and current values, convert

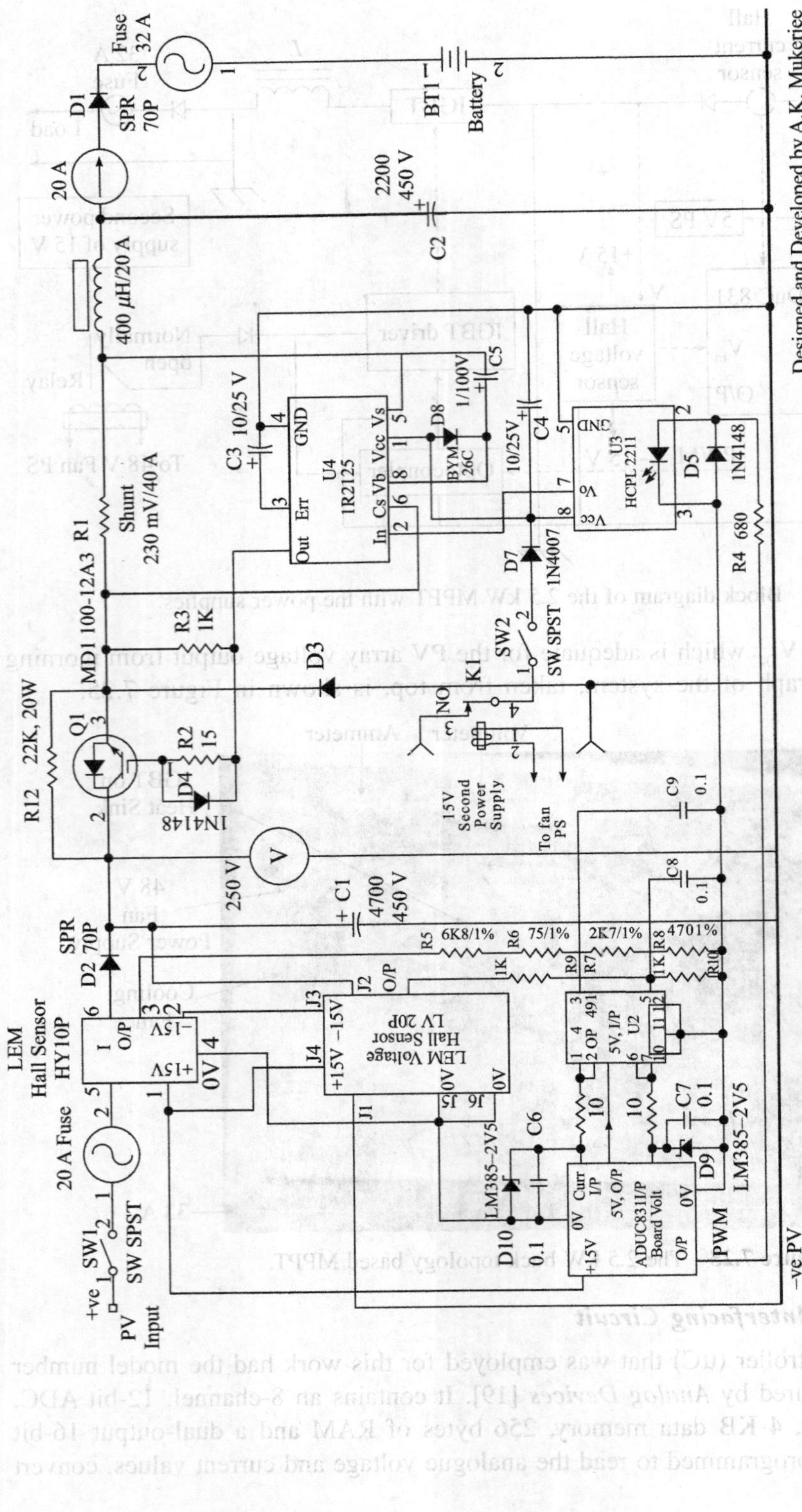

Note : Two Power Supplies are used.
(1) ±15 V for Hall sensors.
(2) 0–15 V for the driver circuit.
(3) The pin Nos. 8 and 9 and 13 and 14 of OP 491 are shorted.

Fig. 7.26 Schematic of 2.5 kW buck based MPPT.

them into usable register values, run the logics coded in *C* and to output the correct PWM signal accordingly. The C-compiler platform for programming was provided by SPJ Systems, Pune, India [20]. This is compatible with the 8051 instruction-set of this particular μC.

The interfacing circuit consists of a Hall-Effect sensor (LEM HY10P) for sensing current and appropriate voltage dividers for sensing voltage. Hall sensor for voltage (LEM LV 20-P) is used for the 2.5 kW system only. IR 2125/2121 MOSFET/IGBT driver IC for triggering the converter switch is connected to on-chip PWM peripheral via HCPL 2211 opto-coupler. Buffer circuits consisting of OP491 operational amplifiers are employed at the input of the ADC channels of the μC. A resistive load R4 was connected to the output port of the converter. The complete circuit diagram of the instrument is given in Figure 7.26.

Experimental results

Table 7.1 show the input and output powers of the MPPT under changing V_{oc} conditions. This will be the case, if the number of solar panels used at the input, is changed randomly or in case of change in atmospheric conditions. In spite of using a controller with relatively low resolution, the system efficiency has been recorded to be more than 92%.

Table 7.1 Table showing the MPPT holding the operating voltage at MPP when fed with different input DC voltages from the PV simulator

V_{oc} (V)	V_m (V)	I_m (A)	P_m (W)	V_o (V)	I_o (A)	P_o (W)	*Efficiency* (%)
50	25.5	1.42	36.21	16.8	1.80	30.24	83.51
100	49.7	2.8	139.16	34.3	3.96	126.56	90.09
150	78.2	4.02	314.36	52.0	5.56	289.52	91.97
200	103	5.45	561.3	69.5	7.36	511.52	91.13
250	130	6.67	867.1	87.6	9.10	797.16	91.93
300	157	7.78	1221.14	104.0	10.84	1127.36	92.29

The interfacing and other circuits were similar to those given in Example 7.1. The oscillogram in Figure 7.27(a) shows that the MPPT attains the MPP after initialization in about 64 seconds. However after reaching the peak value it latches on to the next peak in very fast if the insolation suddenly changes as revealed in Figure 7.27(b).

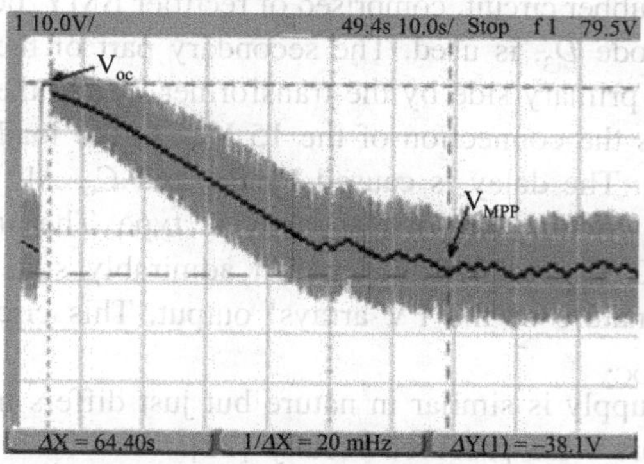

Figure 7.27(a) The MPPT reaches the MPP after almost 64 seconds.

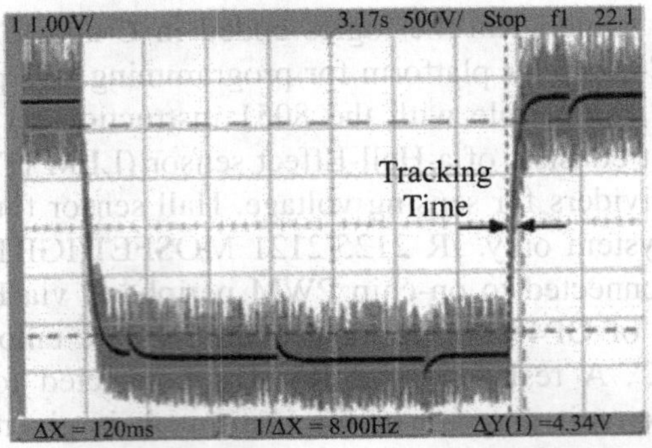

Figure 7.27(b) Voltage waveform for step change of input voltage.

The recovery time has been measured to be 120 milliseconds approximately by the oscilloscope.

The circuit diagrams of the +48 V, +15 V and ±15 V DC power supplies are shown in Figures 7.28, 7.29 and 7.30. The appendix of this chapter gives the details of the transformers used in these circuits. These are called *universal power supplies* because they are designed to work in any part of the world from the local power grid since they cover an rms input voltage range from 85 V to 265 V. For example, a single unit covers the countries that use 110 V, like the United States, to Great Britain and India, where the operating voltage level is 220/240 V. In addition, any voltage excursion up to 270 V and down to 85 V, assuming an erratic grid, will be taken care of. The working of the 15 V circuit is given here. The input AC is first connected to a line filter, explained in Chapter 11, comprising a coupled inductor and capacitor C_1 before being fed to a full wave rectifier made up of four 1N 4007 diodes. These EMI filters prevent the subsequent circuits from polluting the grid with the noise they themselves generate. The DC so produced is filtered by the capacitor C_2 before being supplied to the switching converter TOP 243 Y [21] which converts it to 15 V with the help of the step down transformer T_3.

Since high voltages, coupled with frequencies in the range of hundreds of kilohertz, are involved it is normal to expect very high voltage pulses at pin #7 of the IC which are sufficient to damage the device. To protect the circuit from such effects and increase its operational reliability a snubber circuit, comprised of rectifier BMY26C, capacitor C_3, resistance R_4 and a 200 V zener diode D_3, is used. The secondary part of the circuit generating 15 V is fully isolated from the primary side by the transformer T_3 and the optocoupler U_1. A 15 V relay at the output delays the connection of the 15 V_{DC} to the load at the output unless the circuit is fully functional. The delay is caused by R_{11} and C_9. All the rectifiers involved in the switching part of the circuitry are of fast recovery type. The universal input of 85 V to 265 V works out to 120 V_{DC} to 375 V_{DC} which admirably suits the PV instrumentation because of the variable nature of the PV arrays' output. This circuit worked satisfactorily from 100 V_{DC} to 350 V_{DC}.

The 48 V_{DC} power supply is similar in nature but just differs in component details. The transformers and inductors were procured locally [22].

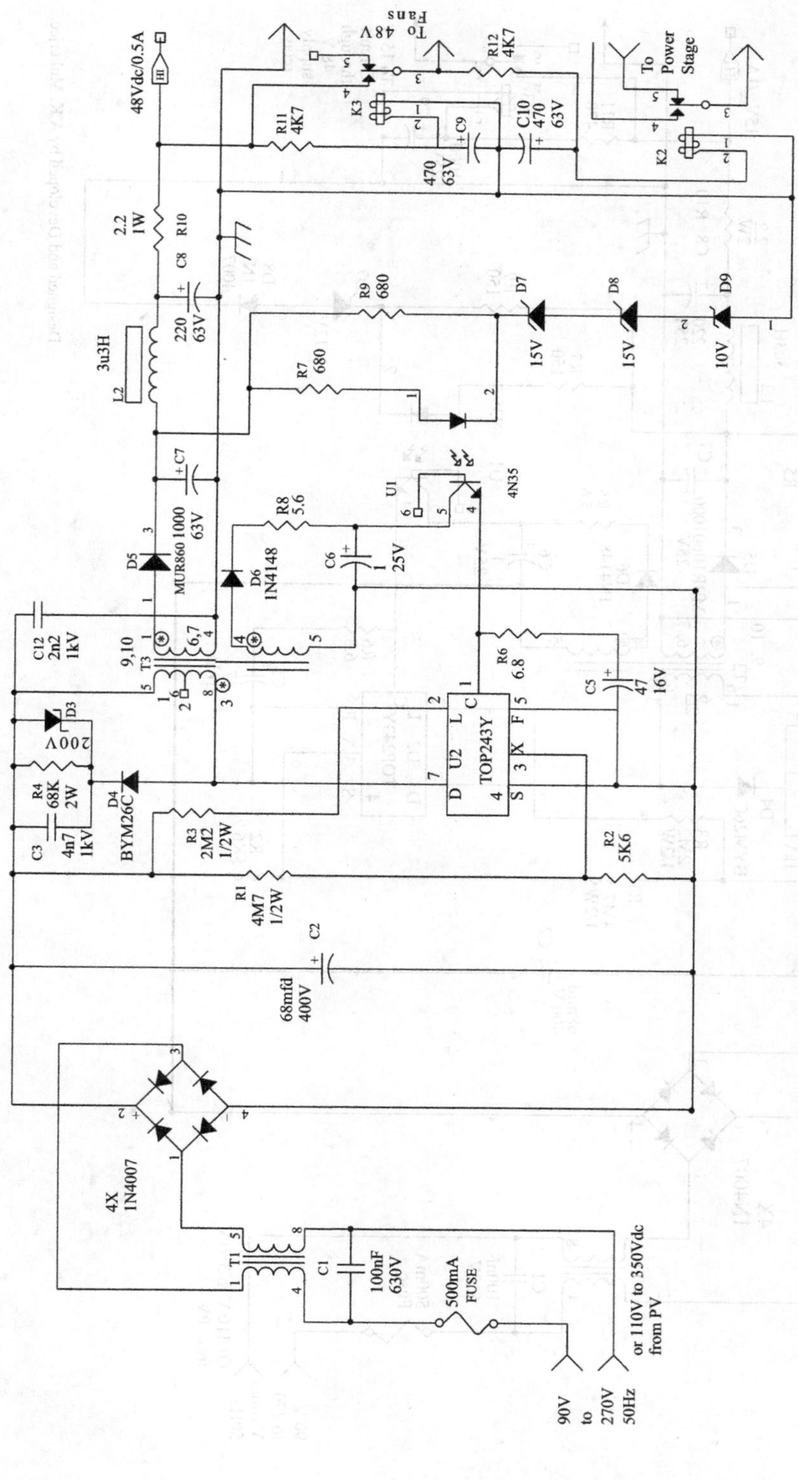

Figure 7.28 48 V fan power supply for MPPT.

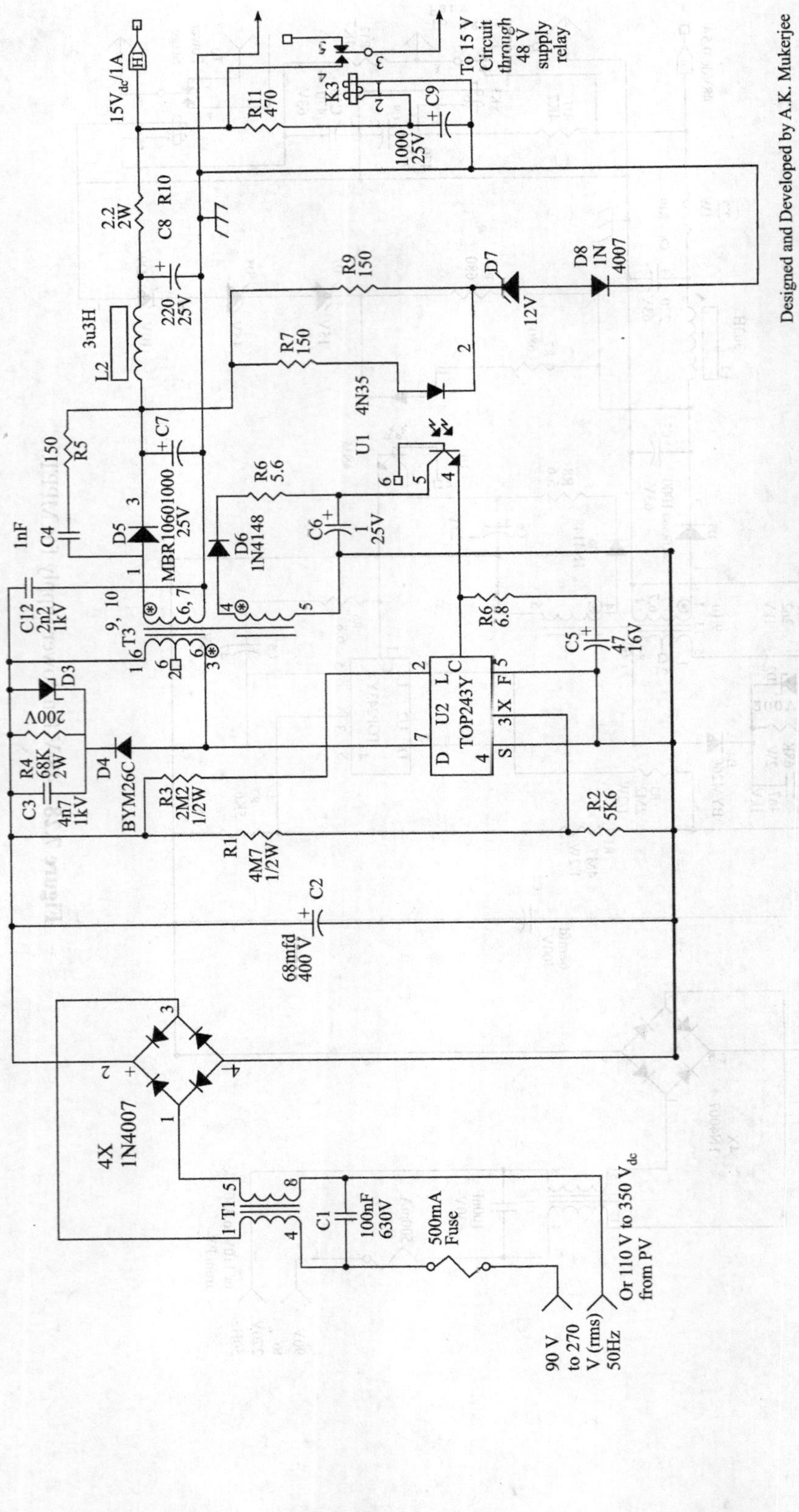

Figure 7.29 Circuit diagram of 15 V universal power supply for MPPT.

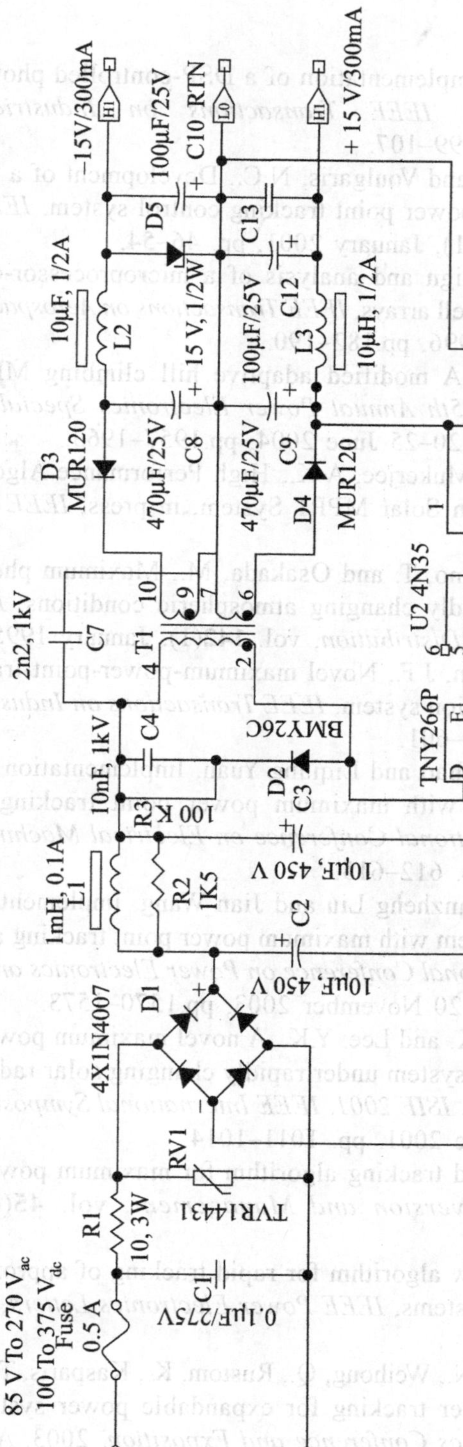

Transformer: COSMO Ferrites, EE 2005S, CF 138, 10 Pin Vertical Bobbin

Figure 7.30 ±15 V universal power supply for MPPT.

References

[1] Hua, C., Lin, J. and Shen, C., Implementation of a DSP-controlled photovoltaic system with peak power tracking, *IEEE Transactions on Industrial Electronics*, vol. **45**(1), February 1998, pp. 99–107.

[2] Koutroulis, E., Kalaitzakis, K. and Voulgaris, N.C., Development of a microcontroller-based, photovoltaic maximum power point tracking control system, *IEEE Transactions on Power Electronics*, vol. **16**(1), January 2001, pp. 46–54.

[3] Huynh, P. and Cho, B.H., Design and analysis of a microprocessor-controlled peak-power-tracking system for solar cell arrays, *IEEE Transactions on Aerospace and Electronic Systems*. vol. **32**(1), January 1996, pp.182–190.

[4] Xiao, W. and Dunford, W.G., A modified adaptive hill climbing MPPT method for photovoltaic power systems, *35th Annual Power Electronics Specialists Conference*, 2004. PESC 04, *IEEE*, vol. **3**, 20–25 June 2004, pp.1957–1963.

[5] Pandey, A., Dasgupta, N. and Mukerjee, A.K., High Performance Algorithms for Drift Avoidance and Fast Tracking in Solar MPPT System, in press, *IEEE Transactions on Energy Conversion*.

[6] Hussein, K.H., Muta, I., Hoshino, T. and Osakada, M., Maximum photovoltaic power tracking: an algorithm for rapidly changing atmospheric conditions, *IEE Proceedings Generation, Transmission and Distribution*, vol. **142**(1), January 1995, pp. 59–64.

[7] Kuo, Y.C., Liang, T.J. and Chen, J.F., Novel maximum-power-point-tracking controller for photovoltaic energy conversion system, *IEEE Transactions on Industrial Electronics*, vol. **48**(3), June 2001, pp. 594–601.

[8] Kunlun Chen, K. Zhengming Zhao and Liqiang Yuan, Implementation of a stand alone photovoltaic pumping system with maximum power point tracking, ICEMS 2001. *Proceedings of the 5th International Conference on Electrical Machines and Systems*, vol. **1**, 18–20 August 2001, pp. 612–615.

[9] Libo Wu, Zhengming Zhao, Jianzheng Liu and Jian Wang, Implementation of a stand-alone photovoltaic lighting system with maximum power point tracking and high pressure sodium lamp, *The 5th International Conference on Power Electronics and Drive Systems*, 2003. PEDS 2003. vol. **2**, 17–20 November 2003, pp.1570–1573.

[10] Kim, T.Y., Ahn, H.G., Park. S.K. and Lee, Y.K., A novel maximum power point tracking control for photovoltaic power system under rapidly changing solar radiation, *Industrial Electronics, 2001. Proceedings. ISIE 2001. IEEE International Symposium on Industrial Electronics*, vol. **2**, 12–16 June 2001, pp. 1011–1014.

[11] Hua, C. and Lin, J., A modified tracking algorithm for maximum power point tracking of solar array, *Energy Conversion and Management*, vol. **45**(6), April 2004, pp. 911–925.

[12] Jain, S. and Agarwal, V., A new algorithm for rapid tracking of approximate maximum power point in photovoltaic systems, *IEEE Power Electronics Letters*, vol. **2**(1), March 2004, pp. 16–19.

[13] Wenkai Wu, Pongratananukul, N., Weihong, Q., Rustom, K., Kasparis, T. and Batarseh, I., DSP-based multiple peak power tracking for expandable power system, 18th Annual *IEEE Applied Power Electronics Conference and Exposition,* 2003. APEC '03, vol. **1**, 9–13 February 2003, pp. 525–530.

[14] Bozetepe, M. and Colak, M., Implementation of Maximum Power Point Tracking Control in Single Phase Utility Interconnected Photovoltaic Inverter, *Compatibility in Power Electronics*, 2007. CPE '07, *IEEE 5th International Conference-Workshop*, May 29 to June 1, 2007, pp. 1–3.

[15] Masoum, M.A.S., Dehbonei, H. and Fuchs, E.F., Theoretical and experimental analyses of photovoltaic systems with voltage and current-based maximum power-point tracking, *IEEE Transactions on Energy Conversion,* vol. **17**(4), December 2002, pp. 514–522.

[16] Masoum, M.A.S, Badejani, S.M.M. and Fuchs, E.F., 'Microprocessor-controlled new class of optimal battery chargers for photovoltaic applications,' *IEEE Transactions on Energy Conversion*, vol. **19**(3), September 2004, pp. 599–606.

[17] Ned Mohan, *Power Electronics—Converters, Applications and Design*, 3rd ed., John Wiley & Sons, Inc., New York, 2003, p. 167.

[18] Kitsum, K., *Switch Mode Power Conversion: Basic Theory and Design*, Marcel Dekker Inc., New York and Basel, 1984, p. 10.

[19] Analog Devices ADuC831 MicroConverter Datasheet. www.analog.com

[20] SPJ Systems Programming Instructions manual supplied with the controller kit and the Integrated Development Environment. SPJ Systems, 101, Beaver Grandeur, Baner Road, Pune–411045, India. www.spjsystems.com.

[21] Power Integrations. www.powerint.com/.

[22] M/S Supreme Systems, 71, phase-IV, Udyog Vihar, Gurgaon-122015, Haryana. www.supremesystems.in

APPENDIX

Transformer for 15 V, 1 A power supply

Description	*Variable*	*Value*	*Units*
Core Type	Core/Bobbin	EE19	
Primary Inductance	LP	399	μH
Primary Number of Turns	NP	64.2	
Bias Winding Number of Turns	NB	7.3	
Primary Wire Gauge	AWG	30	AWG
Primary Winding Current Capacity	CMA	293	Cmils/A
Reflected Output Voltage	VOR	112.02	V
Bobbin Winding Width	BW	9	mm
Safety Margin Width	M	0	mm
Primary Number of Layers	L	2.4	
Core Cross Sectional Area	AE	23	mm^2
Gapped Core Effective Inductance	ALG	97	nH/T^2
Maximum Flux Density	BM	2143	gauss
Peak Flux Density	BP	2600	gauss
AC Flux Density for Core Loss	BAC	1071	gauss
Gap Length	LG	0.28	mm
Primary Leakage Inductance	LL	8	μH
Secondary Trace Inductance	LSEC	30	nH

Secondary winding

Description	*Variable*	*Output 1*	*Units*
Secondary Number of Turns	NSx	9	
Rounded to Integer Secondary Number of Turns	Rounded Down NSx		
Auxiliary Output Voltage for Rounded down to Integer Secondary Number of Turns	Rounded Down Vox		V
Rounded to Next Integer Secondary Number of Turns	Rounded Up NSx		
Auxiliary Output Voltage for Rounded up to Next Integer Secondary Number of Turns	Rounded Up Vox		V
Secondary Wire Gauge Range	AWGSx Range	20–24	AWG
Comment: Secondary wire gauge is less than recommended minimum (26 AWG) Tip: Consider a parallel winding technique (bifilar, trifilar) for > 1.5 A outputs, increase size of transformer (larger BW) or reduce margin (M).			

Transformer for 48 V, 0.5 A power supply

Transformer primary

Core and Bobbin Type	*Core/Bobbin*	*E25/13/7 (EF25)*	*Units*
Primary Inductance	LP	802	μH
Primary Winding Number of Turns	NP	67	
Bias Winding Number of Turns	NB	8.97	
Primary Wire Gauge (Rounded to next smaller standard AWG value)	AWG	27	AWG
Primary Winding Current Capacity	CMA	448	Cmils/A
Reflected Output Voltage	VOR	117.74	V
Bobbin Physical Winding Width	BW	15.30	mm
Safety Margin Width	M	0.0	mm
Number of Primary Layers	L	2.0	
Core Effective Cross Section Area	AE	0.53	cm^2
Gapped Core Effective Inductance	ALG	177	nH/T^2
Maximum Operating Flux Density	BM	1802	gauss
Peak Flux Density	BP	2186	gauss
AC Flux Density for Core Curves	BAC	501	gauss
Gap Length	LG	0.33	mm
Estimated Transformer Primary Leakage Inductance	LL	12.0	μH
Estimated Secondary Trace Inductance	LSEC	30	μH

Secondary winding data

Secondary Number of Turns	*NSx*	*28*	
Rounded to Integer Secondary Number of Turns	Rounded Down NSx		
Auxiliary Output Voltage for Rounded to Integer NSx	Rounded Down Vox		V
Rounded to Next Integer Secondary Number of Turns	Rounded Up NSx		
Auxiliary Output Voltage for Rounded to Next Integer NSx	Rounded Up Vox		V
Secondary Wire Gauge Range	AWGSx Range	24–28	AWG
Comment: Primary wire gauge is less than recommended minimum (26 AWG) and may overheat			
Tip: Consider a parallel winding technique (bifilar, trifilar), increase size of transformer (larger BW) or reduce margin (M)			

CHAPTER 8

BOS Other MPP Tracking Techniques

Chapter 7 dealt with the basic concepts of MPP tracking and the classical algorithms. This chapter discusses the special MPP tracking techniques that are devised to address specific issues or have special characteristics that are different from the classical methods. The algorithms introduced and discussed here include schemes to alleviate problems due to partial shading on PV modules. Further, sensor reduction techniques (that involve the usage of only one sensor instead of the commonly used two sensors for reading the electrical parameters at the PV end), hardware based modifications etc. are also explained. These algorithms have come into being with continuous research towards developing robust, cost-effective and more reliable solutions.

8.1 Partial Shading on PV Modules

Complete or partial shading can occur on photovoltaic modules under various conditions like dirt, dust, different angle of incidence, shading due to poles or nearby cables and by vehicular parts in case of race-cars etc. [1–5]. Under such conditions, the output power of the PV module decreases significantly. In addition, under cases like array damage and partial shading conditions resulting in non-uniform cell illumination, several local maxima arise in the *P–V* curve apart from a global maximum.

Local maxima arise due to the complicated conduction mechanism of partially and unshaded cells and bypass diodes [6]. In such cases, the general classical MPPT algorithms fail to track the global maxima. The resultant sub-optimal tracking impairs the efficiency of the system and defeats the concept of MPP tracking. Figure 8.1(a) shows the effect of partial shading [7]. It can be seen from Figure 8.1(b) more than one local maximum can occur in such cases.

The constant voltage ratio method (discussed earlier in Section 7.5.3) may or may not hold well in such cases. This means that isolating the array from the tracker and measuring the V_{oc} may not help locate V_{MPP} as the ratio may not be in the range in which it is generally

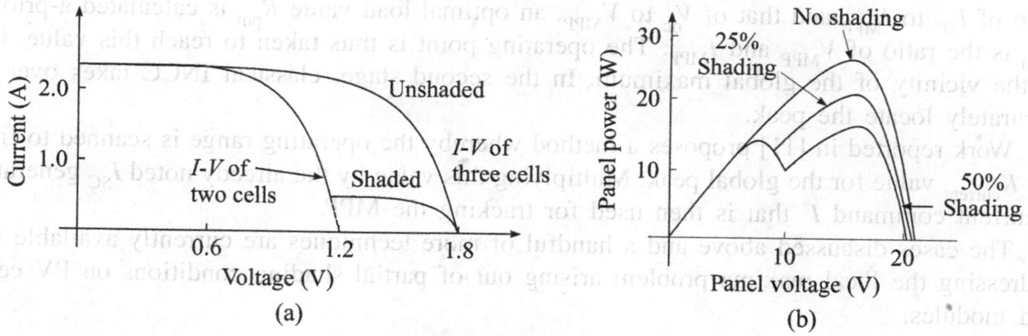

Figure 8.1 Effect of partial shading on *I* vs *V* curves.

used in the constant voltage ratio method. One feasible method that is generally adopted for alleviating this problem is sweeping the entire power curve occasionally. This is generally done by interrupting the main MPPT algorithm. Such hybrid techniques involve sweeping of the power curve and locating the global maximum and then operating the classical algorithm near the global peak to accurately locate it. The only drawback of this method is that this incurs some loss in power when the power curve is swept fully. The background logic that is generally followed for such a method is illustrated in Figure 8.2.

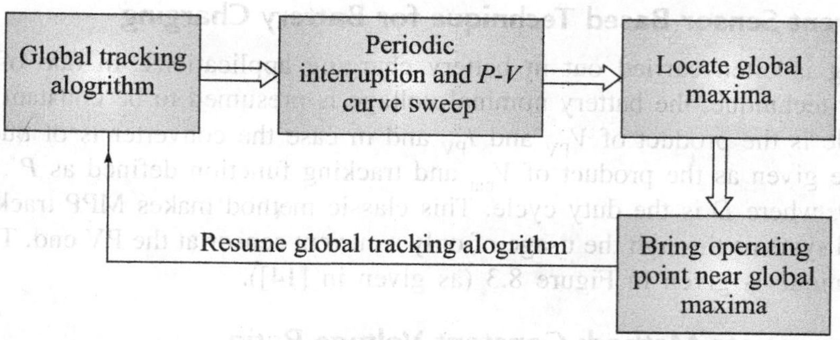

Figure 8.2 Logic of the sweep method for skirting local maxima.

8.1.1 Hybrid Techniques Addressing Local Maxima

A hybrid technique that addresses the problem of local maxima arising from partial shading conditions is discussed here [8]. Here, incremental conductance (INCC) method is used as the global algorithm which is periodically interrupted by P&O method. This is a type of hybrid algorithm having a global and another local algorithm working in synergy to achieve the efficiency and accuracy in tracking the changing MPP in cases of complicated environmental changes that include partial and non-shading conditions. The global algorithm, being INCC, provides better tracking under changing atmospheric conditions. The problem of local maxima is addressed by periodically sweeping the control range of the P&O algorithm.

In papers such as [9–10], two-stage methods are used. In the first stage, V_{oc} and I_{SC} are noted through monitoring cells. The monitoring cells are kept at a place where there is no shading. One monitoring cell is used to determine the V_{oc} and the other for I_{SC}. Knowing the ratio of I_{SC} to I_{MPP} and that of V_{oc} to V_{MPP}, an optimal load value R_{pm} is calculated a-priory. R_{pm} is the ratio of V_{MPP} and I_{MPP}. The operating point is thus taken to reach this value, i.e. to the vicinity of the global maximum. In the second stage, classical INCC takes over to accurately locate the peak.

Work reported in [11] proposes a method whereby the operating range is scanned to find the $K_{current}$ value for the global peak. Multiplying this value by the already noted I_{SC} generates a current command I^* that is then used for tracking the MPP.

The cases discussed above and a handful of more techniques are currently available for addressing the local maxima problem arising out of partial shading conditions on PV cells and modules.

8.2 Sensor Reduction Schemes

The MPP tracking techniques discussed so far generally employ two sensors—one for voltage and the other for current. In several applications, reduction in the number of sensors happens to be desirable because these, especially hall current sensors, are bulky and expensive. Further, the system reliability is enhanced considerably by operating it with lesser number of sensors. Understandably, some truly efficient sensor-reduction techniques have been developed [12–13].

8.2.1 Current Sensor Based Technique for Battery Charging

MPP tracking is often carried out in battery charging applications. In one of the earliest single-sensor technique, the battery nominal voltage is presumed to be constant [14]. Power of PV module is the product of V_{PV} and I_{PV} and in case the converter is of buck type then power can be given as the product of V_{bat} and tracking function defined as P^*, where P^* is simply I_{PV}/D, where D is the duty cycle. This classic method makes MPP tracking easy for battery based systems through the usage of only a current sensor at the PV end. The algorithm for this technique is given in Figure 8.3 (as given in [14]).

8.2.2 Approximate Method: Constant Voltage Ratio

Constant voltage ratio technique has been discussed in Section 7.5.3. This algorithm is based on the assumption that the ratio of the voltage at maximum power point to the open circuit voltage is constant [15]. It is governed by Equation (8.1).

$$\frac{V_{MPP}}{V_{oc}} \cong K < 1 \tag{8.1}$$

here, K is the ratio that is assumed to be constant. The PV panel terminals are momentarily open circuited at regular intervals for measuring V_{oc} and then the V_{MPP} value is calculated. The operating point is then forced to the calculated value. Figure 8.4 gives an overall idea of the tracking mechanism of this method. According to reference [15], the value of *Constant* is not constant but varies with insolation and temperature.

Although this method is extremely simple to implement, the assumption of a constant value of K makes this technique relatively less accurate than search based techniques.

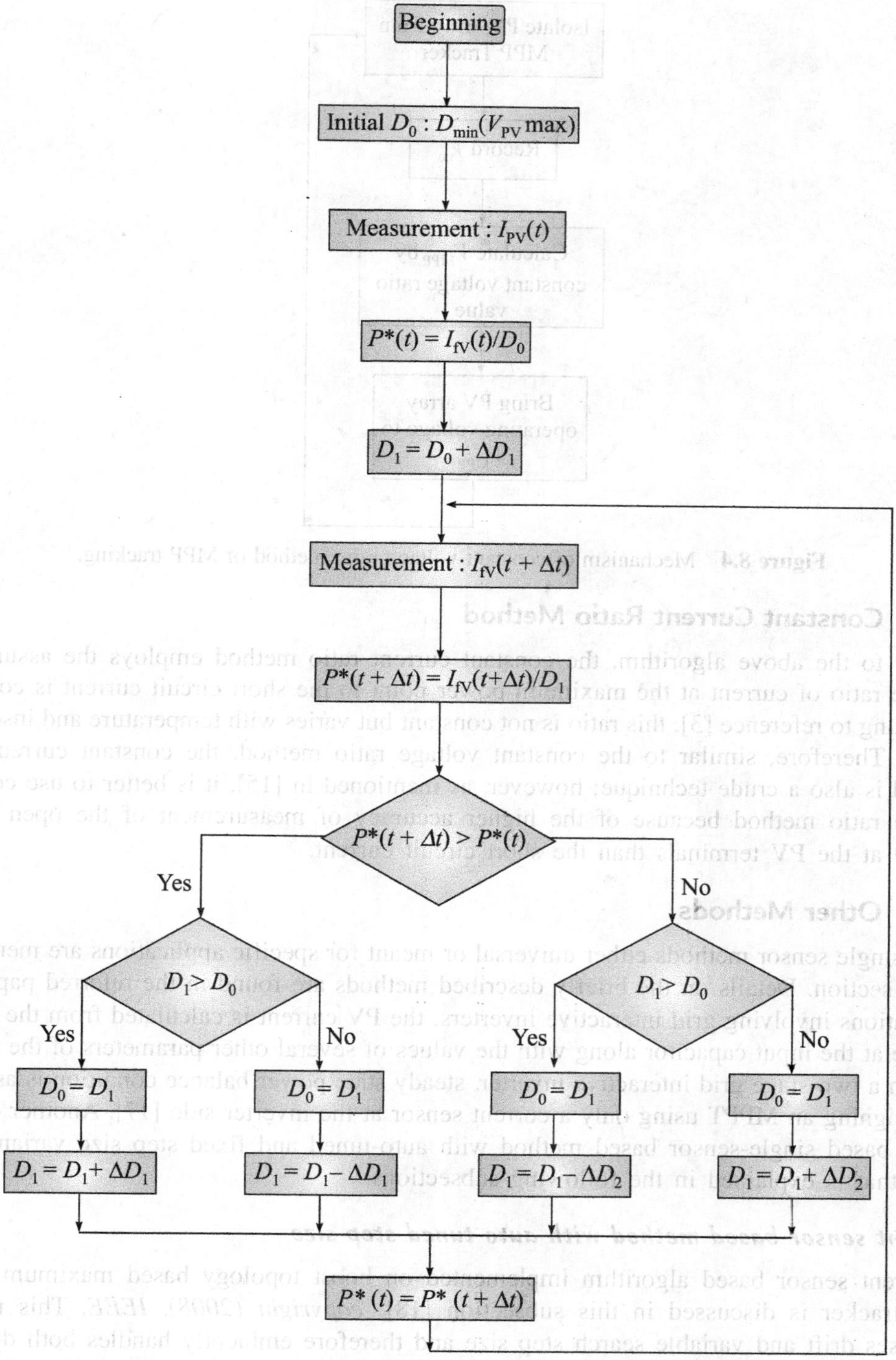

Figure 8.3 Flowchart of single sensor (current sensor) based MPP tracking method.

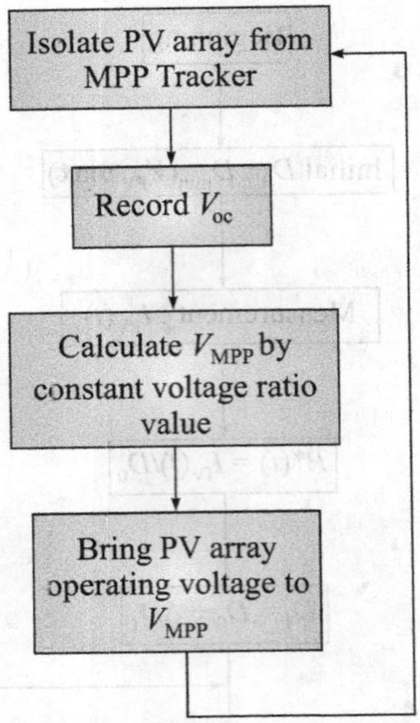

Figure 8.4 Mechanism of constant voltage ratio method of MPP tracking.

8.2.3 Constant Current Ratio Method

Similar to the above algorithm, the constant current ratio method employs the assumption that the ratio of current at the maximum power point to the short circuit current is constant. According to reference [3], this ratio is not constant but varies with temperature and insolation values. Therefore, similar to the constant voltage ratio method, the constant current ratio method is also a crude technique; however, as mentioned in [15], it is better to use constant voltage ratio method because of the higher accuracy of measurement of the open circuit voltage at the PV terminals than the short circuit current.

8.2.4 Other Methods

Other single sensor methods either universal or meant for specific applications are mentioned in this section. Details on the briefly described methods are found in the referred papers. In applications involving grid interactive inverters, the PV current is calculated from the sensed voltage at the input capacitor along with the values of several other parameters of the system [16]. In a two-stage grid interactive inverter, steady state power balance condition is assumed for designing an MPPT using only a current sensor at the inverter side [17]. Another current sensor based single-sensor based method with auto-tuned and fixed step size variants also exists that is explained in the following subsections.

Current sensor based method with auto-tuned step size

A current sensor based algorithm implemented on boost topology based maximum power point tracker is discussed in this subsection [18]; *copyright (2008), IEEE.* This method addresses drift and variable search step size and therefore eminently handles both dynamic and steady state conditions.

Development of fixed-step-size variant

The load power must maximize corresponding to the source peak power. As is known, the output power P_o is given by Equation (8.2).

$$P_o = I_o^2 R_L \tag{8.2}$$

where the load resistance R_L is constant load resistance and assuming no change in efficiency between sampling cycles, maximization of load power reflects peak input power. Therefore, the square of the load current must be maximized to track MPP. Therefore, simply I_o can be maximized. However, in MPPTs, the main endeavour is to closely follow the maximization of power on the array side. Hence, the sensors are always preferred on the source (array) side. Following the boost topology transfer function, thus, the new objective function P^* is given by:

$$P^* = I(1 - D) \tag{8.3}$$

where I is the source current. Experimental curves are obtained by sweeping the entire duty cycle range for the converter. Figure 8.5 presents the normalized plots showing the matching of the real power peak with that of P^*.

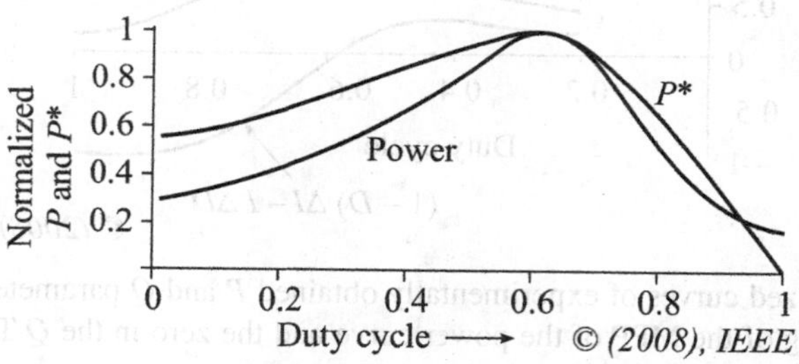

Figure 8.5 Normalized Power and objective-function peaks match at the same maximum power duty cycle.

This objective function P^* can be followed to reach the peak using hill-climbing technique. However, as mentioned earlier, this class of algorithm has an inherent drawback of drifting in wrong direction during fast changing atmospheric conditions. The derivative of this function is thus obtained to derive conditions that definitely indicate which side of the *P-D* curve peak the system is operating at in a particular time. The derivative of P^* w.r.t. D at the peak will be zero, leading to the following condition at MPP

$$(1 - D)\Delta I = I\Delta D \tag{8.4}$$

Further, as observed from the *P-D* curve, the derivative of P^* is positive and negative on left and right side of the MPP respectively and thus, the following conditions are obtained:

$$(1 - D)\Delta I > I\Delta D \quad \text{(on left of MPP on the } P\text{-}D \text{ curve)} \tag{8.5}$$

$$(1 - D)\Delta I < I\Delta D \quad \text{(on right of MPP on the } P\text{-}D \text{ curve)} \tag{8.6}$$

here, I and D values are sampled in two consecutive instants and only two multiplications are involved. The simplicity of computation in logic-code is critical for digital implementation. Since an objective function is not directly maximized here, this method avoids drift during changing conditions. A variable search step size is introduced in the next section for improving the dynamic tracking.

Variable step size parameter

The absolute value of the difference between LHS and RHS of Equation (8.4) (Term 1 and Term 2) is recognized to be suitable as a deciding step size amplitude parameter for this implementation. Using this parameter Q which is in any case computed for the basic tracking, saves the code from involving any extra computation for finding the new step size at every sampling instant. Q is represented by Equation (8.7)

$$Q = (1 - D)\Delta I - I\Delta D \tag{8.7}$$

As can be observed from Figure 8.6, within the usual working range of D, this parameter Q shows a consistent decrease in value as the MPP point is approached. Thus, the controlling equation for the proposed method is given by

$$D_k = D_{k-1} \pm M\,|Q_{k-1}| \tag{8.8}$$

where M is a scaling factor adjusted as per implementation.

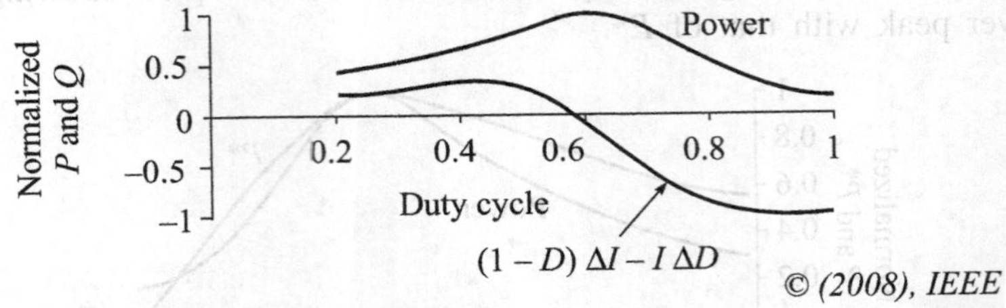

Figure 8.6 Normalized curves of experimentally obtained P and Q parameters showing the close matching of the MPP of the power curve and the zero in the Q Term.

A typical result showing start-up and steady state tracking stages is given in Figure 8.7. This figure shows the power curve for start-up stage (at a very low duty cycle), peak tracking at PV curve 1 (denoted by S_1), step change to another curve 2 (denoted by S_2) and tracking at S_2.

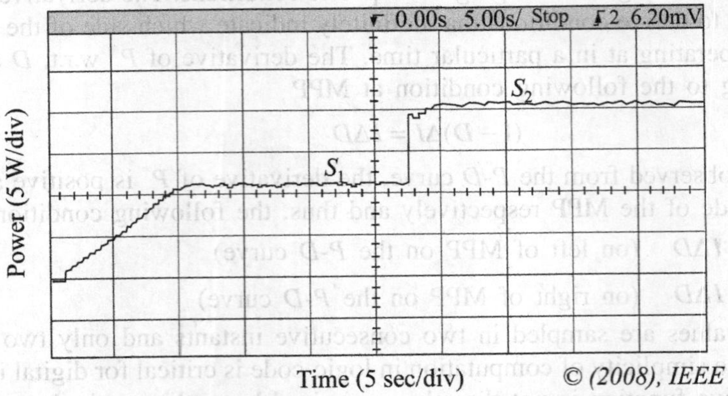

Figure 8.7 Variation in power drawn from the PV source at start-up, steady states and step change conditions when the simple current-sensing based approximate tracking technique is applied.

The final form of the proposed method is therefore represented by the update rule Equation (8.8) with the logic check as given by Equations (8.4–8.6); Q being calculated according to Equation (8.7). Flowchart is shown in Figure 8.8.

Such and other single sensor techniques are being researched upon aggressively with the aim to bring down the cost and enhance reliability of MPP trackers.

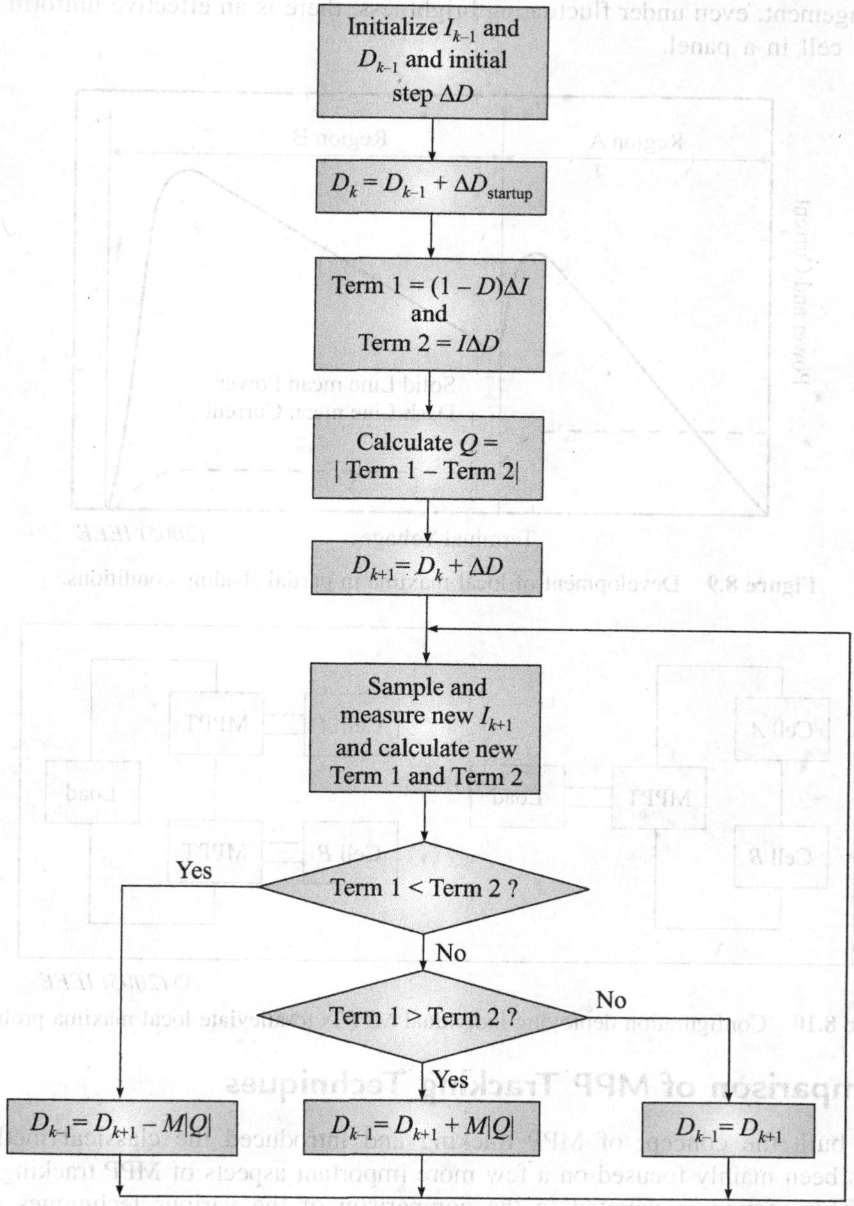

Figure 8.8 Flowchart of the simple current-sensing based approximate tracking technique.

8.3 Hardware-Modification Based Technique

A hardware-modification based solution has been suggested in [19] for mitigating the local maxima problem in partial shading conditions (see Figure 8.9). Unlike the common MPPT based systems where the terminal voltage of the entire array is controlled, this novel method employs dedicated MPPTs connected to each cell individually as illustrated in Figure 8.10. In this arrangement, even under fluctuating brightness, there is an effective uniform intensity within each cell in a panel.

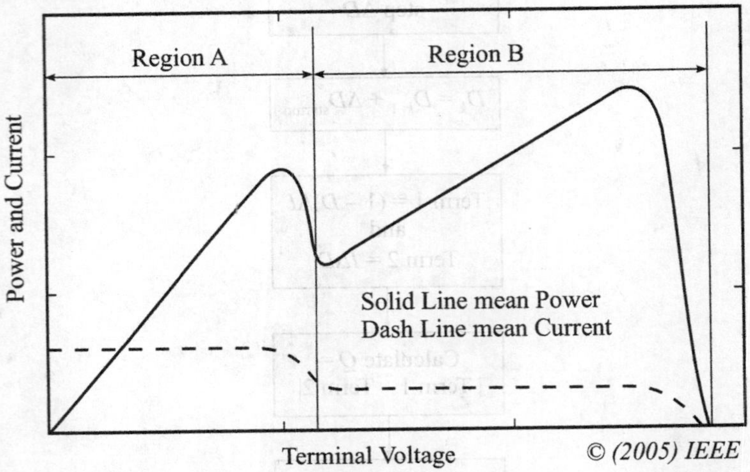

Figure 8.9 Development of local maxima in partial shading conditions.

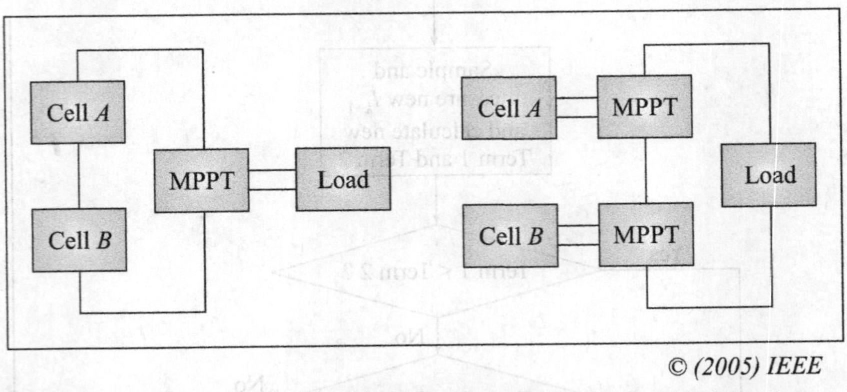

Figure 8.10 Configuration deploying individual MPPTs to alleviate local maxima problem.

8.4 Comparison of MPP Tracking Techniques

Chapter 7 built the concept of MPP tracking and introduced the classical method. This chapter has been mainly focused on a few more important aspects of MPP tracking and their solutions. This section is devoted to the comparison of the various techniques and is an attempt to provide an overview of their applicability. A classical work on the comparison on

MPPTs has presented a test-bed based method for true comparison of a few of the best known MPP tracking algorithms [15]. It also throws light on the quantitative difference between the efficiencies of these methods when compared on one common platform along with defining a suitable way of calculating the efficiency of any MPP tracking technique. This paper is a source that provides detailed description of well-established algorithms such as 'Perturb and observe', 'Incremental Conductance', 'Parasitic capacitance', 'Pilot cell' and 'Constant voltage'. Data were collected on various conditions such as clear, partly cloudy and cloudy days. It says that when suitably optimized, the efficiencies of the two most prevalent algorithms 'Perturb and observe' and 'Incremental Conductance' can prove to be even higher than an appreciable 97%. Both these algorithms prove to be of comparable efficiencies with 'Perturb and Observe' working better during clear sky and cloudy sky conditions as against the 'Incremental conductance' algorithm that shows advantages under partly cloudy conditions. The advantages and limitations of the logics of both these algorithms are critically observed and highlighted. This work suggested that the 'Perturb and Observe' algorithm, if optimized for a given hardware, could serve the purpose of most MPP tracking requirements as compared to other more complicated algorithms. This finding adds to the already well-known advantages of 'Perturb and Observe' algorithm of being a cost-effective option with ease of implementation.

Table 8.1 presents some of the basic MPP tracking algorithms along with their strengths and weaknesses for a ready overview and reference.

Table 8.1 Comparison of various MPP tracking techniques

Algorithm	*Computation***	*Sensors*	*Performance*/Advantages and Limitations*
P&O (with V_{ref} control)	Simple. One multiplication.	Voltage, Current	• Good • Standard method for MPPT. • Fixed step size affects dynamics response and oscillations. Therefore requires trade-off. • Suffers from problem of drift.
Hill-Climbing with D control	Simple	Voltage, Current	• Good • Easy to implement. • Requires trade-off between steady state oscillations (i.e. oscillations at the power peak) and dynamic response (during change in condition and start-up of tracker) • Suffers from problem of drift.
INCC	Complex. Two divisions	Voltage, Current	• Good • Complex to implement • Avoids drift and SS oscillations.
Modified INCC $I+V(\Delta I/\Delta V) = 0$	Average. One division	Voltage, Current	• Good • Computationally somewhat simpler than classical INCC. • Inherits the merits of INCC.

Contd.

Table 8.1 Comparison of various MPP tracking techniques (*Contd.*)

Algorithm	*Computation***	*Sensors*	*Performance*/ Advantages and Limitations*
Variable step method with V_{ref} control	Average. One division.	Voltage, Current	• Good • Involves additional voltage control loop.
P-V curve sweep method for partial shading	Simple	Voltage, Current	• Average • Sweeps entire *P-V* range at intervals leading to power loss.
Fixed Duty Cycle (Data based)	Nil	Nil	• Poor • Open loop control. • Very limited utility.
Constant Voltage Method	Simple. Voltage control loop.	Voltage	• Average • Poor scalability due to wide variations in solar cell characteristics.
Optimizing $P^* = I/D$	Moderate. One Division	Current	• Good • Does not hold for deep discharged battery conditions. • Only for systems with battery at the output of MPPT.
Single (current) sensor with control equation $D_k = D_{k-1} \pm M\,\lvert Q_{k-1}\rvert$	Simple. (No division) Robust (variable step size)	Current	• Good • Avoids drift • Simple computation • Robust because of adjustable step size

*MPP Tracking
** Computation of objective function for maximization

Although no rules exist for identifying the best suited maximum power point tracking algorithm for particular cases, appropriate MPPT algorithm should be judiciously chosen depending upon the characteristics based on the requirement of the applications. For applications that involve low cost, i.e. where cost is a constraint, the constant current ratio method, constant voltage ratio method, empirical observation based single voltage sensor methods, etc. should be chosen. For race vehicles and space vehicles, algorithms addressing partial shading conditions should be preferred. For those requiring simple circuitry yet high performance and reliability, sensor reduction techniques should be applied. For terrestrial, medium cost applications, either INCC or simple hybrid techniques are comparatively more suitable. Understandably, there is no set rule as such but judicious choice of algorithms can significantly influence the performance of a MPPT.

References

[1] Borg, N.J.C.M. and Jansen, M.J., Energy loss due to shading in BIPV application, *3rd World Conference on Photovoltaic Energy Conversion*, 2003, Japan, pp. 2220–2222.

[2] Woyte, A., Nijs, J. and Belmans, R., Partial shadowing of photovoltaic arrays with different system configurations: literature review and field test results, *Solar Energy*, vol. **74**(3), March 2003, pp. 217–233.

[3] Bekker, B. and Beukes, H.J., Finding an optimal PV panel maximum and power point tracking method, *IEEE AFRICON*, 2004, pp. 1125–1130.

[4] Kobayashi Kenji, Takano Ichiro and Sawada Yoshio, A study of a two stage maximum power point tracking control of a photovoltaic system under partially shaded insolation conditions, *Solar Energy Materials and Solar Cells*, vol. **90**(18–19), 23 November 2006, pp. 2975–2988.

[5] Kovach, A., Effects of inhomogeneous irradiation distribution on a PV array in an urban environment, *Photovoltaic Energy Conversion*, 1994, Conference Record of the Twenty Fourth IEEE Photovoltaic Specialists Conference–1994, 1994 IEEE First World Conference on vol. **1**, 5–9 December 1994, pp. 994–997.

[6] Sullivan, C.R. and Powers, M.J., A high-efficiency maximum power point tracker for photovoltaic arrays in a solar-powered race vehicle, *24th Annual Power Electronics Specialists Conference, IEEE 1993*. PESC '93 Record, 20–24 June 1993, pp. 574–580.

[7] Quaschning, V., *Regenerative Energiesysteme, Technologie-Berechnung—Simulation*, Book, www.hauser.de.

[8] Wenkai, Wu, Pongratananukul, N., Weihong, Q., Rustom, K., Kasparis, T. and Batarseh, I., DSP-based multiple peak power tracking for expandable power system, *18th Annual IEEE Applied Power Electronics Conference and Exposition, 2003*. APEC '03, vol. **1**, 9–13 February 2003, pp. 525–530.

[9] Kobayashi Kenji, Takano Ichiro and Sawada Yoshio, op. cit.

[10] Irisawa, K., Saito, T., Takano, I. and Sawada, Y., Maximum power point tracking control of photovoltaic generation system under non-uniform insolation by means of monitoring cells, *Twenty-Eighth IEEE Photovoltaic Specialists Conference, 2000*, 15–22 September 2000, pp. 1707–1710.

[11] Noguchi, T., Togashi, S. and Nakamoto, R., Short-current pulse-based maximum-power-point tracking method for multiple photovoltaic-and-converter module system, *IEEE Transactions on Industrial Electronics*, vol. **49**(1), February 2002, pp. 217–223.

[12] Park, J.H., Ahn, J.Y., Cho, B.H. and Yu, G.J., Dual-Module-Based Maximum Power Point Tracking Control of Photovoltaic Systems, *IEEE Transactions on Industrial Electronics,* vol. **53**(4), June 2006, pp. 1036–1047.

[13] Veerachary, M., Senjyu, T. and Uezato, K., Voltage-based maximum power point tracking control of PV system, *IEEE Transactions on Aerospace and Electronic Systems*, vol. **38**(1), January 2002, pp. 262–270.

[14] Salas, V., Olías, E., Lázaro, A. and Barrado, A., New algorithm using only one variable measurement applied to a maximum power point tracker, *Solar Energy Materials and Solar Cells,* vol. **87**(1–4), May 2005, pp. 675–684.

[15] Hohm, D.P. and Ropp, M.E., Comparative study of maximum power point tracking algorithms, *Program Photovolt: Res. Appl. 2003*, vol. **11**, pp. 47–62.

[16] Kasa, N., Iida, T. and Chen, L., Flyback inverter controlled by sensorless current MPPT for photovoltaic power system, *IEEE Transactions on Industrial Electronics*, vol. **52**, August 2005, pp. 1145–1152.

[17] Kitano, T., Matsui, M. and Xu, D., Power sensorless control scheme utilizing power balance at DC link, *IEEE Industrial Electronics Conference, IECON*, 2001, pp. 1309–1314.

[18] Dasgupta, N., Pandey, A. and Mukerjee, A.K., Current-Sensor-Based Photovoltaic MPP Tracking Algorithm with Efficient Dynamic Response, *IEEE International Conference on Industrial Technology* (ICIT) 2008, China.

[19] Kajihara, A. and Harakawa, A.T., Model of photovoltaic cell circuits under partial shading, *IEEE International Conference on Industrial Technology* (ICIT), 2005, 14–17 December 2005, pp. 866–870.

CHAPTER

9

Design Examples

Having gone through the components of a photovoltaic (PV) system in detail in Chapter 4, it is only logical to apply the principles discussed there to actually determine their quantitative values to complete the system configuration. This will not only enable the reader to install a PV system of desired specifications but also have a feel of the magnitude of the problem, and its cost, to compare with those of the conventional energy systems leading to the feasibility of this renewable energy source for different purposes. Many of the readers will be familiar with a 1 kW domestic inverter used to supply standby power during outages of conventional electric power. Since this is used as a reserved source of power for 24 hours it will be interesting to compare this with a photovoltaic inverter of the same capacity.

9.1 Design of a 1 kW Stand-alone Photovoltaic Power Supply

All PV systems are similar in nature but differ only in details from each other. It is therefore necessary to have a general view of the system design. Once the true nature of the system is understood, any other individual product design becomes a mere corollary. For example, a solar water pump and a solar lantern both have the inputs from a solar module or an array of different power, processed through a maximum power point tracker and followed by a load, which is either a pump or a lamp.

The photovoltaic system cited above can be better appreciated with quantitative values attached to each component. It is then crucial to design a PV power plant, for a stand-alone application, of the following specifications:

1. Average power output = 1 kW into a DC load at a DC voltage of 108 V
2. Duration of operation = 24 hours a day
3. Average time of sunlight available = 8 hours a day
4. Number of sunless days = 2 per week

5. Peak value of insolation in Delhi = 900 W/m^2
6. Maximum depth of discharge of battery = 50%
7. Array should have a fixed tilt.

The block diagram of the PV system is re-drawn in Figure 9.1.

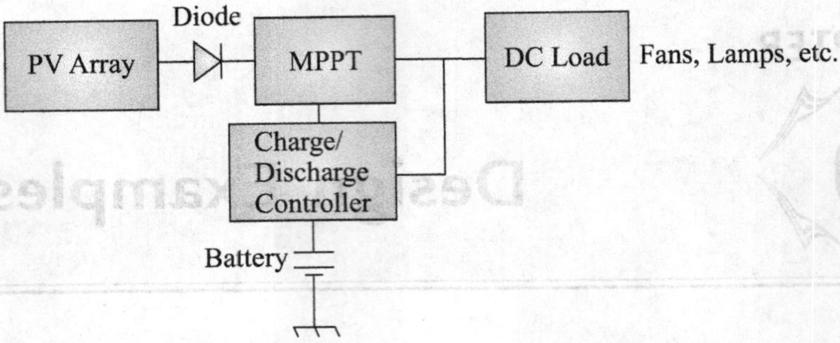

Figure 9.1 Block diagram of the 1 kW PV power supply.

In Figure 9.1, the charge/discharge controller has been separated from the MPPT, which jointly form the power conditioner. To calculate the size of the PV array and that of the battery, the total load on the array has to be calculated which will include the losses and efficiencies of the devices in between the array and the load. The following assumptions are being made:

(a) The electrical efficiency of the circuit of the MPPT = 90%
(b) The charge/discharge cycle efficiency of the battery (assuming new ones) = 90%
(c) The diode is usually a built-in part of the MPPT and therefore neglected. However, it is essential to save the circuit from accidental input voltage reversal.
(d) Wiring and cabling will introduce another 5% loss.

The energy requirement for seven days can be calculated as follows:

Power required = 1000 W

Therefore, the energy needed for seven days is:

$$E = 1000 \text{ W} \times 24 \text{ hr} \times 7 \text{ days} = 168{,}000 \text{ Whr}$$

For eight hours of sunlight per day, available for five days, the energy given directly to the load is:

$$1000 \text{ W} \times 8 \text{ hr} \times 5 \text{ days} = 40{,}000 \text{ Whr}$$

Hence, the rest of the energy must be stored in, and supplied by, the battery bank. This energy is:

$$168{,}000 - 40{,}000 = 128{,}000 \text{ Whr}$$

Since the charge-discharge efficiency of the battery bank is 90% (assumption b) the energy to be supplied to the battery is:

$$128{,}000 \div 0.9 = 142{,}222 \text{ Whr}$$

9.1.1 The Size of the Battery Bank

If the voltage of the battery bank is 108 V, as desired, then its charge capacity is:

$$142{,}222 \text{ Whr} \div 108 \text{ V} = 1316.87 \text{ Ahr}$$

Since it is stipulated that the batteries must retain 50% of the charge after supplying the required energy, their charge holding capacity must be twice this value. That is:

Total charge = 1316.87 × 2 = 2633.75 Ahr

Such high charge capacity (Ahr) cells have an output voltage of 2 V each.

To give a nominal output of 108 V a string of 54 cells, of 2 V each, must be connected in series. That is 54 cells of 2633.75 ampere hours each should be attached in series to give the necessary energy. In case 2633.75 Ahr cells, which are rare, are not available then two strings of 54, 1316.87 Ahr ones may be connected in parallel. These must be the antimony doped deep discharge type. The batteries are to be connected as shown in Figure 9.2.

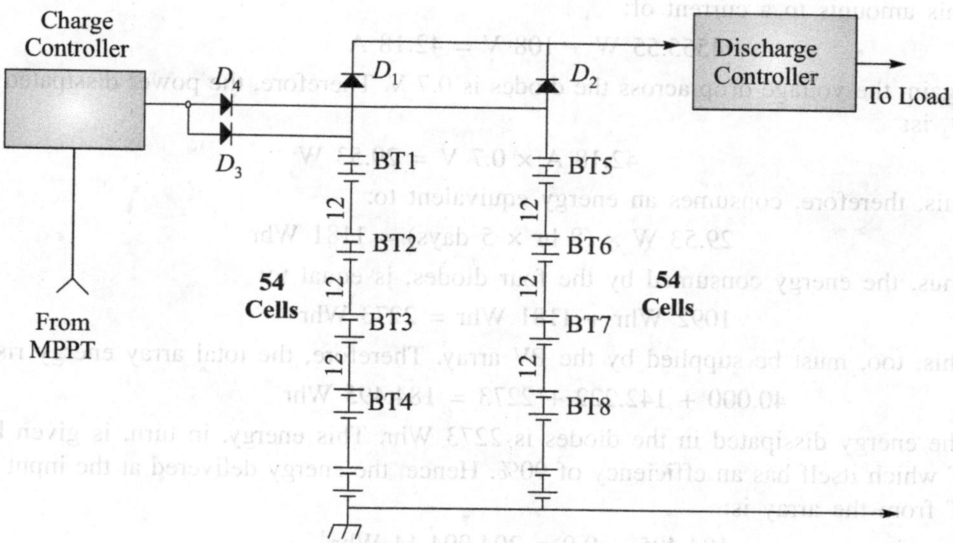

Figure 9.2 Battery bank for the storage.

Diodes D_1 and D_2 ensure that one string of batteries is isolated from the other during discharge into the load whereas diodes D_3 and D_4 isolate them during the charging process. However, it must be noted that these four diodes waste some photovoltaic power which depend on the currents. At 108 V the load current is:

$$1000 \text{ W} \div 108 \text{ V} = 9.26 \text{ A}$$

If the diodes are made of silicon the nominal voltage drop, during conduction, across them is 0.7 V. Therefore, both the diodes will dissipate a power of

$$9.26 \text{ A} \times 0.7 \text{ V} = 6.5 \text{ W}$$

Each diode will conduct half the current of 9.26 A, that is, 4.63 A and the maximum voltage at any condition will be 108 V. Therefore each diode, with 100% overrating, should be capable of conducting 10 A and with a reverse breakdown voltage of 200 V to take care of heating and surges if any. These diodes should be mounted on suitable heat sinks and air cooled with fans if necessary.

In terms of energy the diodes will consume:

$$6.5 \text{ W} \times 24 \text{ hrs} \times 7 \text{ days} = 1092 \text{ Whr}$$

To calculate the dissipation in diodes D_3 and D_4, it should be remembered that the energy passing through them is:

$$40{,}000 \text{ Whr} + 142{,}222 \text{ Whr} = 182{,}222 \text{ Whr}$$

This energy is passed in:

$$8 \text{ hr} \times 5 \text{ days} = 40 \text{ hr}$$

Hence, the power going through them is:

$$182{,}222 \text{ Whr} \div 40 \text{ hr} = 4555.55 \text{ W}$$

This amounts to a current of:

$$4555.55 \text{ W} \div 108 \text{ V} = 42.18 \text{ A}$$

Again, the voltage drop across the diodes is 0.7 V. Therefore, the power dissipated in D_3 and D_4 is:

$$42.18 \text{ A} \times 0.7 \text{ V} = 29.53 \text{ W}$$

This, therefore, consumes an energy equivalent to:

$$29.53 \text{ W} \times (8 \text{ hr} \times 5 \text{ days}) = 1181 \text{ Whr}$$

Thus, the energy consumed by the four diodes, is equal to:

$$1092 \text{ Whr} + 1181 \text{ Whr} = 2273 \text{ Whr}$$

This, too, must be supplied by the PV array. Therefore, the total array energy rises to:

$$40{,}000 + 142{,}222 + 2273 = 184{,}495 \text{ Whr}$$

The energy dissipated in the diodes is 2273 Whr. This energy, in turn, is given by the MPPT which itself has an efficiency of 90%. Hence, the energy delivered at the input of the MPPT from the array is:

$$184{,}495 \div 0.9 = 204{,}994.44 \text{ Whr}$$

As assumed earlier, there is a 5% loss in wiring and cabling. Hence, the output of the array would be:

$$204{,}994.44 + \frac{5}{100}\, 204{,}994.44 = 215{,}244.16 \text{ Whr}$$

9.1.2 Array Size

1. An approximate calculation

This energy of 215,244.16 Whr is to be generated by the array in 5 days with 8 hours of sunlight on each day. Hence, the power of the array becomes:

$$215{,}244.16 \text{ Whr} \div (8 \text{ hr} \times 5 \text{ days}) = 5381.10 \text{ W}$$

The modules available in the market are calibrated at an illumination of 1000 W/m². But the insolation received on earth, at Delhi, never exceeds 900 W/m² with almost zero illumination in early morning and at night. The intensity varies approximately in a sinusoidal way as shown in Figure 9.3. For this reason if modules of 5381.1 W are

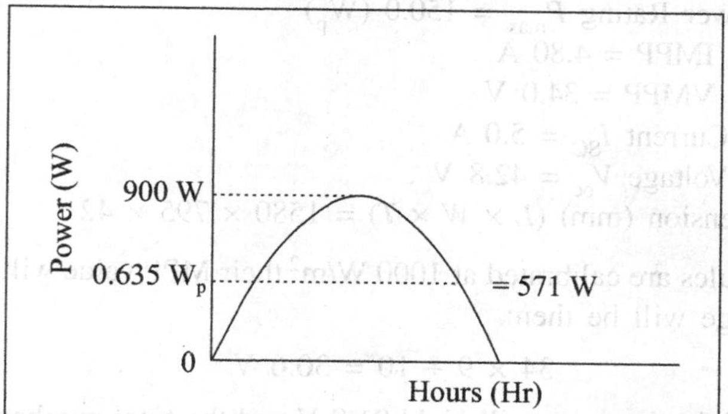

Figure 9.3 Solar intensity varies as an approximate sine function over the day.

procured from the market they will not produce the required power. To compensate for the lower radiation an array of higher power has to be installed. If the peak radiation is 900 W/m² and the nature of its variation is assumed to be sinusoidal, to a first approximation, then the average works out to:

$$900 \text{ W} \times 0.635 = 571 \text{ W}$$

Therefore, the total size of the array becomes

$$5381.1 \text{ W} \times 1000/571 = 9424 \text{ W}$$

To determine the number of modules to be put in series and parallel, the array MPP voltage must be defined and this is dependent on the operating point of the MPPT. For low MPP voltages and power levels at the input a boost switching converter based MPPT is preferable because it can raise low input to higher output operating voltages. The transfer function of the boost converter is:

$$V_o = V_{in} \div (1 - D)$$

where V_o is the output voltage of the converter, V_{in} is its input voltage and D is the duty cycle. Thus, if D is 0.9, then on one extreme:

$$V_o = 10\, V_{in}$$

And on the other extreme if $D = 0.1$

$$V_o = V_{in} \div (1 - 0.1) = V_{in} \div 0.9 = 1.11\, V_{in}$$

This means that if the output is fixed at 110 V, to account for the drop in the diodes D_1 to D_4, the input to the MPPT can vary from 11–99 V which will indeed cover almost the entire time from morning to evening as shown in Figure 9.3. So the peak MPP of the array can be set at above 90 V when the insolation is 900 W/m². Since quite a high power is needed the CEL PM 150 module is reasonably attractive and the modules should be arranged in a way shown in Figure 4.5. It should also be noticed from the same figure that the diodes R_1 to R_4 will consume some of the power. The specifications of the solar module are given below:

Electrical Parameters of module type PM 150 under standard test conditions

1. Maximum Power Rating P_{max} = 150.0 (W_p)
2. Rated Current IMPP = 4.80 A
3. Rated Voltage VMPP = 34.0 V
4. Short Circuit Current I_{SC} = 5.0 A
5. Open Circuit Voltage V_{oc} = 42.8 V
6. Physical Dimension (mm) ($L \times W \times T$) = 1580 × 795 × 42

Since these modules are calibrated at 1000 W/m² their MPP value will reduce at 900 W/m². The real MPP voltage will be then:

$$34 \times 9 \div 10 = 30.6 \text{ V}$$

Therefore 3 modules in series will yield 91.8 V and the total number of modules required for the array is:

$$9424 \div 150 \text{ (W}_p) = 62.83 = 63$$

Hence the number of strings, with each strings containing three modules is 63/3 = 21. Thus, the array will have a similar formation as shown in Figure 4.5 with 21 strings instead of 4 and 21 diodes placed at R_1 to R_{21}. Since each string will have a rated current IMPP of 4.80 A, de-rated by a factor of 0.9, that is 4.32 A, the specifications of the diodes must be 6 A, 150 V and mounted on heat sinks. This shall also be the ratings of the bypass diodes. The heat dissipated by each diode;

$$4.32 \text{ A} \times 0.7 \text{ V} = 3 \text{ W}$$

Therefore, the power lost in 20 diodes is equal to:

$$3 \text{ W} \times 20 = 60 \text{ W}$$

This is about 0.64% of the total array power of 9424 W and can therefore be considered negligible.

2. A more accurate calculation

Another way to calculate the array size is to proceed according to Section 4.8.2. If the array is set to collect the maximum energy over one year, then the tilt angle should be fixed at

$$\phi = (\text{Latitude} \times 0.9)° = 28.58 \times 0.9 = 25.7°$$

The radiation data for Delhi at a tilt angle of 25° gives a value of 2131.2 kWhr/m² annually [1]. This amounts to an average of

$$2131 \div 365 = 5.84 \text{ kWhr/m}^2\text{/day}$$

Therefore, the equivalent hours of full sunlight

$$\text{EHFS} = 5.84 \text{ kWhr/m}^2\text{/day} \div 1 \text{ kW/m}^2 = 5.84 \text{ hr}$$

It means that if a solar module is placed under an illumination of 1 kW/m^2 for 5.84 hr, it will receive this much energy. It is very convenient because all the modules are rated at this solar intensity. The power of the solar array will now be calculated. The energy required as calculated earlier is 215,244.16 Whr and has to be generated in five days. Therefore, the power of the array required is

$$215{,}244.16 \text{ Whr} \div (5.84 \text{ hr} \times 5 \text{ days}) = 7371.4 \text{ W}$$

This value is 2.053 kW lower than the one calculated earlier and is more accurate since it is based on the radiation collected over one year whereas the value of 900 W peak insolation was approximate. A further correction for temperature has now to be applied to the value just calculated. The temperature on the surface of the modules kept in the sun at the peak summer months is roughly 40°C. The temperature of the solar cells inside a module is another 25°C higher than that at the surface. Then it can be safely assumed the temperature of the cells is approximately 40°C higher than the STC temperature of 25°C at which the modules are rated. The decrease in power of the modules is 0.5% per degree centigrade [2]. Therefore, the de-rating of a 150 W module when the inside cell temperature is 40°C higher is

$$(0.5 \div 100) \times 40 \times 150 = 30 \text{ W}$$

This means that each 150 W module will deliver a power of 120 W that is a reduction of 20% power. Extending the same argument to the entire array the total array power must be raised by 20%. Therefore, the required power of the array becomes

$$7371.4 \text{ W} + (20 \div 100) \times 7371.4 \text{ W} = 8846 \text{ W}$$

Therefore, the number of 150 W modules needed = 8846 ÷ 150 = 59

Making this a multiple of three 60 modules will be sufficient to form the array. As discussed earlier 20 parallel strings of modules are needed with each string having three modules in series. The next problem is to place the components of the array such that one module does not cast its shadow on the one behind it.

9.1.3 Distance between Modules in the Array Formation

If one string of three modules is kept such that one module is behind another, the shadow of the first one will cast its shadow on the second one just behind it while the second one will cover the third one with its shadow. Therefore, an adequate distance should be kept between the first and the second and between the second and the third module so that under the worst condition, when the sun is over the tropic of capricorn, all the modules are exposed to light.

Figure 9.4 is a reproduction of Figure 4.14 for facilitating the calculation of the safe distance between two successive modules. Although two modules are shown here at M and C a third one is placed beyond the one at C.

The length of the shadow made the module placed at M is given by Equation (4.6) is written below:

$$L = D \sin \beta \tan (\phi - \delta)$$

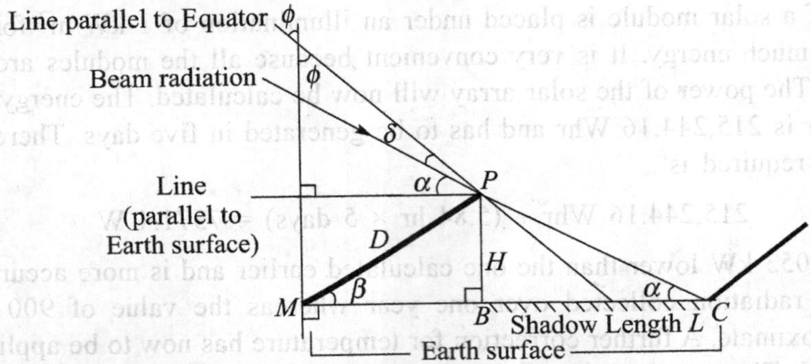

Figure 9.4 Two modules of a string of three shown for calculation of inter module distance.

When the sun is over the tropic of Cancer, the angle of declination δ is taken to be negative as described in Section 4.8.1 so that the Equation (4.6) becomes

$$L = D \sin \beta \tan (\phi + \delta) \tag{9.1}$$

The parameters to be inserted in the Equation (9.1) are:

1. Length of the PM 150 module, $D = 1.58$ m
2. Latitude of Delhi, $\phi = 28.58°$
3. Inclination of the module to earth, $\beta = \phi \times 0.9 = 28.58° \times 0.9 = 25.7°$
4. Declination angle, $\delta = 23.45°$

Substituting these values in Equation (9.1), we get,

$$L = 0.88 \text{ m}$$

The length of the shadow under the module is

$$MB = D \cos \beta = 1.58 \times \cos 25.7° = 1.58 \times 0.9 = 1.424$$

Therefore, the distance between the points where the two modules touch the earth is

$$MB + L = 1.424 + 0.88 = 2.3 \text{ m}$$

Conclusions

1. Three rows of twenty modules each should be placed on the ground from east to west and facing south.
2. The next row should be placed 2.3 metres behind the one in the front. This way no shadow will fall from the row of modules in front to the one just behind. This is shown in Figure 9.5. Although the modules are kept on the ground in this diagram, it is suggested that the array be raised at least twenty centimetres from the ground to avoid coming in contact with water during rains.

Example 9.1: It will be instructive to calculate the area covered by this array and find the cost of the land required to place it.

If a space of 2 cm is kept between adjacent modules, the total space between 20 modules is

$$2 \times 19 = 38 \text{ cm.}$$

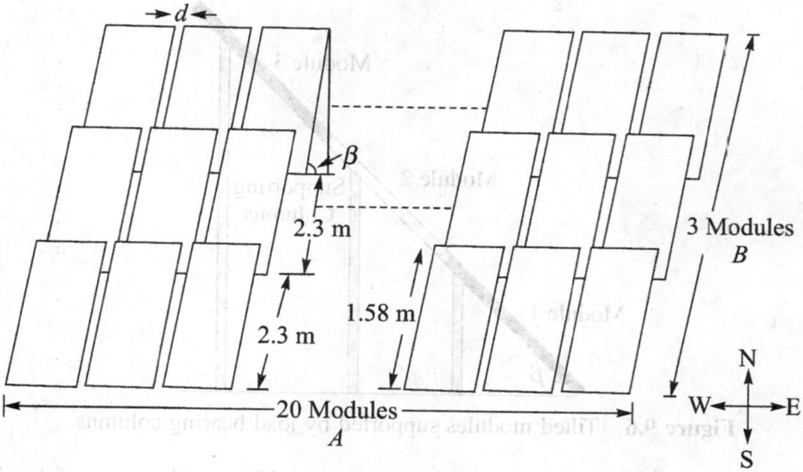

Figure 9.5 The complete array with dimensions.

Width of one module = 79.5 cm. Therefore, the total space occupied by the modules = 79.5 × 20 cm = 1590 cm. Hence, the total length of the array is

$$A = 1590 + 38 = 1628 \text{ cm} = 16.28 \text{ m}$$

Depth of the array, $B = (MB + L) + (MB + L) + MB = 2.3 \text{ m} + 2.3 \text{ m} + 1.42$

$= 6.02 +$ thickness of the three modules $= 6.02 + 0.042 \times 3 = 6.15$ m

Therefore, the area covered by the array $= A \times B = 16.28 \times 6.15 = 100 \text{ m}^2$.

The cost of land in South Delhi is ₹ 200,000 per m^2. Therefore, the cost of the land is: 200,000 ×100 m^2 = ₹ 20,000,000. This exorbitant amount definitely cannot be spent for the intended purpose as it would make the PV power plant most uneconomical. There are two solutions to the problem. First the modules can be mounted on rooftops which are not possible for domestic use as most of the households are lodged in multistoried apartments. However, large office buildings and factories can accommodate them. Such arrangements are possible in building integrated systems where the modules are located on the sides of the structures.

The second method is to take the large arrays out of the costly and congested cities to cheaper places like deserts and load the generated solar electricity on power grids and send it for consumption in the cities. Such methods are called *grid interactive systems* and have their own synchronization and islanding problems, discussed in Chapter 10.

Another way to reduce the land area is to put one module over another at the same angle with respect to the earth as shown in Figure 9.6.

The length of the array will remain the same but its depth will reduce. In Example 9.1 the length of the array $A = 16.28$ m.

The depth of the array B has now to be calculated. Since the three modules are kept in series, their total length = 3 × 1.58 m = 4.74 m. Therefore, depth

$$B = 4.74 \times \cos 25.7° = 4.27 \text{ m}.$$

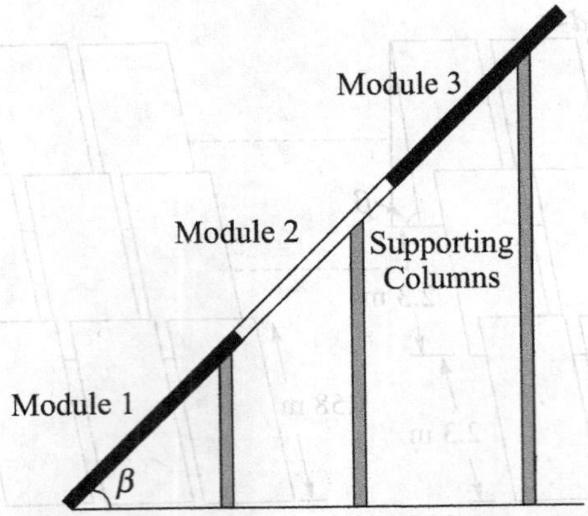

Figure 9.6 Tilted modules supported by load bearing columns.

This is smaller than 6.15 m of the earlier example. Hence, the area of land covered by the array

$$4.27 \text{ m} \times 16.28 \text{ m} = 69.53 \text{ m}^2.$$

This area is more than 30% lower than that in the earlier case.

9.1.4 Costing

The cost of the components is tabulated below:

The current price is ₹ 120.00 per W.

1. Solar module @ ₹ 160.00/W = 9415.52 W × 160 = ₹ 1,506,483.00
2. 54 cells, 2633.75 Ahr, 2 V each @ ₹ 15/Ahr = ₹ 2,133,337.00
3. MPPT and Charge/Discharge controller = ₹ 50,000.00

Total = ₹ 3,689,820.00

If a lifetime of 10 years is taken for the array and the electronics and five years for the battery bank, then the cost amounts to: ₹ 5,823,157.00. In 10 years the electricity produced is equal to

$$1 \text{ kW} \times 24 \text{ hr} \times 365 \text{ days} \times 10 \text{ years} = 87,600 \text{ kWhr}.$$

Therefore, the cost of this energy is

$$₹\ 5,823,157/87,600 = ₹\ 66.47 \text{ per kWhr}$$

That is, the cost of the electricity produced is now dependent on the cost of the battery, which has become the dominant factor. At present the cost of domestic electricity from the grid is ₹ 4.60 per kWhr (unit). However, the cost of electricity produced from diesel generating sets is ₹ 12.50 per unit. This figure was taken when the global price of oil was $ 100 per barrel. Till now, the diesel oil is subsidized in India even though its global price has gone up to $ 135 a barrel. If the life span of the PV array is taken to be 20 years, the PV generated power will be approximately ₹ 57.59 per unit, as the cost of the battery will double, and will still be costlier than that from diesel. Therefore, for any PV application the storage system has to be reduced in size to make PV generated power comparable to diesel generated power. From the above calculations, it is clear that the battery may be used as a voltage stabilizing device rather than a storage facility and that for large energy applications the PV source can be effectively used during sunlight hours to augment the electricity produced by diesel generators. The additional benefits of PV power are:

- Carbon credits
- Lack of emission of polluting gases.

All photovoltaic systems are covered by some standards among which those of the IEEE and the International Electro-Technical Commission (IEC) dominate. The following standard must also be adhered to in this case.

IEC 61194 Ed. 1.0 b: Characteristic parameters of stand-alone photovoltaic (PV) systems. Another one that deals with the efficiency measurement of the power conditioners is given in **IEC 61683 Ed. 1.0 en.**

9.2 PV Based Lighting: CFL and LED Based Solar Lanterns and Streetlights

Solar lanterns were made as a viable alternative to the hurricane lantern for use in the remote areas of India, where conventional electric power was not available [3]. Subsequently a model was designed and fabricated [4] (*Copyright (2000), Reprinted with Permission from Elsevier*) to be marketed as desired, in a project financed by the Ministry of Non-Conventional Energy Sources (MNES), Government of India, in 1992, with an object to supplying the basics lighting needs of people living in those areas and were dependent on the Hurricane lantern for illumination [5]. Since sunlight is freely and abundantly available at all places, not covered with clouds, it can be easily converted into electricity by solar cells; a string of such cells make a solar panel. The details of solar panels are given in the catalogue of Central Electronics Limited (CEL) [6] and other manufacturers. This electricity can then be stored in suitable batteries, normally lead-acid maintenance-free type, to be used when needed. To produce light, the options were limited to either a general lighting service (GLS) incandescent filament lamp, the ordinary household electric bulb, or the compact fluorescent lamp (CFL) since the other devices such as the 4 feet tube light and the low pressure sodium lamp (SOX 18) were either too bulky or too power consuming or both. The objective was simply to produce enough light to equal or slightly exceed that of the kerosene lantern, hence the name solar lantern. Since the cost of a solar panel is about ₹ 200 ($4) per peak watt, the light output of the lamp per unit power input, defined as its *efficacy*, had to be the highest to keep the cost of the solar lantern within the reach of the buyer. The clear choice was therefore the CFL with an efficacy of 50 to 60 lumens per watt [7] over the GLS lamp with its efficacy of about 10–14 lumens per watt [8].

CFLs are gas discharge tubes and therefore require high voltages, of the order of hundreds of volts, to start glowing. For this reason, an oscillator circuit with high output voltage and the required power to operate the tube, has to be constructed. This also must have a high electrical efficiency, exceeding 80%, so that most of the solar electricity actually reaches the tube instead of being wasted in the circuit. The block diagram of the solar lantern is shown in Figure 9.7. A solar panel generates the electricity, which is given to the oscillator and the control circuit. The control circuit, in addition to operating the oscillator, also contains a charge controller which saves the battery from being damaged by preventing it from getting either over-charged or over-discharged. The ballast, in this case a capacitor, ensures that excessive current does not flow through the lamp and thus prolongs the lamp's life.

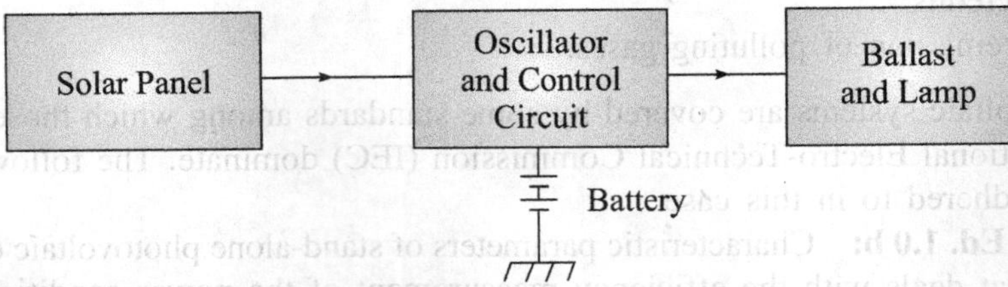

Figure 9.7 Block diagram of a solar lantern.

The CFL is the miniature version of the normal 4 feet tube light called the *fluorescent lamp*. The fluorescent lamp must be described first to understand the CFL.

A breakthrough in more efficient lighting technology came with the invention of the fluorescent lamp which is commonly referred to as the tube light. The normal 4 feet tube light is a low pressure gas discharge tube with two filaments at the ends. The finished lamp is filled with an inert gas such as argon, to a pressure of 1.5 to 5 torricelli (millimetre of mercury). A tiny droplet of mercury is introduced which is used in the discharge to create light. The discharge produces ultraviolet radiation, in addition to visible light, which then is converted to visible light by a coating of phosphor on the inside of the tube [9]. This conversion of wavelength is responsible for the increased emission of visible light resulting in a higher efficacy.

The actual working diagram of the lamp is shown in Figure 9.8. When the power is switched on, the tube does not ignite and a starter is needed to initiate this process. The starter contains a bimetallic strip which bends on heating owing to different coefficients of linear expansion of the two constituents. It is usually filled with an inert gas such as neon

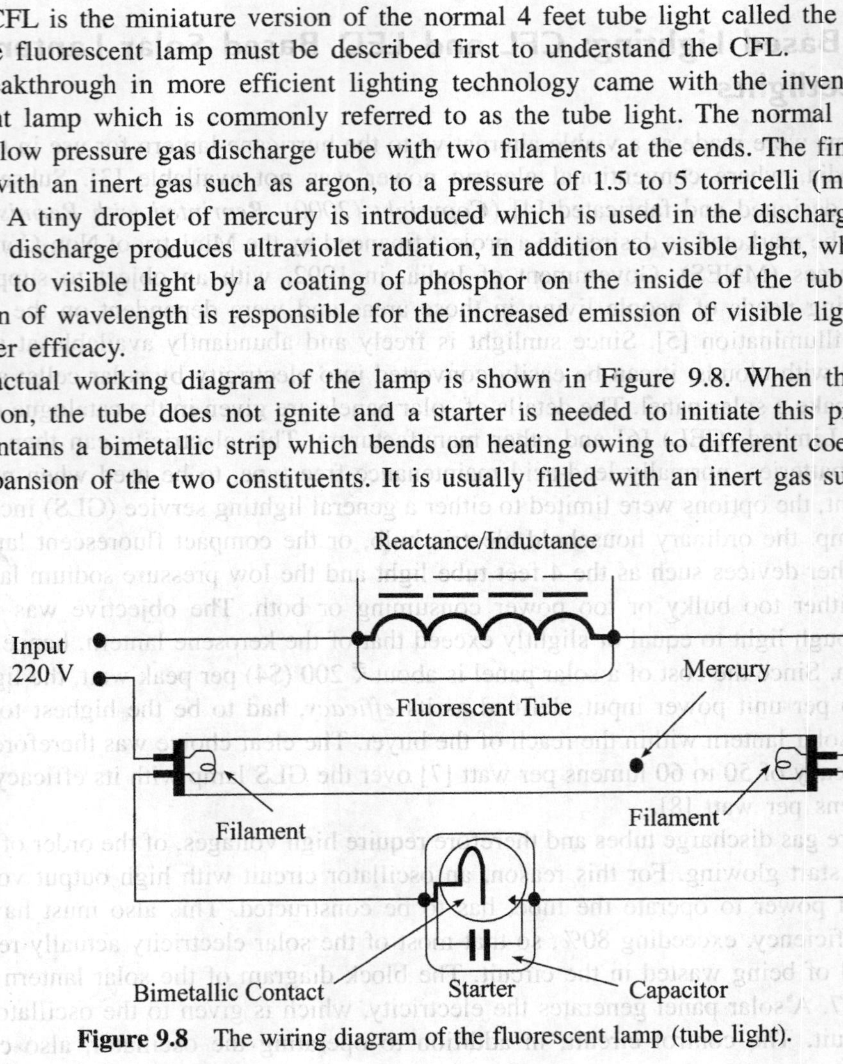

Figure 9.8 The wiring diagram of the fluorescent lamp (tube light).

with an ionizing agent such as tritium, which ignites when the voltage is applied. This in turn heats up the bimetallic strip which closes the switch. A current, limited only by the reactance of the inductance called the *choke*, flows through the two filaments of the tube, which get heated and release electrons. Almost at the same time the gas inside the starter stops glowing thus making the bimetallic strip to cool down so that it comes back to its original position.

This suddenly breaks the circuit so that the current is interrupted. This sudden reduction of current causes a sharp increase in the voltage, to about 600 V, at the inductor according to the equation:

$$V = L\,(di/dt) \tag{9.2}$$

where L is the value of the inductance and di/dt is the rate of change of current.

This voltage of 600 V across the tube is sufficient to cause an arc discharge through it. Once the discharge is established the voltage across the 4 ft tube comes down to about 100 V and the current is then limited by the inductor to 400 mA approximately. The function of the starter then ceases.

However, the power factor of the combination is about 0.55, which is poor. To improve the power factor, several methods are employed. One is to put a capacitor of a suitable value at the input to bring up the value to 0.9 while another is to use electronic ballast, which is costlier.

The CFL

The compact fluorescent lamp is a miniature version of the 4 ft tube light and essentially works on the same principle but when used in a similar circuit configuration gives a very poor power factor, typically 0.3 for a 9 W tube. It is always advisable to use electronic ballasts with these. Most of the electronic ballasts sold in India generally have a poor power factor of 0.5 to 0.8 with a harmonic distortion factor of 120%, which is harmful to the mains if large numbers of these ballasts are used. The power factor corrected ballasts are few and far between since they are very expensive. Therefore, it is incorrect to advocate the use of electronic ballasts without all these improvements. In a solar lantern electronic ballast is used for this reason and being a low voltage device, typically 12 V DC, a step up transformer is needed.

The use of LEDs

At present light emitting diodes (LEDs) have come up as an energy efficient device alternative to the CFL at low level illumination, with a considerable long life. Being solid state devices they can be operated with direct current at low voltage levels of 2 to 4 V. Hence, it was an interesting exercise to compare lanterns made with these two light sources.

The efficacy of LEDs has steadily increased over the last few years. An interesting comparison among efficacies of different light sources, and the progress the LEDs have made with time, has been made in an article in the Scientific American [10]. It shows a steep rise in efficacy of the LEDs from the year 1985 to 2000. The white light LED was in its infancy with 20 lumens per watt (lm/W) in 2000. Now, it is 107 lm/W [11] and is a strong rival of the CFL.

The basic LED consists of a junction between two different semiconductor materials, one *n-type* and the other *p-type*, illustrated in Figure 9.9, in which an applied voltage produces a current flow, that generates light when charge carriers injected across the junction recombine.

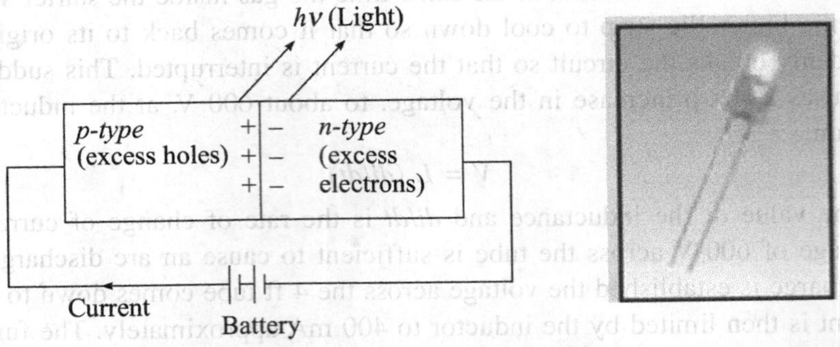

Figure 9.9 A *p-n* junction (left) as a LED (right).

This *p-n* junction in the form of an integrated circuit chip is placed in a concave reflector with two wires attached to carry current to it, with an epoxy based lens in front [12]. The reflector sends the backward going light to front whereas the lens focuses it in the desired angle, which can range from 15 degrees to 100 degrees. The frequency, and perceived colour, of emitted photons is characteristic of the semiconductor material, and consequently, different colours are achieved by making changes in the semiconductor composition of the chip. Reference [13] gives figures to indicate that there is a large variation in the efficacies of the LEDs available in the market. Therefore, care should be taken to select suitable products while fabricating the lanterns.

The dependence on fossil fuels such as kerosene and the resulting polluting green house gases had also to be reduced for which the solar lantern was the obvious answer. Besides, the cost of transporting the oil to such locations is high and the buyers, especially in the hilly regions, have to travel long distances over difficult terrain to fetch it. Fuel-based lighting was shown to be significantly costlier than solar-powered compact fluorescent lamps (CFLs) and solar-powered light emitting diode (LED) alternatives per unit of lighting services delivered. It was calculated [14] that the running cost of a Hurricane-style kerosene lamp is approximately \$0.40 per thousand lumen hours (klmhr) or \$2.89 per thousand lux hours (klxhr), while that of a solar-CFL lantern costs \$0.17/klmhr and \$1.20/klxhr while a solar-LED device costs \$0.15/klmhr and \$0.03/klxhr. Furthermore, as LED efficiencies continue to improve, solar-LED products will become even more economical. Dutt [15] gives the data about the output of different sources as given in Table 9.1 and Table 9.2. The two tables show difference in the light output from the kerosene wick since the material, the thickness and the burning conditions may not have been the same in the two cases. However, an idea of the light outputs can be had by comparing the two. Rajvanshi [16], [17] has given additional data on the output from different sources but the data of interest, the light output from kerosene lanterns, do not differ much from those given in Table 9.2. The 1 W white LED from Cree Inc., type XR7090WT, gives an output of 47 lm/W [11] and can be placed as the second best source of light next to the 16 W CFL. This device costs \$3 per unit in strip package and \$3.50 if reflow mounted on ¾″ round or star printed circuit board. This also has a built-in lens that can diverge light at an angle of 100 degrees.

Table 9.1 Light characteristics for different light sources (1)

Light source	*Flux* (lumens)	*Colour* (2)	*Luminous efficacy* (lumens/W)	*Spec. Fuel Consumption* (kg/klum-hr)	*Fuel Consumption Rate* (kg/hr)
Candle	12	Excellent	0.2 (3)	0.5 (4)	0.006
Kerosene wick	40	Excellent	0.1	0.8	0.032
Kerosene mantle	400	Poor	0.8	0.1	0.040
60 W incandescent	730	Excellent	12	0.21 (5)	0.016
16 W CFL	900	Good	56	0.005	0.004
1.2 W LED (6)	56.4	Good	47	–	–

Notes:

1. Source: van der Plas, R., 1988, "Domestic lighting", Working Paper WPS 68, Industry and Energy Department, World Bank; original includes additional non-electric sources—carbide lamp, butane lamp, and gasoline mantle lamp; compact fluorescent data calculated here from data presented earlier in the chapter.
2. Colour Rendering Index (CRI) is difficult to measure for non-electric sources, according to van der Plas; thus colour quality is shown here on a qualitative scale.
3. For non-electric lamps, the power input in watts has been calculated using the higher heating value of the fuel.
4. Specific fuel consumption is expressed as kg of fuel use per 1000 lm hr of light (kg/klum-hr).
5. For electric sources, we assume a conversion efficiency of 0.30 from kerosene to electricity, and express specific fuel consumption and fuel consumption rates as equivalent amounts of kerosene.
6. The LED data has been taken from reference [11].

Table 9.2 Another comparison electric and non-electric light sources (1) [5]

Light Source	*Flux* (lumens)	*Luminous efficacy* (lumens/W)	*Spec. fuel Consumption* (kg/klum-hr)	*Fuel Consumption Rate* (kg/hr)
Kerosene				
Chirag (diya) (2)	21	0.26 (3)	0.299 (4)	0.006
Hurricane lantern	69	0.38	0.203	0.014
Petromax (mantle)	1303	1.25	0.056	0.073
Electricity				
25 W incandescent	225	9	0.029 (5)	0.007
60 W incandescent	660	11	0.023	0.016

Notes:

1. Selected from Rajvanshi (1987) [18] as quoted in Sinha and Kandpal (1991) [19].
2. The 'Chirag' or 'Diya' is a simple lamp made using a wick and a glass bottle. 3, 4 and 5. Same as notes 3, 4 and 5 of Table 9.1.

Two lanterns were, therefore, constructed for comparison. One was made from the white LEDs, of 47 lumens per watt, and the other was made from a 5 watt CFL with a soft/cold-start, the most efficient firing mechanism, to give the maximum electrical circuit efficiency [20]. The design and experimental results are given below for the two lanterns.

Experiments

The comparisons between the two circuits for their electrical efficiency and cost were made. Both these circuits were designed and fabricated by the authors; one for the MNES, as mentioned earlier, but subsequently modified to compensate for the changes in the battery voltage with temperature and the other to operate a white LED from a constant current source. The first circuit, for the CFL, has been widely manufactured and sold all over India and is very reliable. The basis for this comparison is to find the most cost effective solution for the replacement of the Hurricane lantern, which should provide light for about three hours with the required protections for the battery [21].

The CFL based circuit

Design parameters:

1. Power of lamp = 7 W
2. Time of use = 5 hours per day
3. Conversion efficiency of inverter = 80%
4. Charge remaining in the battery after use = 50% of the full charge
5. Number of sunless days per week = 2 days

If it is desired to operate a 7 watts CFL for five hours per day after sunset in the non-electrified area, the energy required would be

$$E = 7 \times 5 \text{ Whr} = 35 \text{ Whr}$$

Taking the efficiency of the inverter, which converts 12 V_{DC} to 220 V_{rms}, to be 80% the DC energy required is

$$35 \div 0.8 = 43.75 \text{ Whr}$$

If 50% of the charge has to be retained in the battery after a five hour's use, the energy stored should be

$$43.75 \times 2 = 87.5 \text{ Whr}$$

If a 12 V_{DC} battery is used, its charge capacity will be

$$87.5 \text{ Whr} \div 12 \text{ V} = 7.29 \text{ Ahr}$$

It must be remembered that each night a charge equal to 43.75 Whr ÷ 12 V = 3.65 Ahr is removed from the battery by the lantern and stored back the next day by the solar panel. If this goes on uninterrupted for all the seven days the battery size as calculated will suffice. However, this is not so and after five days there will not be any sun to charge the battery which will have only 50% of the total charge left. This charge will be exhausted on the sixth day. Such a condition will have two repercussions: first the battery will be fully discharged and thus may be permanently damaged and secondly, of course, the lantern will not work on the seventh day. It is imperative therefore that the charge holding capacity of the battery must increase to store the required charge generated in five days, to last for seven days. This means that the size of the battery is

$$(7.29 \times 7) \div 5 = 10.2 \text{ Ahr}$$

However, the standard battery available close to this value is the 12 V, 7 Ahr sealed maintenance–free type used in the uninterruptible power supplies (UPS) needed as a backup power source for personal computers. Apart from its easy availability it is also cheap due to its mass production. Hence it was chosen for the solar lantern.

Module size: The average daily radiation in Delhi is 5.84 kWhr/m^2/day calculated earlier. Therefore the equivalent hours of full sunlight

$$\text{EHFS} = 5.84 \text{ kWhr/m}^2\text{/day} \div 1 \text{ kW/m}^2 = 5.84 \text{ hr}$$

Since 43.75 Whr of energy is consumed by the lantern everyday then in seven days the energy consumed is:

$$43.75 \text{ Whr} \times 7 = 306.25 \text{ Whr}$$

It is assumed here that there are two sunless days in a week. That means 306.25 Whr of energy has to be gathered in five days and that in turn means that the total sunshine hours at 1 kW is 5.84 × 5 hours. Hence the power needed to operate the lantern becomes

$$306.25 \text{ Whr} \div (5.84 \times 5) = 10.5 \text{ W}_p$$

A 10 peak watts module is used for both the CFL and LED based lanterns. With the advent of higher efficacy LEDs a smaller panel can do the job for the same light output which will make the instrument cheaper. With a 5 W CFL along with the same battery and PV module the lantern will work for more number of hours. It must be decided whether the solar lanterns should produce just the same amount of light as its Hurricane equivalent or more. The following discussion will clarify the issues.

Circuit Design: The circuit of the whole lantern is split into two parts: (i) the charge-discharge controller as shown in Figure 9.10 and the (ii) oscillator plus the lamp as shown in Figure 9.11. These are discussed below.

(i) *The charge-discharge controller:* Figure 9.10 shows the circuit of the charge-discharge controller of the CFL based lantern, which can operate either a 4-pin 5 W, or a 7 W CFL of any standard make. This circuit is needed to prevent the battery, a sealed maintenance-free lead-acid variety [22], from being over-charged or over-discharged thus prolonging its life. This also controls the switching of the oscillator, which in turn drives the CFL as shown in Figure 9.11. The description of the oscillator follows later.

An LM 324 14-pin quad operational amplifier (four OPAMPS in one envelope) is used to operate the controller. The four sections of the OPAMP are indicated by four triangles with their accompanying components. One section U3C, with its associated components, makes a stable 5 volts regulator used as a reference source. U1A uses this voltage to form a 14.5 V feedback regulator with the *p*-channel MOSFET MTP 2955E as the series pass element. Diodes D_1, D_2, and D_3 are low voltage drop Schottky diodes, which protect the circuit against

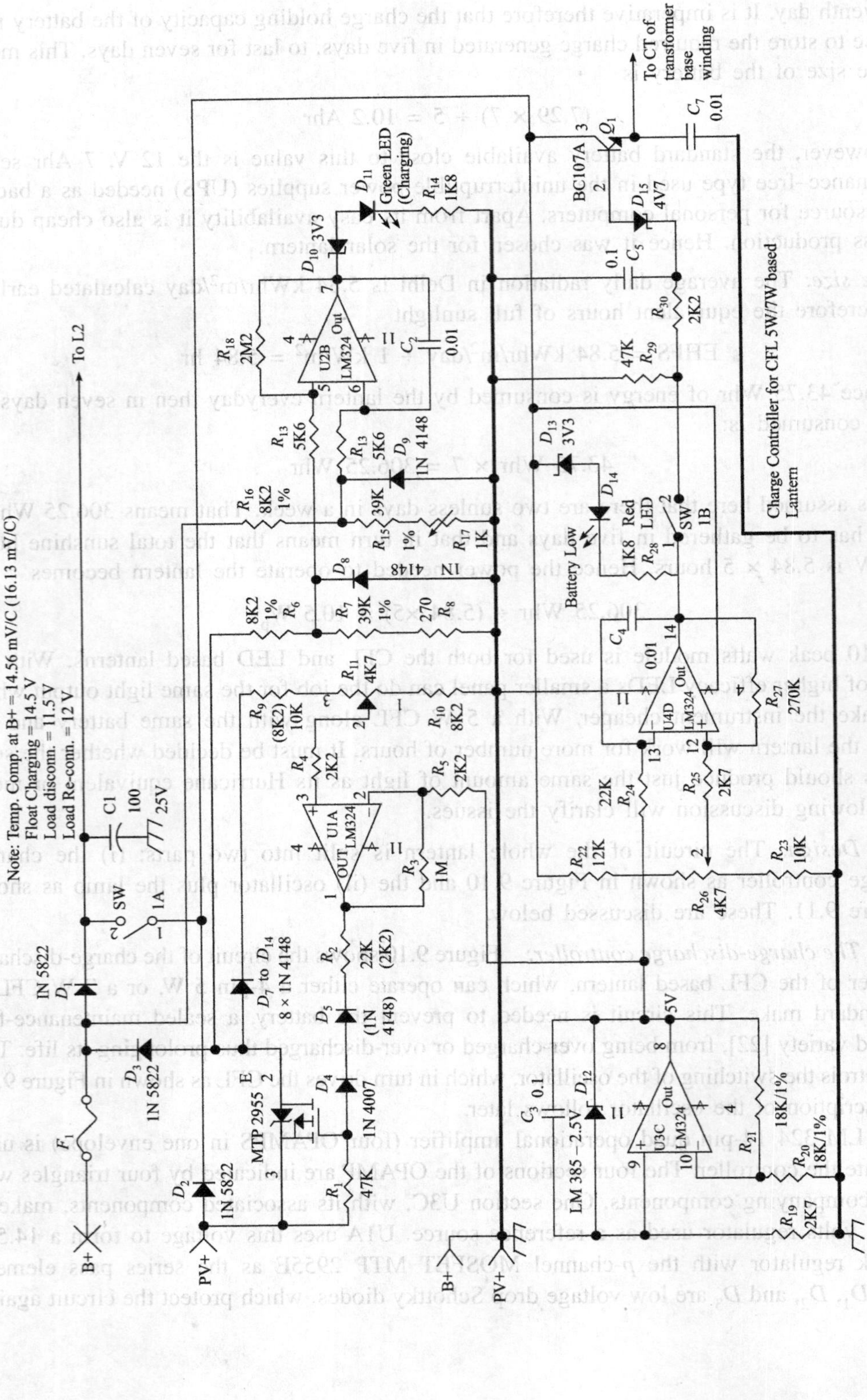

Figure 9.10 Charge-discharge controller of the CFL based solar lantern.

accidental reversal of photovoltaic panel and battery polarities. Diodes D_7 to D_{14} are 1N 4148 silicon diodes in the feedback loop of the regulator, used to provide temperature compensation for the battery during the process of charging. The 12 V lead-acid maintenance-free battery is connected between B+ and the B– terminals to store the charge generated by the PV panel during daytime while the PV panel is connected between the PV+ and PV– terminals. U2B acts as a comparator by detecting the voltage across diode D_3. When the battery is being charged the terminal 5 of U2B is more positive the terminal 6 and the green light emitting diode, D_{11}, glows to indicate it.

When the battery voltage goes lower than 11.5 V, the output voltage of operational amplifier U4D goes low and the transistor Q_1 stops conducting. This shuts down the CFL driving oscillator, shown in Figure 9.11, while the red LED D_{14} gives an indication of 'Battery Low'. When the battery is charged from a partially discharged condition, the series resistance of the PV panel reduces its output voltage to latch on to that of the battery plus some voltage drop across the MOSFET and diodes that come in series. When the battery is fully charged, the output of the regulator settles down to a fixed value of 14.5 V so that overcharging does not damage the battery. However, the regulator continues to furnish a small current to compensate the loss of charge, an inherent property of lead-acid cells, which is known as *trickle charging*. A double pole, double throws switch, SW_1, turns ON/OFF the lantern manually.

A 10 W solar panel is used to charge the 12 V battery. Although this is the nominal battery voltage its actual value really depends on its state of charge. This may vary from 10.8 V, for deep discharge, to 14.5 V for a fully charged state for both lead-acid and VRLA types [21], [22] and [23]. Therefore, the open circuit voltage of the panel should be 12/0.8 = 15 V so that it operates at its maximum power point. Since there are some diodes and a MOSFET in the charging path the panel voltage is usually kept between 16 to 17 V to also account for the temperature rise in the tropics.

(ii) *The oscillator and the lamp circuit:* The circuit diagram of the oscillator used to drive the CFL is shown in Figure 9.11. This is a current-fed, class-D sinusoidal circuit in a push-pull configuration, described by Yarrow [24] and Baxandall [25] that has a theoretical efficiency of 100%, which is the reason for its selection. Diode D_1 ensures that a negative polarity bus voltage from a solar panel or a battery does not damage the circuit while the inductor L_2 maintains a current fed topology. The value of L_2 should be five times, or more, the primary inductance of the transformer T_1 from its centre tap. The primary voltage is then stepped up, to about 125 V, to ignite the CFL, which is either a 5 W or a 7 W device through a ballast capacitor C_2. To help start the discharge in the tube a shield, a thin strip of copper, is placed between the two bent columns of the CFL and tied to its horizontal upper part with a few turns of thin copper wire. This effectively reduces the length of the tube for creating a glow.

A Hewlett-Packard digitizing oscilloscope model 54501A was used to measure the voltage across the lamp at F_3 and ground and the signal at point 4 of the secondary of the transformer that is across R_2, which was proportional to the current through the lamp. A built-in multiplier gave the product of these two quantities to display the power waveform. And since it is point-to-point multiplication, the power factor was automatically taken into account. The power was then averaged over sixteen cycles to get a stable power value. A 3n3F capacitor,

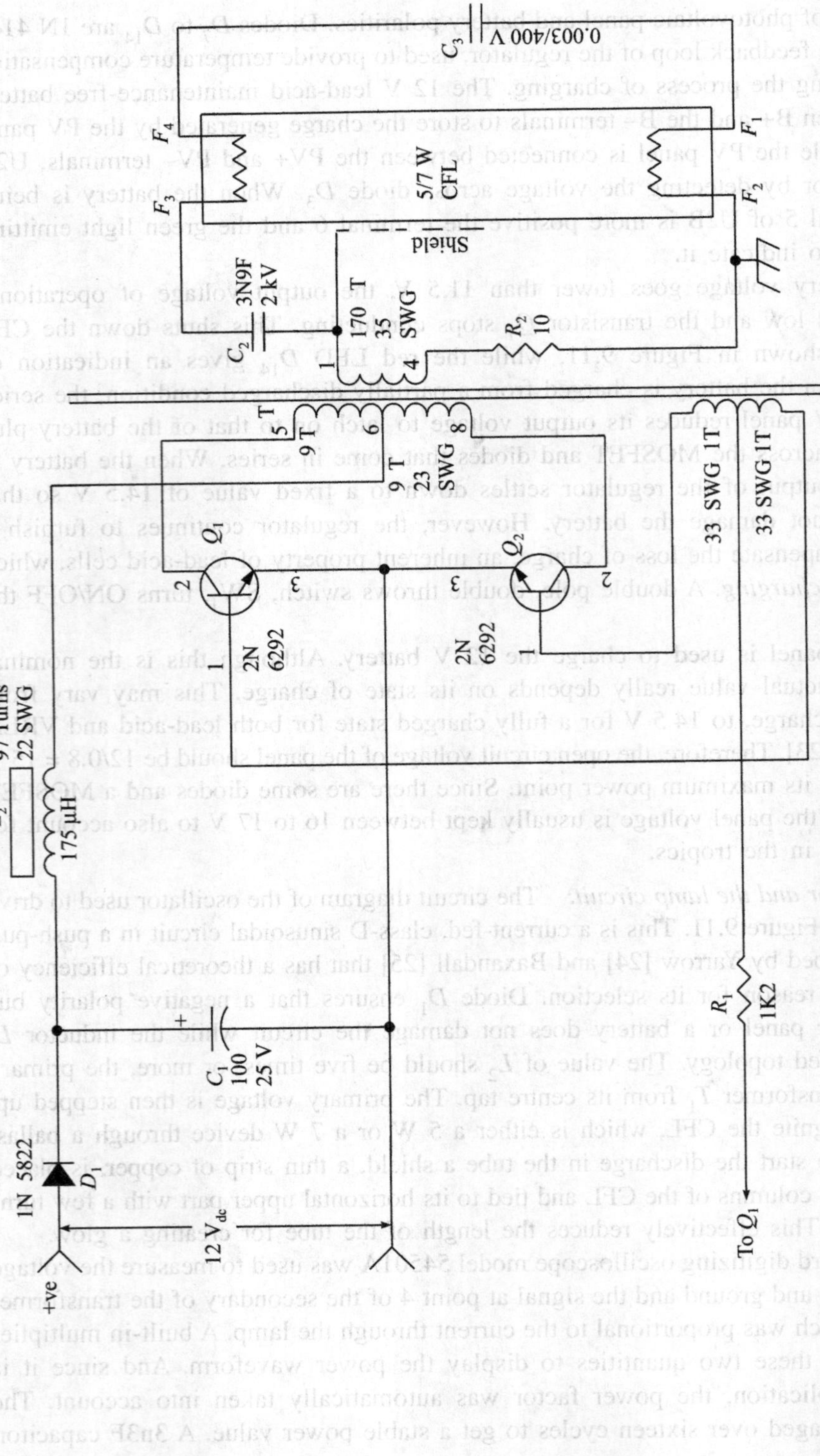

Notes:

1. All resistor values are in Ohms and 250 mW, 5% tolerance unless stated.
2. All capacitor values are in microfarad unless stated.

Figure 9.11 Circuit diagram of the current-fed oscillator.

C_3, is placed between the two filaments of the tube such that it provides a current path through them when the inverter is switched on. Although this current of approximately 200 mA is adequate for pre-heating the filaments, it also reduces the voltage across the tube because of the capacitive divider formed by C_2 and C_3. Initially this low voltage fails to start a discharge but when the hot filaments emit enough electrons the voltage is sufficient to cause a glow. The slow heating of the filaments thus delays the ignition and allows for a gradual increase in the intensity of the emitted light, which is called a *soft start*. Once the tube attains full conduction the voltage across it comes down to its sustaining value, which is about 30 V for a 5 W CFL, and therefore the current through C_3 at this lower voltage reduces.

Another factor that affects the current flow through the filaments is the reduction of the oscillator frequency when the CFL lights up. The ballast capacitor C_2, in series with the tube effective resistance of about 170 Ω, comes in parallel with the secondary winding of the transformer. The shift in frequency is quite drastic as can be seen from Table 9.3. The two effects combined reduce the current through the capacitor C_3 since its reactance increases at low frequency while the voltage across it drops to 30 V from the initial value of 125 V.

Table 9.3 Key performance parameters shown for soft and cold start CFLs

Types of circuit	*4-pin* (5 W) *Soft Start*	*4-pin* (5 W) *Cold Start*
DC input voltage	12.06	12.06
DC input current	474	470
Output voltage ON LOAD	29.66	29.66
Output voltage WITHOUT LOAD	212	212
Output current	156	150
Power output	4.58	4.59
Efficiency %	80.12	80.96
Frequency ON LOAD	25.8	26.5
Frequency WITHOUT LOAD	41.0	80.6
Short circuit current DC (mA) ON LOAD	82.0	83.0
Short circuit current DC (mA) WITHOUT LOAD	34.0	34.0
Power Factor	0.995	0.995
Intensity (klux)[a]	4.3	4.3

[a]Measured on the surface of the lamp.

For cold starting of the CFL capacitor C_3 is removed from circuit. The circuit was cold started and switched off 5000 times without any appreciable blackening of the CFL near its filaments, which was earlier observed for a limited number of cold start operations [26]. The oscillator is started when the winding of the base terminal of the transistors 2N6292 is raised to 11 V by Q_1 of Figure 9.10, the output transistor of the controller. When the battery voltage falls below a certain value, corresponding to a level of discharge, the voltage at the emitter of Q_1, goes to zero thereby shutting down the oscillator.

The LED based circuit

Figure 9.12 shows the circuit diagram of the 12 V LED based solar lantern complete with the charge-discharge controller. The circuit is similar to that given in Figure 9.10 except that a constant current source comprising transistors Q_1 and Q_2 has been added. Resistor R_{31} has to be adjusted to give a current of 350 mA through the LEDs D_{17} to D_{19}, which are placed in series.

This circuit has a poor efficiency as shown in Table 9.4 as compared to the one with the CFL. This is because of the power dissipated in the transistors and the resistance R_{31}.

Table 9.4 Parameters when the LEDs are operated from the circuit of Figure 9.12

Input DC Voltage (V)	*Input DC Current (A)*	*DC Voltage Across Lamp (V)*	*DC Current Through Lamp (A)*	*DC Voltage Across Lamp and Q_1 (V)*	*Voltage Drop Across R_{31} & Q_1*	*Efficiency %*
15.1	0.34	10.0	0.327	14.2	4.2	63.7
12.2	0.288	9.72	0.25	11.5	1.78	69.2

Result

The solar lantern with a 5 W CFL gives a total light output of about 250 lumens whereas the one with the LEDs gives 150 lumens. The light output data are the nominal values supplied by the manufacturers in their data sheets. This is because the lowest power CFL available is 5 W, which yields 250 lumens whereas the circuit of Figure 9.10 can support only 3 white LEDs of 1 W each, which together emit 150 lumens.

The overall electrical efficiency of the CFL based lantern exceeds 80% with a 12 V DC input. The corresponding figure for the LED based lantern is 69.2% because 0.445 (1.78×0.25) watts of power is wasted in the constant current generator. This wastage increases to (4.2×0.327) = 1.373 W thereby reducing the efficiency to 63.7%. A switch mode design will yield a better efficiency figure. The suggested part is LM 3402/LM 3402HV by National Semiconductor [27]. There are other parts which perform exceedingly well and can give up to 90% efficiency. Some of these are CAT 4201 and CAT 4240 by catalyst semiconductors [28].

Conclusion

A replacement for a hurricane lantern needs a light source that can give an output of 40 to 69 lumens as given in Table 9.1 and Table 9.2. 5 W CFLs were used earlier, since lower power tubes were not available, which gave a lot more light than required. However, with the arrival of high intensity LEDs in the market only one or two are needed to give the required light. With one LED in circuit the battery and the solar panel size also will reduce and the whole system will cost between \$20 and \$25.

The circuit given in Figure 9.12 has three LEDs but costs as much as that given in Figures 9.10 and 9.11 because the cost of the oscillator is eliminated. This also requires a solar panel of 8-W as against the 10-W panel used for the CFL based lantern, which amounts a reduction of 20% in the panel cost. However, the life of a white LED is very sensitive to temperature [29]. A complete description of this work will be found in [30]; *Copyright (2007), Reprinted with permission from Elsevier.*

Thus, it is concluded that:

1. For a direct replacement of the hurricane lantern, with a light output of 47 lumens, a single LED based lantern is cheaper than a CFL based lantern. This whole system will cost between \$20 and \$25.
2. For a higher illumination lantern a 5 W CFL, because it is the minimum power available, based lantern is cheaper. For solar powered street lights that should give higher light output CFL based designs are still cheaper than their LED counterparts.

Solar Street Lights

These are more powerful than the lanterns although their basic circuits are similar. The additional feature incorporated in this is an automatic system that switches the lights ON when the sun sets and switches them OFF when the sun rises. This is easily done by monitoring the module current. When the charging indicator green LED of Figure 9.12 turns off, during the nightfall, the street light is turned ON by an additional circuit.

9.3 Large Area Lighting Using High Intensity Discharge (HID) Lamps

Lighting is one of the most common and important application areas in PV technology. For illuminating parking lots and interiors of warehouses, high intensity discharge (HID) lamps are extensively used with conventional electric power. Effort has therefore been underway to replace conventional electricity in this application with PV power. Since PV sources are not only expensive but also have a variable output voltage, a voltage stabilizing circuit is first necessary. After utilizing the maximum power available, with maximum power point trackers, the voltage is often stabilized to various conventional values using battery banks of corresponding rated nominal voltages. A buck or boost regulator is then employed to produce a stable output voltage to be fed to the lamp ballast. Ballasts are electrical or electronic circuits used for limiting the current passing through the discharge lamp. The block diagram of a general PV based lighting system is shown in Figure 9.13.

The following discussion illustrates a PV based HID lamp lighting system that uses a buck based MPPT, a power factor controller (PFC) chip based circuit used as boost stage for stabilization and a current-fed half bridge (CFHB) oscillator and capacitor ballast for feeding a 400 W HID lamp. The laboratory prototype of the regulator and oscillator circuit was

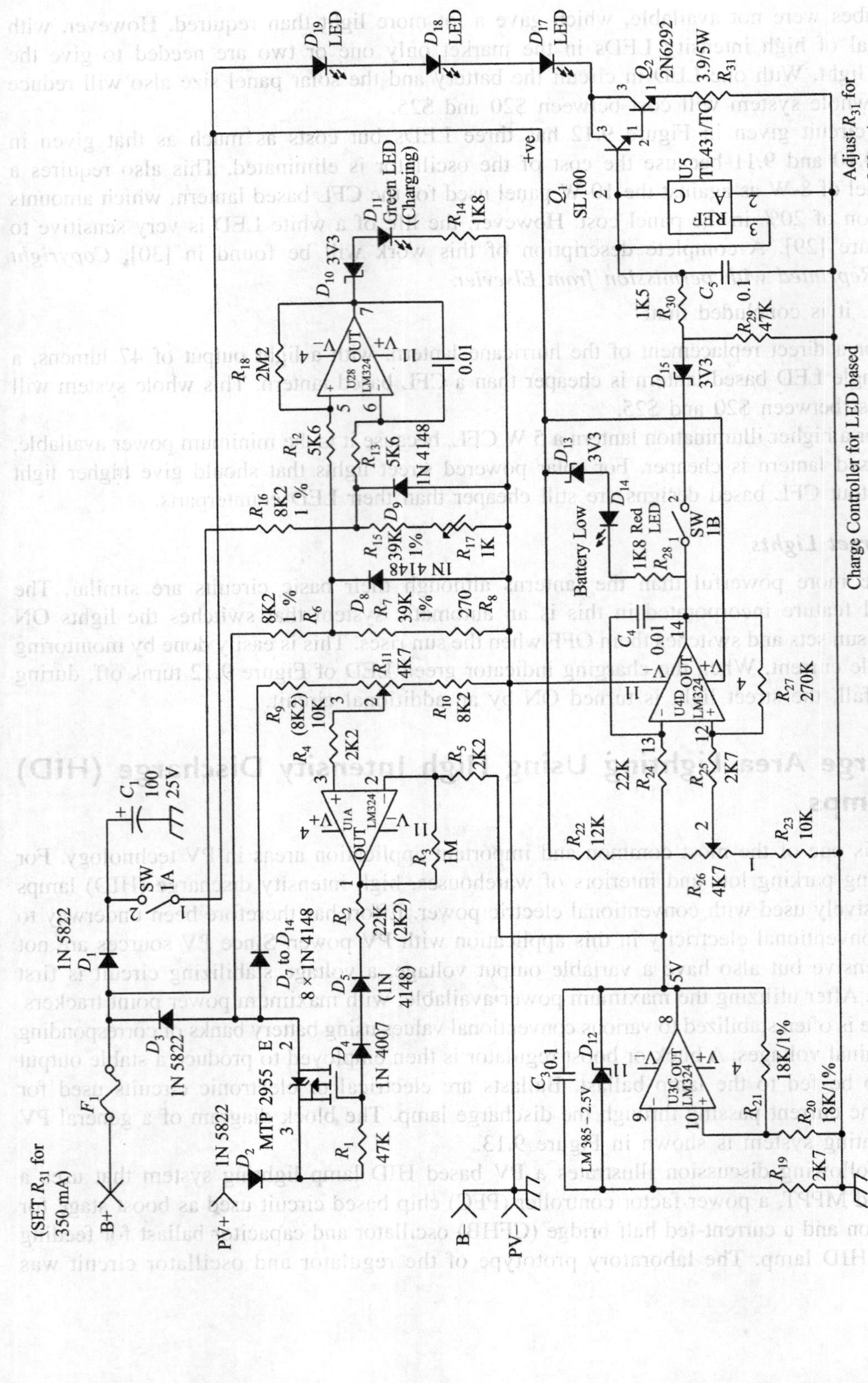

Note: Temp. Comp. at B+ = 14.56 mV/C (16.13 mV/C)
Float charging = 14.5 V, Load discon. = 11.5 V, Load re-conn. 12 V

Figure 9.12 Circuit diagram of charge-discharge controller for LED-based lantern.

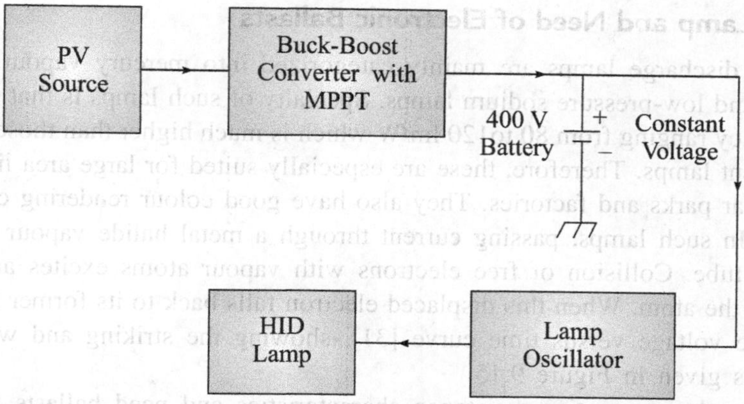

Figure 9.13 Block diagram of a lamp ballast system with a boost based MPPT.

designed, fabricated and tested with various input voltages from a DC power supply improvising various nominal DC voltages normally obtained from PV charged battery banks. The efficiency of the PFC stage was 98% and that of the CFHB oscillator stage was 82% (approx.). The efficiency of the overall circuit exceeded 80% at 400 W.

9.3.1 System Overview

A battery is not only necessary to store energy during sunlight hours to deliver the required power at night but also to give the essentially stable DC voltage. As the calculations will later show, an oscillator that drives the HID lamp used here requires, DC, an input of 400 V, which has to be supplied by a battery bank. Such a battery bank will be unwieldy and risky for a 400 W lamp although it can be charged by a boost converter-based maximum power point tracker. This, however, will obviate another boost converter to raise the voltage to the required level. The present system in Figure 9.14, on the other hand, provides a simpler solution with a lower voltage battery, charged by a buck-based MPP tracker described in Chapter 7, and followed by a boost converter, which can be an integral part of the oscillator circuit board. The analysis is divided into two parts: (i) boost converter and (ii) the driver oscillator for the HID lamp, which together constitute the electronic ballast.

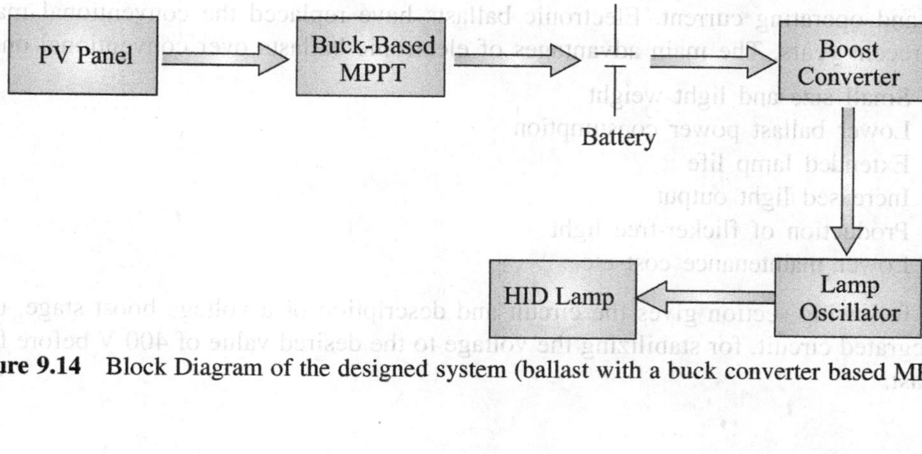

Figure 9.14 Block Diagram of the designed system (ballast with a buck converter based MPPT).

9.3.2 HID Lamp and Need of Electronic Ballasts

High intensity discharge lamps are mainly categorized into mercury vapour, metal halide, high-pressure and low-pressure sodium lamps. Specialty of such lamps is that they have high luminous efficacy ranging from 80 to120 lm/W which is much higher than those of fluorescent and incandescent lamps. Therefore, these are especially suited for large area lighting such as gymnasiums, car parks and factories. They also have good colour rendering capabilities and compact size. In such lamps, passing current through a metal halide vapour produces light inside the arc tube. Collision of free electrons with vapour atoms excites an electron into higher orbit of the atom. When this displaced electron falls back to its former level, radiation is emitted. The voltage versus time curve [31], showing the striking and working voltage characteristic is given in Figure 9.15.

These lamps have negative resistance characteristics and need ballasts to regulate the current flow through the tube. For the present experiment, a 400 W metal halide lamp HQI-BT 400/D manufactured by OSRAM was chosen. The efficacy of this lamp is 76 lm/W. The lamp specifications are given in OSRAM manual.

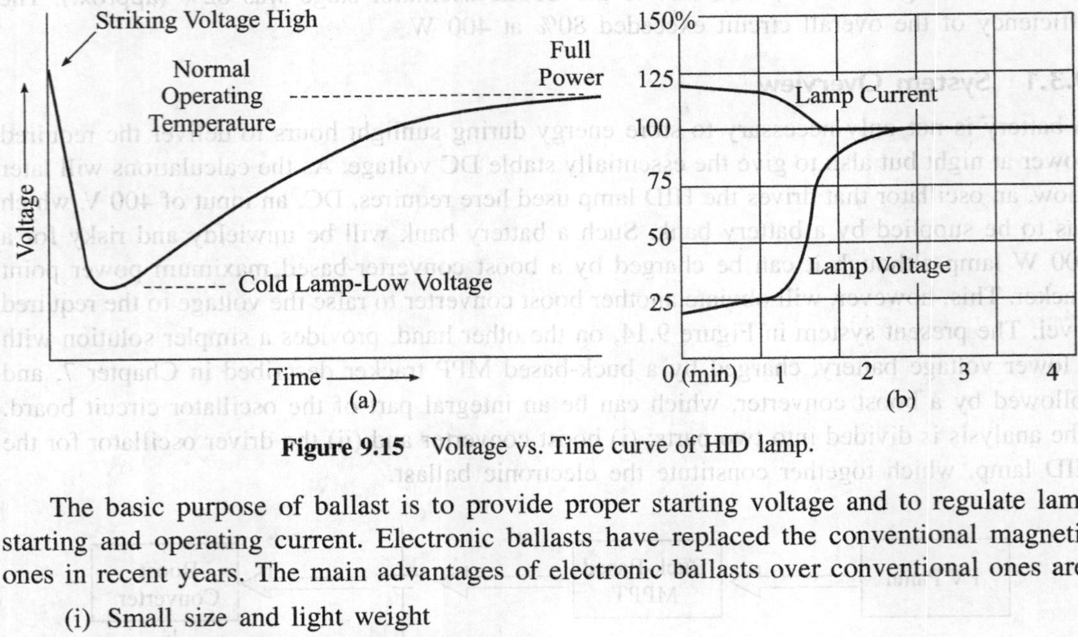

Figure 9.15 Voltage vs. Time curve of HID lamp.

The basic purpose of ballast is to provide proper starting voltage and to regulate lamp starting and operating current. Electronic ballasts have replaced the conventional magnetic ones in recent years. The main advantages of electronic ballasts over conventional ones are:

(i) Small size and light weight
(ii) Lower ballast power consumption
(iii) Extended lamp life
(iv) Increased light output
(v) Production of flicker-free light
(vi) Lower maintenance cost etc.

The following section gives the circuit and description of a voltage boost stage, using a PFC integrated circuit, for stabilizing the voltage to the desired value of 400 V before feeding the ballast.

9.3.3 Method of Using a PFC as DC to DC Boost Converter [32]

Discontinuous mode DC-DC boost converters are well described in literature (*Copyright (2008), Reprinted with permission from Elsevier*). Another type of converter, in a bridge configuration, has been utilized to stabilize varying DC input voltages [33]. This is a relatively complicated circuit and is used for high power requirements. A need was felt for a circuit for low and medium power applications. The DC to DC boost converter described here is inexpensive, in as much as it is unusual, because it uses a power factor controller (PFC) integrated circuit normally used in AC applications in a discontinuous mode.

Literature survey showed that some manufacturers [34], [35] provide integrated circuits for active AC power factor correction to be used as pre-converters in electronic ballasts and also in off-line power converter applications. Motorola [34] provides such a solution for an AC to DC power factor corrected pre-converter for a 450 W load, with an output voltage of DC, 400 V, using the integrated circuit MC 34262. For this kind of operation the line AC voltage is first converted to a haversine, by full wave rectification, then chopped 10 to 20 thousand times and processed to get the desired result. Since the haversine has a significant DC component [36], it stands to reason that the power factor corrector (PFC) can work from a pure DC photovoltaic source also. An experiment was therefore carried out, with a slight modification to the circuit, to determine the characteristics of the circuit and use it for lighting a 400 W HID lamp. The modifications were necessary to stabilize the circuit and to protect it from an accidental reversal of photovoltaic source polarity while keeping the device loses, due to dissipation, low.

The following equation shows the DC component ($I_m \times 2/\pi$) in the haversine current expression

$$i = I_m \left[\frac{2}{\pi} - \frac{4}{\pi} \sum_{\substack{k=\text{even} \\ k \neq 0}} \frac{\cos k\omega t}{(k+1)(k-1)} \right] \tag{9.3}$$

where

$$I_m = \frac{E_m}{r_p + R_l} \tag{9.4}$$

The quantities involved above are:

E_m = peak AC voltage

R_l = load resistance

r_p = rectifier resistance

$\omega = 2\pi f$, f = frequency.

The circuit was operated first for a typical AC operation. Then, this was tried with pure DC from a laboratory power supply. The description and results presented in this thesis are with the DC inputs.

9.3.4 Experimental Arrangement

The circuit diagram of the PFC is given in Figure 9.16. The notable differences with this and the one given in [34] are:

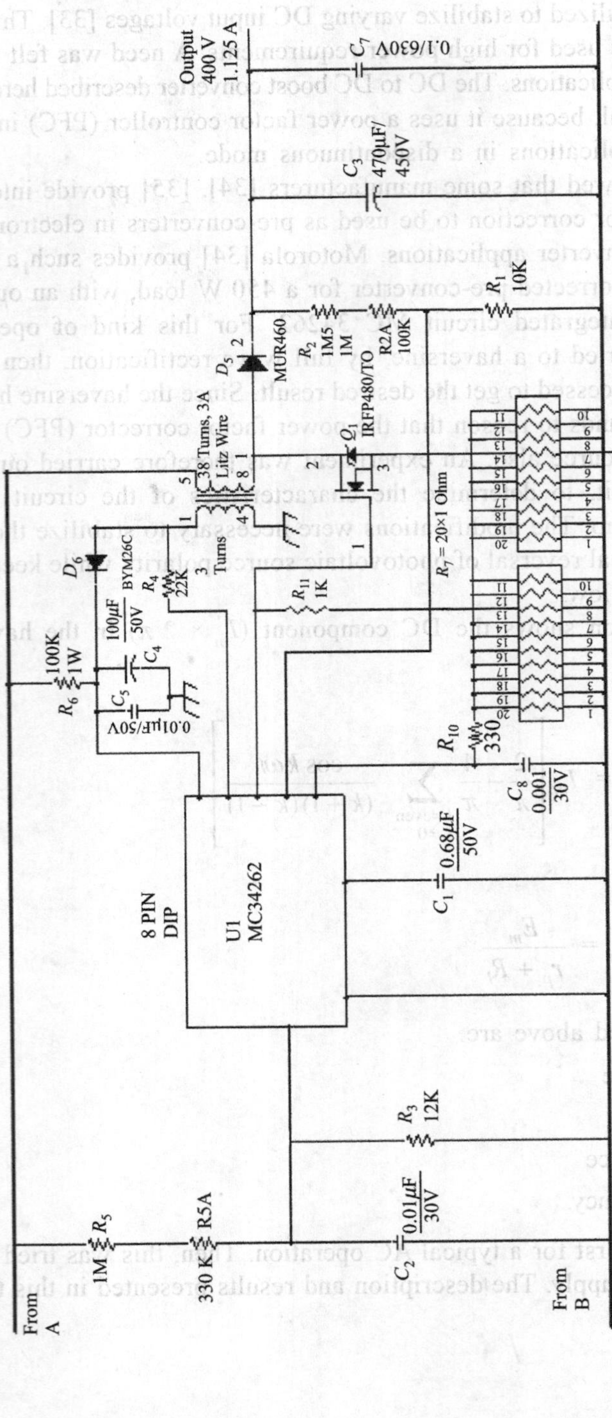

Figure 9.16 Circuit diagram of the Power Factor Controller.

(i) MOSFET IRFP 460 was used instead of MTW 20N50E.

(ii) The resistance R_7 is a parallel combination of 20 units of 1 Ω resistances to reduce the effect of soldering resistance. This is difficult if one 0.05 Ω resistance is used.

(iii) A resistance R_{11} of 1 K has been added. This helps to stabilize the circuit. The transformer uses an EE 4220 ferrite core of AL = 200 nH/T² with Litz wire winding. The Litz wire was made in-house from 180 strands of 42 SWG wire. This was necessary as the recommended transformer was not available.

Figure 9.17 shows the circuit diagrams of the rectifier and EMI/RFI filter preceding the PFC for DC operation while Figure 9.18 shows the block diagram of the experimental set-up. During the measurements with DC three and a half digit DC voltmeters and ammeters were used, while a DC power supply was used in the input to supply the required power.

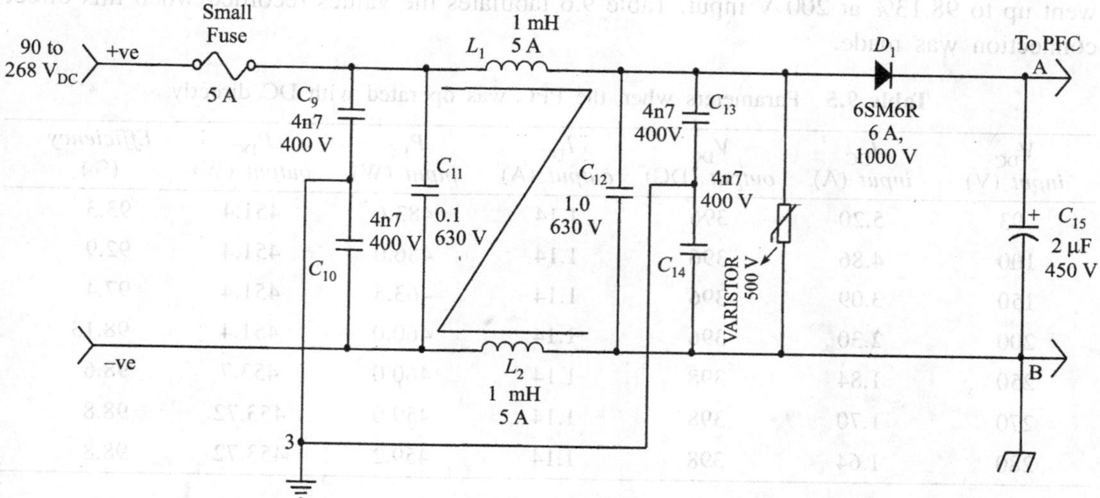

Figure 9.17 Rectifier and EMI filter for the PFC for DC Operation.

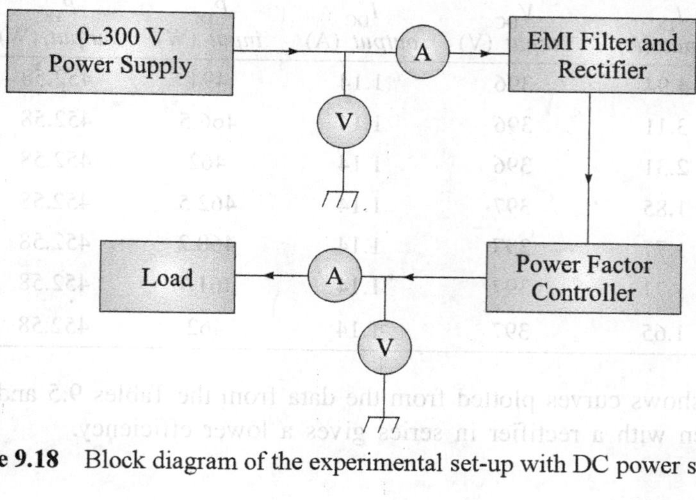

Figure 9.18 Block diagram of the experimental set-up with DC power supply.

Results

The circuit was operated with DC under the two following conditions:

When operated directly from the source and when a rectifier was placed between the source and the PFC as shown in Figure 9.17, to prevent damage to the circuit if the source voltage is accidentally reversed. Table 9.5 shows the efficiency values of this converter stage when the system was tested with various input DC voltages. This table shows that at lower operating voltages the efficiency figures are lower than those at higher voltages. This is so because the currents are considerably higher at lower voltages giving higher I^2R losses. Also the device switching and conduction losses increase with higher currents. When the circuit was used with a rectifier 6SM6R (600 V, 6 A), the efficiency of this converter stage at 200 V was recorded to be 97.96%. This was employed to protect the circuit against accidental voltage reversal. However, when the circuit was run directly, without the rectifier, the efficiency went up to 98.13% at 200 V input. Table 9.6 tabulates the values recorded when this direct connection was made.

Table 9.5 Parameters when the PFC was operated with DC directly

V_{DC} *input* (V)	I_{DC} *input* (A)	V_{DC} *output* (DC)	I_{DC} *output* (A)	P_{DC} *input* (W)	P_{DC} *output* (W)	*Efficiency* (%)
93	5.20	396	1.14	483.6	451.4	93.3
100	4.86	396	1.14	486.0	451.4	92.9
150	3.09	396	1.14	463.5	451.4	97.4
200	2.30	396	1.14	460.0	451.4	98.13
250	1.84	398	1.14	460.0	453.7	98.6
270	1.70	398	1.14	459.0	453.72	98.8
280	1.64	398	1.14	459.2	453.72	98.8

Table 9.6 Parameters when the PFC was operated with DC through a rectifier

V_{DC} *input* (V)	I_{DC} *input* (A)	V_{DC} *output* (V)	I_{DC} *output* (A)	P_{DC} *input* (W)	P_{DC} *output* (W)	*Efficiency* (%)
100	4.94	396	1.14	494	452.58	91.62
150	3.11	396	1.14	466.5	452.58	97.02
200	2.31	396	1.14	462	452.58	97.96
250	1.85	397	1.14	462.5	452.58	97.86
260	1.77	397	1.14	460.2	452.58	98.34
270	1.71	397	1.14	461.7	452.58	98.02
280	1.65	397	1.14	462	452.58	97.96

Figure 9.19 shows curves plotted from the data from the Tables 9.5 and 9.6. They show that the one taken with a rectifier in series gives a lower efficiency.

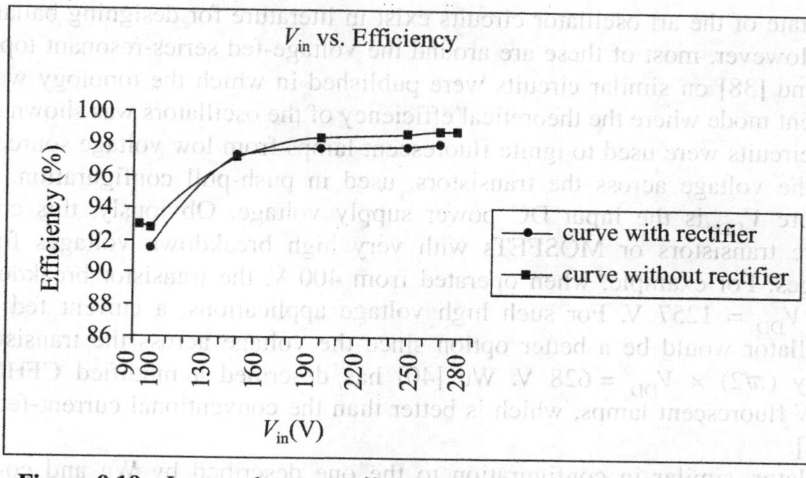

Figure 9.19 Input-voltage vs. efficiency curves with and without rectifier.

There is, however, a serious shortcoming in this circuit. This converter if used directly from a PV array will result in instability because the *I-V* curve at the input of the PFC has a negative slope as shown in Figure 9.20(a), which is plotted from experimental values. If this negative resistance curve is superimposed on an *I-V* curve of a solar array a situation will arise, as shown in Figure 9.20(b), where two points of intersection will result. This will lead to instability in the circuit. Hence it is necessary to operate this circuit from a stable DC source like a battery. For this reason the converter was tested with several DC voltage values in the laboratory to simulate different battery nominal voltage conditions.

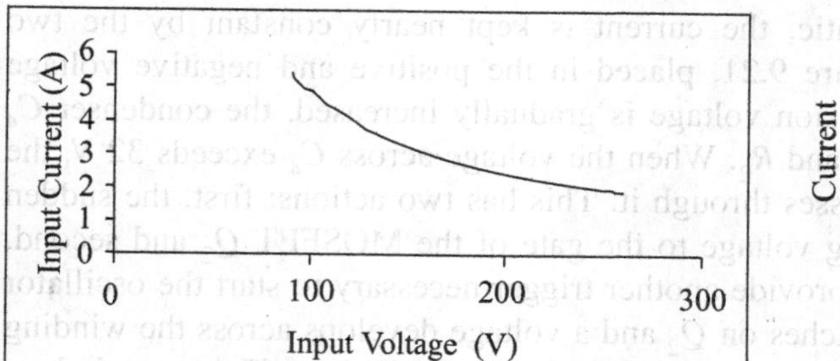

Figure 9.20(a) Experimental *I-V* curve of the PFC showing a negative resistance.

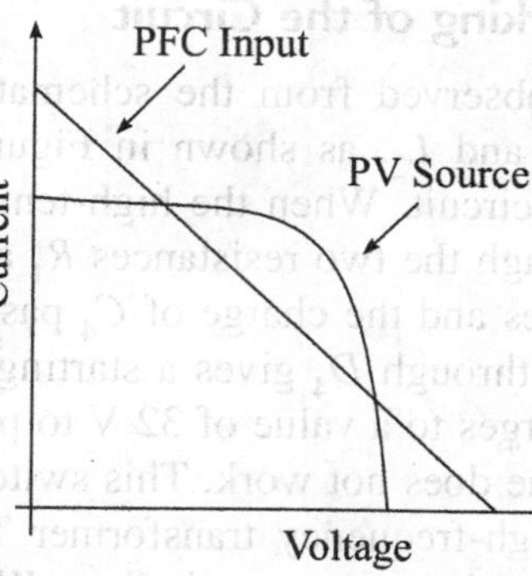

Figure 9.20(b) Intersection points of hypothetical *I-V* curves of a typical PFC and a PV array.

9.3.5 The Oscillator Stage

After stabilizing the voltage at 400 V, the output of the converter stage was fed to an oscillator for producing the appropriate voltage and power for the lamp operation ([45]; *Copyright (2006), IEEE*). The complete description and design of the oscillator and ballast stage is given in the following discussion.

Several state of the art oscillator circuits exist in literature for designing ballasts for HID lamps [31]. However, most of these are around the voltage-fed series-resonant topology. Two papers [37] and [38] on similar circuits were published in which the topology was based on constant current mode where the theoretical efficiency of the oscillators was shown to approach 100%. Such circuits were used to ignite fluorescent lamps from low voltage sources [39]. It is known that the voltage across the transistors, used in push-pull configuration, is equal to $\pi \times V_{DD}$, where V_{DD} is the input DC power supply voltage. Obviously, this configuration would require transistors or MOSFETs with very high breakdown voltages for high DC supply voltages. For example, when operated from 400 V, the transistor breakdown voltage will be $\pi \times V_{DD}$ = 1257 V. For such high voltage applications, a current fed half bridge (CFHB) oscillator would be a better option since the voltage across the transistors will be approximately $(\pi/2) \times V_{DD}$ = 628 V. Wu [40] has described a modified CFHB oscillator lighting 40 W fluorescent lamps, which is better than the conventional current-fed oscillators [41] and [42].

An oscillator, similar in configuration to the one described by Wu and co-workers, is explained here to light the 400 W HID lamp that has an efficiency of approximately 89%. This circuit is not damaged even if the HID lamp is destroyed, or is left unconnected or the output of the oscillator is short circuited accidentally. Here MOSFETs are used instead of bipolar junction transistors (BJTs). This is done keeping in mind the high voltage requirements for the present purpose. The only shortcoming of this oscillator is that it does not begin to oscillate when connected to a resistive load, but once it starts to oscillate it can be connected to the load thus making this circuit suitable for igniting fluorescent and HID lamps.

The circuit diagram of this current-fed parallel resonant half-bridge oscillator is shown in Figure 9.21.

9.3.6 Working of the Circuit

As can be observed from the schematic, the current is kept nearly constant by the two inductors L_1 and L_2, as shown in Figure 9.21, placed in the positive and negative voltage paths of the circuit. When the high-tension voltage is gradually increased, the condenser C_4 charges through the two resistances R_1 and R_2. When the voltage across C_4 exceeds 32 V, the DIAC D_4 fires and the charge of C_4 passes through it. This has two actions: first, the sudden high current through D_4 gives a starting voltage to the gate of the MOSFET Q_2 and second, C_4 again charges to a value of 32 V to provide another trigger necessary to start the oscillator if the first one does not work. This switches on Q_2 and a voltage develops across the winding W_1 of the high-frequency transformer TR_1. If the voltage across W_1 is sufficient to induce high enough voltage across windings W_3 and W_4, the circuit starts oscillating. Otherwise, the high-tension voltage has to be raised further. This particular oscillator starts to operate from a high-tension DC voltage of 150 V.

Once the oscillations begin, the voltage across C_4 is discharged through R_3 and the fast-recovery diode D_1 so that the DIAC is not able to fire. The zener diodes D_2, D2A, D_3 and D3A prevent high voltage signals at the gates of the MOSFETs Q_1 and Q_2 and thus protect them from damage. Resistances R_5 and R_6 limit the currents through the zener diodes.

The 10 nF capacitor C3A prevents very high voltages from being generated by the $L\,(di/dt)$ mechanism of the inductors L_1 and L_2 during the switching actions of the transistors.

This also generates sinusoidal oscillations. However, if C3A has a large value, then the circuit acts as a constant-voltage oscillator while a very low value of C3A generates high voltages across it, which can lead to breakdown of the MOSFETs. The secondary winding W_2 of TR_1 delivers current to the HID lamp through the ballast capacitors C_7, C_8 and C8A. For initial firing of the lamp the voltage multiplier, comprising diodes D_5 to D_{10} and capacitors C_{10} to C_{12}, generates the required DC voltage while resistances R_7 to R_9 limit the current.

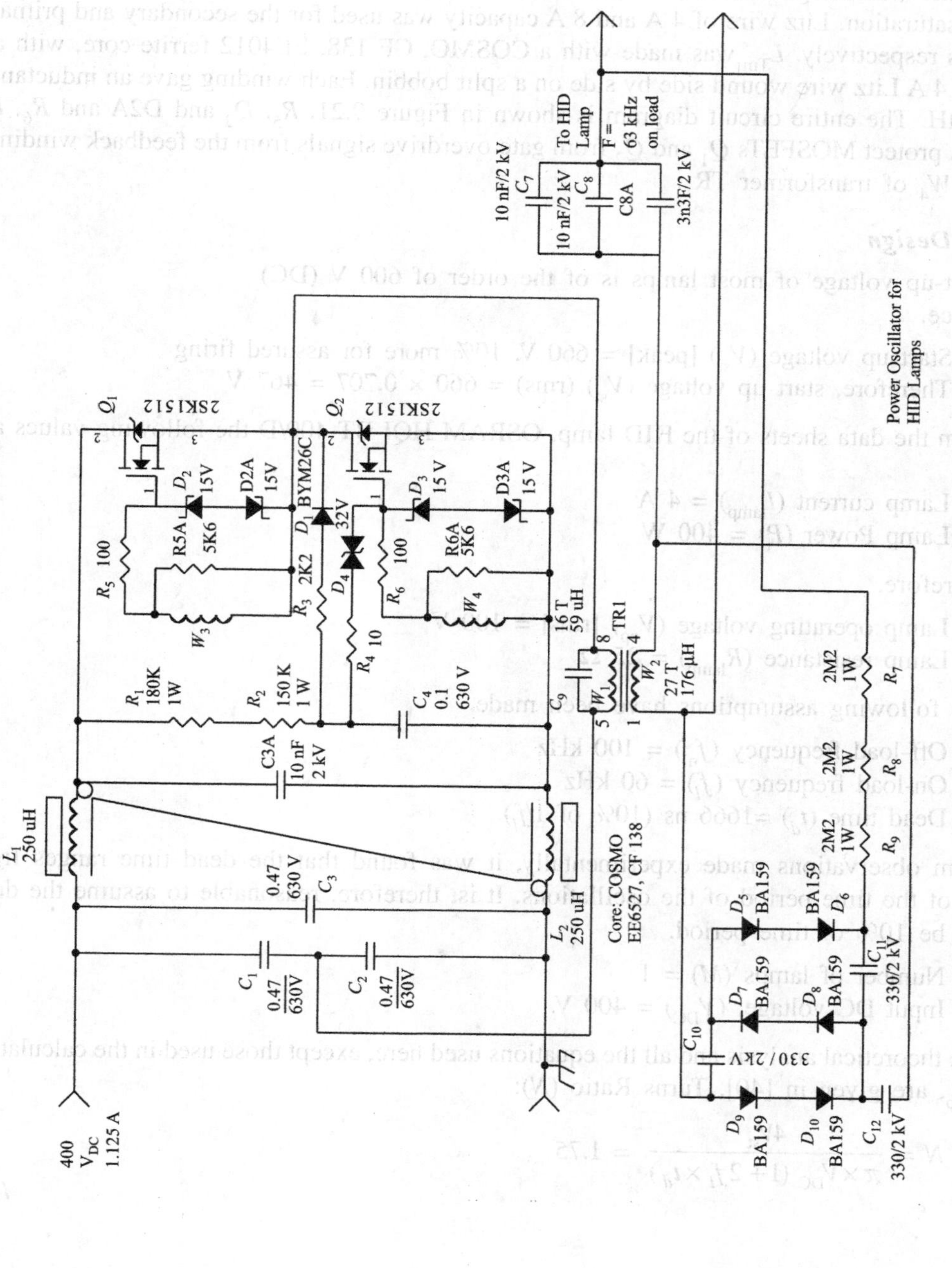

Figure 9.21 Circuit diagram of the oscillator.

The frequency of oscillation is determined by C3A and the primary inductance of TR_1. L_1 and L_2 also play a part. For further tuning, a capacitor C_9 may be connected across the primary winding W_1 of TR_1. However, C_9 has been omitted in the present circuit.

9.3.7 Construction Details of Some Components

The transistor selected for this case was the MOSFET 2SK1512 with anti parallel diode. The transformer was designed around a COSMO EE6527, CF138 ferrite core with an air gap to prevent saturation. Litz wire of 4 A and 8 A capacity was used for the secondary and primary windings respectively. L_{Tm} was made with a COSMO, CF 138, EE4012 ferrite core, with air gap, and 4 A Litz wire wound side by side on a split bobbin. Each winding gave an inductance of 250 μH. The entire circuit diagram is shown in Figure 9.21. R_5, D_2 and D2A and R_6, D_6 and D3A protect MOSFETs Q_1 and Q_2 from gate overdrive signals from the feedback windings W_3 and W_4 of transformer TR_1.

Circuit Design

The start-up voltage of most lamps is of the order of 600 V (DC).

Hence,

(a) Start up voltage (V_{st}) [peak] = 660 V, 10% more for assured firing
(b) Therefore, start up voltage (V_{st}) (rms) = 660 × 0.707 = 467 V

From the data sheets of the HID lamp, OSRAM HQI-BT 400/D the following values are taken:

(c) Lamp current (I_{lamp}) = 4 A
(d) Lamp Power (P) = 400 W

Therefore,

(e) Lamp operating voltage (V_{op}) [rms] = 100 V
(f) Lamp resistance (R_{lamp}) = 25 Ω

The following assumptions have been made.

(g) Off-load frequency (f_{nl}) = 100 kHz
(h) On-load frequency (f_l) = 60 kHz
(i) Dead time (t_d) =1666 ns (10% of $1/f_l$)

From observations made experimentally, it was found that the dead time ranges from 7–13% of the time period of the oscillations. It is, therefore, reasonable to assume the dead time to be 10% of time period.

(j) Number of lamps (M) = 1
(k) Input DC voltage, (V_{DC}) = 400 V.

The theoretical analysis and all the equations used here, except those used in the calculation of C_{lamp}, are given in [40]. Turns Ratio (N):

(a) $$N = \frac{4V_{st}}{\pi \times V_{DC}\,(1 + 2f_l \times t_d)} = 1.75$$

(b) Ballast capacitance (C_{lamp})

(c) $$C_{lamp} = \frac{V_{op}}{2\pi f_l R_{lamp}\sqrt{V_{st}^2 - V_{op}^2}} = 23.1 \text{ nF}$$

(d) C_{eff} and L_{rm} are calculated from: $f_{nl} = \dfrac{1}{2\times\pi\times\sqrt{C_{eff}\times L_{rm}}}$ and (9.5)

(e) $$f_l = \frac{1}{2\times\pi\times\sqrt{(C_{eff} + MN^2 C_{lamp})\times L_{rm}}} \tag{9.6}$$

where C_{eff} is the effective capacitance across the primary of the output transformer TR_1 and L_{rm} is its magnetizing primary inductance.

Therefore, C_{eff} = 39.6 nF and L_{rm} = 63.9 μH

Now,

(f) $C_{eff} = (4C_7 + C_9)$ (9.7)

$C_9 = 0$ in this case, therefore,

(g) $C_7 = C_{eff}/4 = 9.9$ nF.

(h) C_7 refers to the component in [40]. C_7 is denoted as C3A in Figure 9.21.

(i) $V_{primary}$ (Peak) = $V_{secondary}$ (Peak) ÷ N = 660 ÷ 1.75 = 377 V

(j) $$V_{transistor} = \frac{\pi(1 + 2f_l \times t_d)}{2} V_{DC} = 753 \text{ V}$$

(k) Magnetizing inductance of the coupled inductor = $L_{Tm} > 8*L_{rm}$ (9.8)

This has been chosen to be 250 μH, which is $4L_{Tm}$, to keep the size of the inductor small.

(l) Current through C_7 is $$I_{C_7} = \frac{C_7 V_{transistor}}{(4C_7 + C_9 + MN^2 C_{lamp})\pi f_l L_{rm}} = 5.6 \text{ A}$$

(m) Current through one transistor = $I_Q = I_{C_7} + \dfrac{2P}{V_{DC}}$ (5.6 + 2.2) = 7.8 A

Results

A comparison of the theoretical and experimental values of components and parameters (at 400 V) is shown in Table 9.7. It can be seen that the peak-to-peak voltage across the primary of the transformer, $V_{pp(primary)} = 2 \times V_{p(primary)} = V_{p(transistor)}$, the peak voltage across one transistor.

Table 9.7 Comparison between theoretical and experimental values at 400 V_{DC}

Parameters	*Theoretical Values*	*Experimental Values*
f_{nl} (kHz)	100	83.8
f_l (kHz)	60	62.4
N	1.75	1.69 (27T/16T)
C_{lamp} (nF)	23.0	23.0
C_7 (nF)	9.9	10.0
L_{rm} (μH)	63.9	59.0
L_{TM} (μH)	500 (250,250 Coupled)	500 (251,261 Coupled)
V_{peak} (transistor)	753 V	745 V

Table 9.8 gives the $V_{transistor}$ and frequency variations with 'no load' condition. Table 9.9 gives the values when the oscillator was loaded. From Table 9.8, it is observed that the 'no load' frequency is 83.2 kHz instead of the assumed value of 100 kHz. This deviation can be attributed to stray capacitances and the tolerances of components. The current through the HID lamp was measured as voltage across a carbon resistance combination of 0.275 Ω/8 W. This voltage was multiplied with the voltage across the lamp in a digitizing oscilloscope HP 54501A, which was then averaged over 32 cycles to give the AC power. Several measurements yielded efficiency values between 88–89% as given in Table 9.9.

Figure 9.22 shows a curve, drawn from Table 9.8, where the voltage across one transistor V_p is plotted against the input DC voltage V_{DC}. The V_p is 745 V, which compares well with the theoretical value of 753 V.

Table 9.8 Variation of measured parameters during off-load condition

V_{DC} (V)	I_{DC} (A)	f_{nl} (kHz)	$V_{p\,(transistor)}$ (V)	*Transistor ON Time* (μs)	*Transistor OFF Time* (μs)
125	0.03	93.2	194	5.40	5.32
150	0.03	90.9	244	5.56	5.44
175	0.03	89.0	294	5.76	5.44
200	0.05	87.3	344	6.00	5.44
250	0.06	85.7	444	6.20	5.44
300	0.08	84.2	544	6.40	5.44
350	0.10	83.8	638	6.58	5.44
400	0.21	83.2	745	6.78	5.44

Table 9.9 Values of parameters at different DC voltages when oscillator was loaded

V_{DC} (V)	I_{DC} (A)	P_{DC} (W)	f_l (kHz)	V_{lamp} (V)	I_{lamp} (A)	P_{lamp} (W)	*Efficiency* η (%)
350	1.07	375.57	63.02	104.7	3.25	333.45	88.8
400	1.21	480.3	62.4	104.7	4.08	426.5	88.9

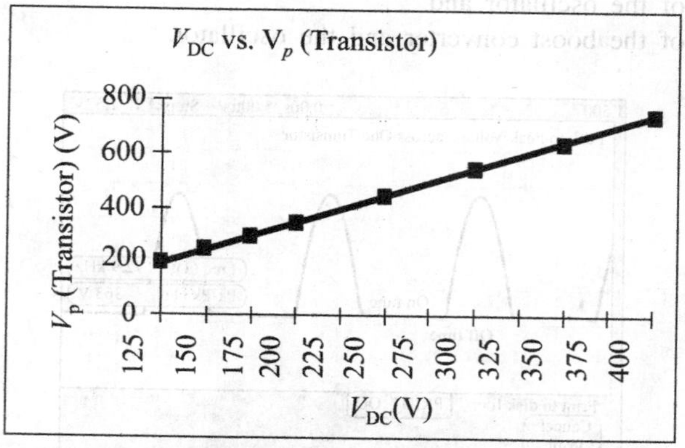

Figure 9.22 Graph between input DC voltage and peak voltage across one transistor.

Converter and Oscillator Combined System Results

The results for the PFC and the oscillator circuit combination, that is when this circuit was used with the boost converter circuit described earlier, are shown in Table 9.10. The overall efficiency of the boost converter and the lamp oscillator exceeds 80%. The voltage waveform (DSO result) across one transistor is shown in Figure 9.23. The OFF time is smaller than the ON time. Figure 9.24 shows the 'dead time' in the waveform across the primary of the power transformer. The circuit is self-oscillating, unlike the one given by Ben-Yakov and co-workers [43], and has a voltage multiplier as the igniter circuit so that increasing the multiplication factor can ignite higher voltage lamps too.

Table 9.10 Parameters for PFC and lamp oscillator when the PFC was operated with DC

V_{in}[1] (DC) (V)	I_{in}[2] (DC) (A)	P_{in}[5] (DC) (W)	V_o[3] (DC) (V)	I_o[4] (DC) (A)	P_o[6] (DC) (VA)	*Efficiency*[7] (PFC) (%)	V_o[8] (AC) (V)	I_o[9] (AC) (A)	P_o[10] (AC) (W)	η[11] (O_{SC}) (%)	*Efficiency*[12] *(Overall)* (%)
100.7	5.2	523.6	397	1.21	480.3	91.7	113	3.4	394.76	82.19	75.38
150	3.32	498	397	1.21	480.3	96.4	113	3.4	394.76	82.19	79.26
200.4	2.44	488.9	397	1.21	480.3	98.2	113	3.4	394.76	82.19	80.74
220	2.22	488.4	397	1.21	480.3	98.3	113	3.42	394.76	82.19	80.82
250	1.96	490	397	1.21	480.3	98	113	3.41	394.76	82.19	80.56
269.7	1.82	490.8	397	1.21	480.3	97.8	113	3.43	394.76	82.19	80.43

[1] input DC voltage to PFC boost circuit
[2] input DC current to PFC boost circuit
[3] output DC voltage of PFC boost circuit

[4] output DC current of PFC boost circuit
[5] input DC power in watts of PFC boost circuit
[6] DC power output in watts of PFC boost circuit
[7] electrical efficiency of the boost converter
[8] rms voltage across the lamp
[9] output rms current through the lamp
[10] power in watts delivered to the lamp
[11] electrical efficiency of the oscillator and
[12] electrical efficiency of the boost converter and the oscillator.

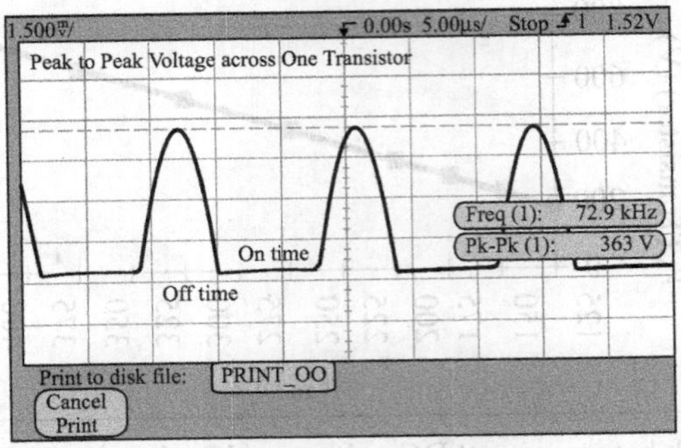

Figure 9.23 The drain-source voltage waveform of one MOSFET.

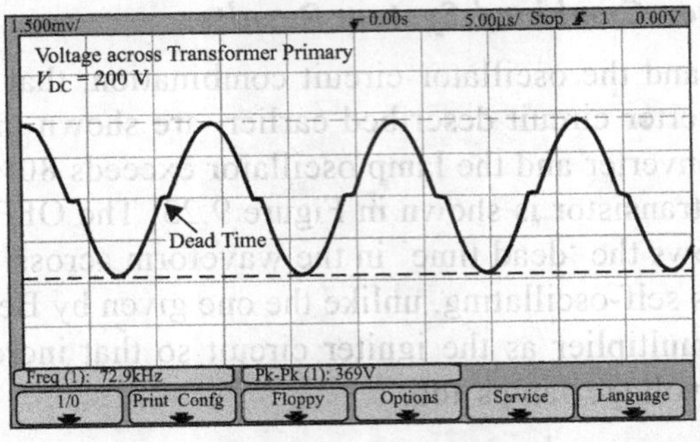

Figure 9.24 'Dead time' in voltage across primary of transformer.

Conclusions

The power factor controller works from DC supplies. It maintains an almost constant output voltage of 397 V when the input varies from 90 to 270 V with an efficiency exceeding 98%. Thus, this circuit is eminently suitable for use with batteries charged from photovoltaic sources. Since the discontinuous mode PFC depends on cut-in and cut-out of current, higher power converters can be made using the same topology by changing the MOSFET, transformer and the resistance R_7.

The CFHB power oscillator fabricated and reported here has lit a 400 W HID lamp, with an efficiency of 88.9%, for lighting factory interiors or parking lots for vehicles and can be used from non-conventional energy sources. Reference [44] shows a set of curves, which indicate that the lamps ignite at very low voltages. In the present case it lit up as soon as the oscillator started oscillating at 150 V. The circuit can now be suitably modified to use a smaller and less expensive transformer TR_1 with a 1:1 primary to secondary ratio. The details of the transformer design are given in Chapter 11. This will reduce the primary current and allow the use of lower current MOSFETs. However, the ballast capacitor will increase in value. When used with the DC to DC converter stage, the total ballast efficiency exceeded 80%. A detailed description of the circuit will be found in reference [45]. The array size and the battery size, if used can be calculated as given in the design of Section 9.1.

The level and the distribution pattern of light on roads are prescribed by the Indian standard IS: 1944 (Parts I and II) – 1970, and is called the Indian Standard Code of Practice for Lighting of Public Thoroughfares [46]. For any large area illumination design only the illumination produced by a device is not sufficient. The light pattern generated by the housing containing the light source is also very important. This depends on the reflector of the housing, the transparency of the outer cover which lets out the light, the ingress protection it offers, the height of the lamp from the ground and its tilt from the horizontal. The IP 66 housing offers excellent insect ingress protection without which insects crawl into the housing and reduce the transmission of light over a period of time. A totally enclosed high power lamp generates a lot of heat which must be taken out of the lamp housing in hot climatic conditions that exist in the tropics to prevent damage to the lamps. IP 66 housings made by several manufacturers accomplish both these tasks admirably in the roads of Delhi.

The required intensity of light varies from place to place. To set a standard the Illuminating Engineering Society (IES) of North America has prescribed minimum maintained levels of illuminance for most of the areas normally used [47]. Some of the values are given in Table 9.11. Thus, the power of the lamp is dictated by the illumination needed at a particular place. In complex cases like the roads simulation is done using appropriate software and the power and location of the lamp are found but in simpler cases like a bedroom or kitchen it is easier to use an adequately powerful lamp. However, in both the cases the intensity of the light on the surface is measured by a luxmeter which should match with the values given in Table 9.11. A comprehensive discussion on interior and exterior lighting is given in reference [8]. Several proprietary software are available but the one that has often yielded good results is cited in [48].

Table 9.11 IES* Recommended Illumination Levels in Lux

Place	Lux	*Place*	Lux	*Place*	Lux
Reading handwriting	750	Garage Entrance	540	Living rooms general	50
Reading rooms, newspaper and magazines	1100	Garage Storage	54	Corridors and Stairways	220
Classrooms	750	Library Study Areas	750	Sewing darning desk etc.	1600
Laboratories	1100	Conference Areas	320	Bedroom general	110
Courtroom Activity Area	750	Gymnasium Match Area	540	Kitchen working Areas	540

*Illuminating Engineering Society (IES) of North America

For PV home lighting a suitable CFL based luminaire can be chosen from [48] and the rest of the circuit may be designed around it. A suitable housing for LEDs, with a predictable radiation pattern, is yet to be made.

9.4 A PV Powered Fan and Pump

An electric motor is needed to drive a fan or a water pump. The power requirement varies with the capacity of either. For instance, for a 12 W fan a small motor will do but for 300 litres per minute capacity pump, with a high water head, a more powerful motor is essential. To start with consider the *I-V* curve of an electric motor shown in Figure 9.25 where the motor is connected to a variable voltage source shown as the battery.

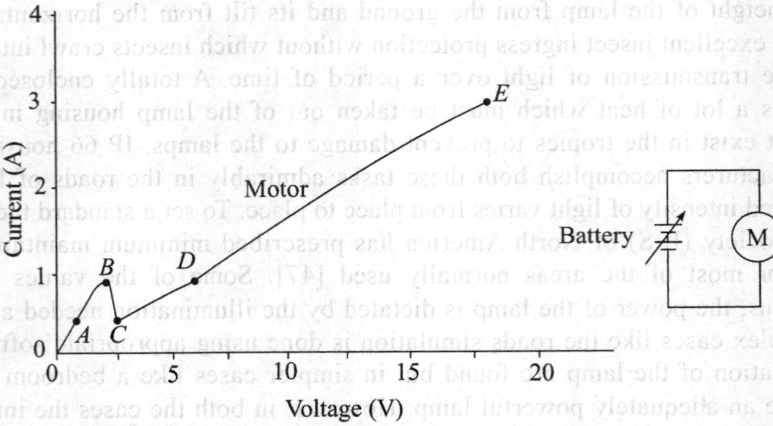

Figure 9.25 The *I-V* curve of a motor.

The nature of the *I-V* curve of the motor needs some explanation. When the voltage increases from 0 V, the current increases rapidly until the point *B* is reached because the motor is in a stalled or non-rotating state. At *B* there is sufficient voltage to rotate the motor which creates a back EMF in the coil that pushes the current down to *C*. A further increase in the applied voltage raises the current to *D*. From there the current increases almost linearly with the voltage until the point *E*. This behaviour changes somewhat when the motor power if fed from a PV module. In such a case the motor power has to be matched with the maximum power point of the PV array for the best performance in terms of the power transfer. Since the *I-V* curve of a solar module or array shifts with the light received from the sun so does its MPP. Thus, the power fed to the motor also changes thereby altering its speed which, in turn, modifies the water or air flow. Hence, it is necessary to superimpose the *I-V* characteristics of the motor on that of the array to find whether the two intersect at the MPP of the array. An ideal situation would be if the two intersect at the MPP of the array over the entire range of the insolation.

Figure 9.26 shows a fan motor *I-V* curve superimposed on that of a PV module with the points of intersection marked as *A*, *B*, *C*, *D* and *E*. The former does not intersect the latter at its MPP except at the point *E* resulting in power loss at other points. The MPP shifts from

E to *F* and then to *G* and *H* as the insolation decreases which indicates that the power available at *D*, for instance, is very small as compared to the MPP at *G* on the same curve. There are two ways to overcome this:

1. Use a maximum power point tracker so that it always ensures that the *I-V* curve of the motor falls on *E*, *F*, *G* and *H* regardless of the light intensity. This makes the MPPT a matching network between the module and the motor.
2. The second method is to use a battery whose voltage is almost equal to the V_{MPP} of the module so that the module always works at its MPP. This method was mentioned earlier while discussing the solar lantern. It is needless to point out that the motor should have the same operating voltage as that of the battery.

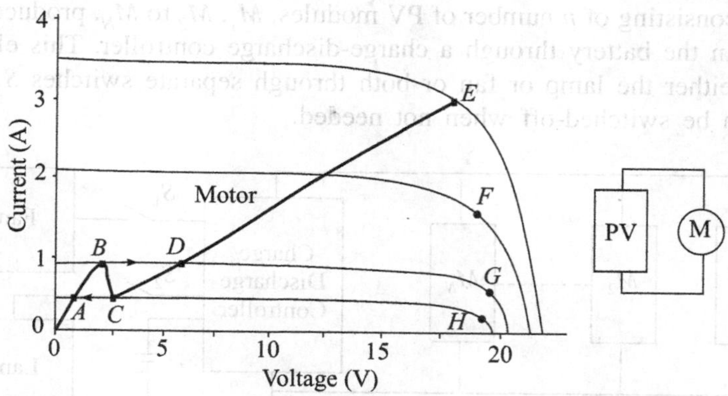

Figure 9.26 *I-V* of a pump superimposed on that of a module.

In absence of either of the above options, that is when the motor is directly connected to the module, the following processes occur. It has been mentioned earlier that the PV module is essentially a constant current source at lower output voltages but soon assumes the characteristics of a constant voltage source near its V_{oc} as can be seen from Figure 9.26. The motor therefore is now being driven by a current source instead of a voltage one. Early in the day when the incident light intensity is weak the current through the motor increases sharply from zero to *B*, as it is in a stalled position, but the voltage of the module does not increase because it is near a short circuit condition. Once the motor starts rotating the back EMF generated reduces the current. Since the module is a constant current source, in this region, it forces the same current through the motor and in order to do so its voltage rises to a value corresponding to the point *D*. This is shown by the arrow from *B* to *D*. Thereafter the current follows an almost linear path to *E*, as the solar radiation intensifies. When the sun's power declines towards the evening, the path traced by the motor current follows the ascending one until the current reaches *C*, where the motor stalls again and its current tries to shoot up but is confined to the same value as at *C* because the module is unable to supply the current. So the path now taken by the motor current is *CA*. Hence there is a hysteresis which would have been absent if the motor was driven by a voltage source.

In India a fan is required for more number of hours than a light. The light often works for three hours after dark whereas a fan is needed for almost 24 hours depending on the occupancy of the room, the place and the time of the year. In northern India the fan is must from early April to late September while in the central and southern parts of the country it is needed for twelve months. Thus, the concept for home or domestic lights has come up based on the demand from the market. Therefore, in the tropical countries the working hours of a fan and a light have to be found first to determine the total load leading to the energy requirement for seven days, with the desired number of sunless days, to calculate the size of the battery and the PV array. It must also be remembered that a very high cost will discourage customers from using PV based systems. A commercially viable combination of fan and light consists of a 7 W CFL based light source which works for five hours and a 12 W DC fan that can give a good air flow from 10 p.m. to 6 a.m. that is eight hours, when people go to sleep. The calculation of the components will follow the routine process outlined in Section 9.1. The block diagram of the solar home light is given in Figure 9.27. The solar array of a suitable voltage consisting of *n* number of PV modules, M_1, M_2 to M_N, produces the electricity which is stored in the battery through a charge-discharge controller. This electric energy is used to operate either the lamp or fan or both through separate switches S_1 and S_2 so that either device can be switched-off when not needed.

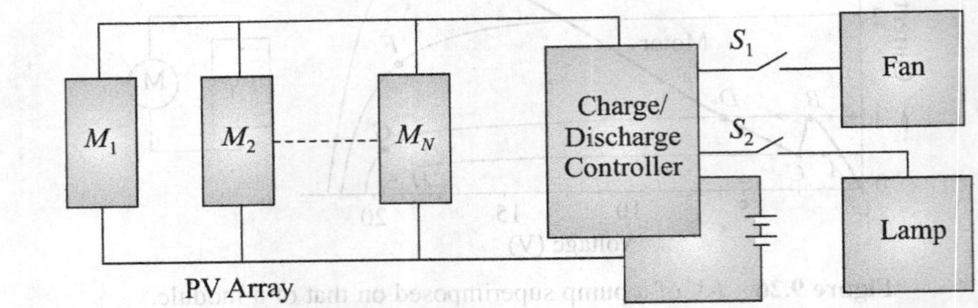

Figure 9.27 Block diagram of a solar home light and fan system.

Figure 9.28 shows two water output curves when a pump is operated straight from the solar array and when it is operated by a battery charged through a solar array. Of course the motor of the pump has the characteristics shown in Figure 9.26 and so its speed also

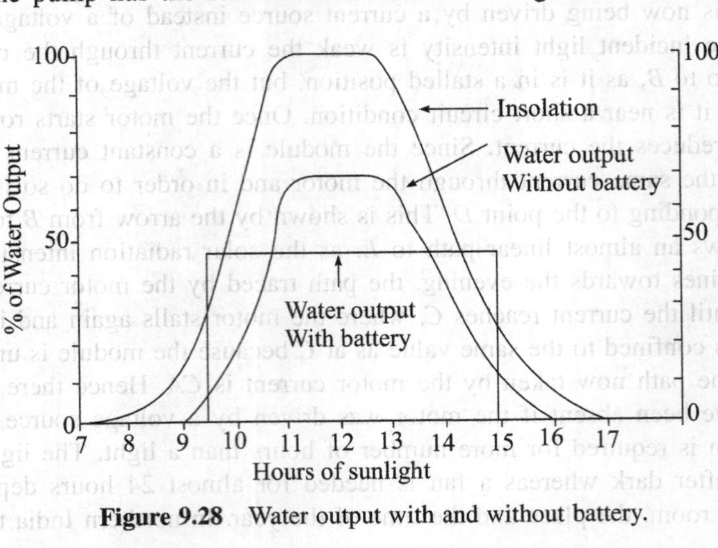

Figure 9.28 Water output with and without battery.

varies almost linearly with the rise in insolation which makes the water output almost follow the insolation curve. However, when a battery is employed to drive the pump, the water output rises to a maximum value and stays there until the pump is switched-off. In the latter case the price of the battery comes into the cost considerations of the system but it also comes in handy during shadowy but rainless conditions. Usually, a battery is used for voltage stabilization and not for storage for long periods.

9.4.1 Determination of the Motor Horsepower for a Pump [1]

First it is necessary to find out the pump power needed to push water at a certain rate to a platform or tank located at a predetermined height. The source of water may be on the surface or deep underground as it is in most of the cases. Water has to be often drawn from subsoil sources and pumped to the surface for irrigation or stored in overhead tanks.

To calculate the horse power of the pump motor the work done to raise the water to a required height has to be found. Suppose 1 L of water is lifted by one metre, the work done to do this is

$$W = m \times g \times h \text{ (joules)} \tag{9.9}$$

To calculate the horse power (HP) of a pump required to lift 1 litre of water by 1 metre the specific gravity of water is taken at 25°C to be 0.997 kg/L.

Therefore, the work done to lift vertically 1 litre of water by 1 metre is

$$W = m \times g \times h = 0.997 \text{ (kg)} \times 9.81 \text{ (m/s}^2) \times 1 \text{ m} = 9.78 \text{ J}$$

If this work is done in one hour, then the required power becomes

$$P = 9.78/(60 \times 60) = 2.72 \times 10^{-3} \text{ W}$$

Now one horse power (HP) = 746 W. Therefore

$$1 \text{ W} = \frac{1}{746} \text{ HP}.$$

Hence

$$2.72 \times 10^{-3} \text{ W} = (2.72 \times 10^{-3})/746 = 3.64 \times 10^{-6} \text{ HP}$$

This means that 3.64×10^{-6} HP is needed to pump one litre of water in one hour to a height of one metre. Therefore, to raise X litres of water to a height of H metres will require 3.64×10^{-6} (X) (H) HP. However the pump, used for this purpose, will also consume some power because of its frictional and other losses. If η be the pump efficiency, called wire to water efficiency, then the horse power of the motor of the pump needed is

$$(3.64 \times 10^{-6})\,(X)\,(H)/\eta \text{ HP} \tag{9.10}$$

If the pump is operated for P_T hours, called the pumping time measured in terms of 'peak sun hours' which is same as EHFS, then the required power is

$$P = (3.64 \times 10^{-6}) \frac{(X) \times (H)}{P_T \times \eta} \text{ HP} \tag{9.11}$$

Wire to water efficiency is defined as the water output of the pump in $m \times g \times h$ J/s, divided by the electric power input in watts. It is not unusual to find fractional horse power motor driven pumps with an efficiency of 65%. When connected to a PV array, the pump works with changing speed depending on the array voltage starting from morning to evening as shown in Figure 9.26. This time is called the *pumping time* P_T, measured in peak sun hours, and is taken as 1 in Equation (9.11) and as explained earlier the pump rarely works at the MPT of the array so that lower power is fed into it. However, if an MPPT is introduced, about 20% excess energy is available to the pump. Therefore another term, called the pumping time factor P_{TF}, is introduced in Equation (9.11) which assumes a value of 1.2 in this case with an MPPT. The resulting expression is given in Equation (9.12) [49].

$$P = (3.64 \times 10^{-6}) \frac{(X)\times(H)}{P_T \times P_{TF} \times \eta} \text{ HP} \tag{9.12}$$

If a battery is used, P_{TF} is the ratio of the actual running time of the pump T and the actual sun hours P_T, that is,

$$P_{TF} = T/P_T \tag{9.13}$$

Substituting Equation (9.13) in Equation (9.12) the expression changes to

$$P = (3.64 \times 10^{-6}) \frac{(X)\times(H)}{T \times \eta} \text{ HP}$$

This equation is same as Equation (9.12) if $P_{TF} = 1$, that is, if the pump were operated directly from a PV panel.

Example 9.2: Calculate the horse power of the motor driven pump required for the following work:

(i) 4000 L of water have to be lifted to the surface from an underground source located at a depth of 50 m.
(ii) A 90% efficient MPPT has to be used to utilize the full power of the solar array.
(iii) The pipe frictional losses are assumed to be 5%.
(iv) The duration of peak solar radiation = 6 hr
(v) Wire to water efficiency of the pump, $\eta = 60\%$

Solution: To calculate the power of the motor assume that an MPPT is used so that $P_{TF} = 1.2$. The frictional losses of 5% effectively increases the height of water so that the effective height is {50 + 5% (50)} = 52.5 m. Substituting these values in Equation (9.12) the power of the pump,

$$P = (3.64 \times 10^{-6}) \frac{4000 \times 52.5}{6 \times 1.2 \times 0.6} = 0.177 \text{ HP} = 132 \text{ W}$$

A 1/6 HP motor is available as this is a standard value but this amounts to only 124.3 W. The concept of service factor can now be applied. The service factor of a motor, in simple terms, is its overload capacity which the designers keep at 25%. This means that a 124.3 W

(1/6 HP) motor will not be damaged even if it were operated at {124.3 + 25% (124.3)} = 155.4 W suggesting that this very motor is suitable for the present application. Software for calculating the pump horse power is available on the internet [50] which can simplify many calculations.

Some other conditions must also be taken into account while designing a pumping system. As shown in Figure 9.29, pipes of larger diameter but of equal lengths, require less force to transfer the same quantity of water. Obviously, very large diameter pipes cannot be used on every occasion not only because of higher cost but also to pump water to higher places will need pumps designed for such work. It is evident that the energy required to pump water will depend on the quantity of water, that is, the same pump will lift larger amount of water

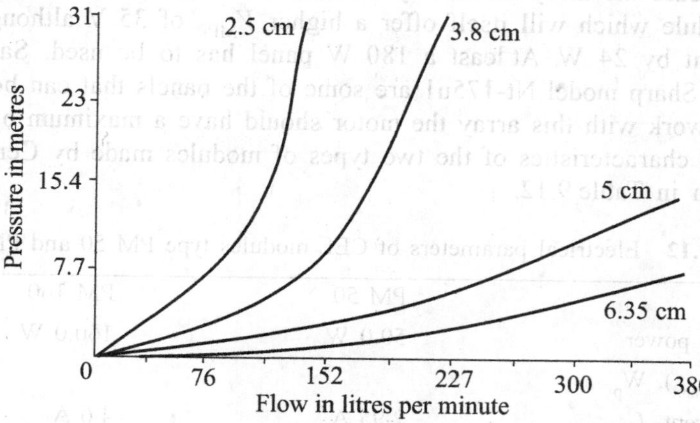

Figure 9.29 Flow curves plotted against pressure for pipes with different diameters.

to lower levels but will lift comparatively smaller volume to greater heights. These are given in detail in any fundamental book on fluid mechanics. The size of the PV array will now be calculated. Although the MPPT increases the quantity of water, but it consumes some power itself, its efficiency being typically 90%. In all the calculations the cost of the MPPT and its electricity consumption must be weighed against the advantage it provides in terms of water output it augments. However, confining the calculations to energy only the power input to the MPPT becomes

$$132/0.9 = 146.7 \approx 147 \text{ W}$$

A PV array must generate 147 W to operate the pump. The value found above needs at least one correction, that is, for the temperature.

The temperature on the surface of the modules kept in the sun at the peak summer months is roughly 40°C. The temperature of the solar cells inside a module is another 25° higher than that at the surface. Then it can be safely assumed the temperature of the cells is approximately 40° higher than the STC temperature of 25°C at which the modules are rated. The decrease in power of the modules is 0.5% per degree centigrade [2]. Therefore, the de-rating of a 147 W array, if used for this purpose, when the inside cell temperature is 40° higher is

$$(0.5 \div 100) \times 40 \times 147 = 29.3 \text{ W}$$

It means that a 147 W array will deliver a power of 117.7 W, that is, a reduction of 20% power. The array power will have to be increased to accommodate the loss.

Incorporating the correction the array power becomes 183.8 ≈ 184 W.

An array of 184 W, when kept at a tilt angle of 25.7° in the North-South direction, at Delhi will adequately drive the pump at 40°C. There are two options open to design the array. The first one is to use four CEL PM 50 panels in a series-parallel combination to give a total of 200 W. This is better by way of cost, being closer to the calculated power of 184 W. There will be certain losses associated with the isolation diodes as shown in Figure 4.5 which will be taken care of by the excess capacity of 16 W. The series connection of the two panels will give an MPP of 34 V while the second series connected pair, in parallel, will increase the array current by a factor of two. The other alternative is to use one PM 160 module which will itself offer a higher V_{MPP} of 35 V although it falls short of the requirement by 24 W. At least a 180 W panel has to be used. Sanyo model HIT Power 205N and Sharp model Nt-175u1 are some of the panels that can be considered for this purpose. To work with this array the motor should have a maximum operating voltage of 34 V DC. The characteristics of the two types of modules made by Central Electronics Limited are given in Table 9.12.

Table 9.12 Electrical parameters of CEL modules type PM 50 and PM 160

Parameter	PM 50	PM 160
Maximum power	50.0 W	160.0 W
Rating (P_{max}), W_p		
Rated current, I_{MPP}	2.95 A	4.6 A
Rated voltage, V_{MPP}	17.0 V	35.0 V
Short circuit current, I_{SC}	3.30 A	5.0 A
Open circuit voltage, V_{oc}	21.4 V	43 V
No. of cells	36	72
Physical dimension, ($L \times W \times T$)	845 × 550 × 35 (mm)	1580 × 795 × 42 (mm)
Weight (kg)	5.2	14

Another method is given here to determine the array size directly from the energy required to lift the same quantity of water as cited above. The total energy for the water lifted to a height of 50 metres is

$$E = (4000 \times 0.997) \times 9.81 \times 50 = 1,956,114 \text{ J}$$

Since 1 joule = $\dfrac{1}{3.6\times10^3}$ Whr therefore, the energy expressed in watt-hour becomes 1956.114/3.6 = 543 Whr

Hence the actual power required to compensate for the frictional losses in the pipe is 5% more, the total being 570 Whr. This energy must be further enhanced to compensate for the wire to water efficiency of the pump, taken as 0.6, and 0.9 for the electrical efficiency of the MPPT. Thus, the compensated energy fed into the MPPT is

$$\frac{570}{0.9\times 0.6} = 1055.8 \text{ Whr}$$

However, an MPPT increases the water output by 20%. Therefore, the energy required is reduced to

$$1055.8/1.2 = 880 \text{ Whr}$$

Further analysis is similar to that given in the beginning of this chapter. It is repeated here for the sake of totality. If the solar array is kept at a fixed tilt, then the tilt angle for Delhi is

$$\phi = (\text{Latitude of Delhi} \times 0.9)^\circ = 28.58 \times 0.9 = 25.7^\circ$$

The radiation data for Delhi at a tilt angle of 25° gives a value of 2131.2 kWhr/m^2 annually [1]. This amounts to an average of

$$2131 \div 365 = 5.84 \text{ kWhr/m}^2\text{/day}$$

Therefore, the 'Equivalent Hours of Full Sunlight' or the peak sun hours is

$$\text{EHFS} = 5.84 \text{ kWhr/m}^2\text{/day} \div 1 \text{ kW/m}^2 = 5.84 \text{ hr}$$

It means that if a solar module is placed under an illumination of 1 kW/m^2 for 5.84 hours it will receive this much energy. The power of the solar array will now be calculated. The energy required, as calculated earlier is 880 Whr. Dividing this by the EHFS the module size is obtained, that is, the power

$$P = 880 \text{ Whr}/5.84 \text{ hr} = 150.7 \text{ W}$$

The temperature compensation will have to be applied here too. The temperature on the surface of the modules kept in the sun at the peak summer months is roughly 40°C. The temperature of the solar cells inside a module is another 25° higher than that at the surface. Then it can be safely assumed the temperature of the cells is approximately 40° higher than the STC temperature of 25°C at which the modules are rated. The decrease in power of the modules is 0.5% per degree centigrade [2]. Therefore, the de-rating of a 150.7 W array, if used for this purpose, when the inside cell temperature is 40° higher is

$$(0.5 \div 100) \times 40 \times 150.7 = 30.14 \text{ W}$$

The value calculated above means that a 150.7 W array will deliver a power of 120.56 W, that is, a reduction of 20% power. The array power will have to be increased to accommodate such a loss. Incorporating the correction the array power becomes

$$P = 188.4 \text{ W}$$

An array of 188.4 peak watts when kept at a tilt angle of 25.7° at Delhi will drive the pump. In conclusion, it can be said that a pump of 151 W used with a solar array of 188 peak W approximately can lift four thousand litres of water from a source 50 m underground.

There is some difference between the sizes of the array found by the two methods. 184 W was found from the first method while 188.4 W was calculated by another. The reason is that the latter one is based on actual data amounting to 5.84 kWhr/m²/day while the former takes a pumping time of six hours which is the EHFS itself.

Data given in Table 9.13 for New Delhi, given earlier in Table 2.1, is reproduced here for convenience. For a ±15° tilt in addition to the latitude, the data show the average daily radiation throughout the year [51]. The average insolation is slightly more than 6 kWhr/m²/day and therefore will yield an EHFS of approximately six hours. Also the insolation is more in the months from October to February, when the array tilt is 15° more than the latitude, than those during the summer. This fact is useful for solar water heaters for they are used for generating hot water in winters but not particularly helpful for water pumping.

Table 9.13 Average Daily Peak Sun Hours (EHFS × 1 kW/m²) in kWhr/m² for New Delhi, India, Latitude: 28° 58′ N, Longitude: 77° 20′ E, 210 m

	Array tilted at latitude		*Array tilted at latitude −15°*		*Array tilted at latitude +15°*		
Month	*Fixed Array*	*One-Axis Tracking*	*Fixed Array*	*One-Axis Tracking*	*Fixed Array*	*One-Axis Tracking*	*Two-Axis Tracking*
Jan	5.83	7.38	5.04	6.38	6.28	7.87	7.92
Feb	7.04	8.97	6.37	8.09	7.31	9.23	9.24
Mar	7.31	9.02	7.05	8.6	7.18	8.83	9.05
Apr	6.94	9.17	7.12	9.23	6.42	8.5	9.3
May	6.87	9.36	7.38	9.83	6.08	8.25	9.86
June	6.19	8.53	6.76	9.15	5.38	7.32	9.23
July	4.20	5.94	4.5	6.31	3.75	5.17	6.34
Aug	5.30	7.27	5.53	7.44	4.83	6.6	7.46
Sep	5.70	7.44	5.66	7.23	5.46	7.13	7.45
Oct	6.57	7.99	6.09	7.34	6.69	8.09	8.13
Nov	6.43	8.56	5.62	7.49	6.88	9.05	9.08
Dec	5.73	7.11	4.87	6.06	6.26	7.68	7.77
Yearly Avg.	6.18	8.06	6	7.76	6.04	7.81	8.4

Modern pumps mostly use brushless DC motors because they can be directly used with PV generators and also for their high electrical efficiencies. These are particularly useful in submersible versions, where the problem of carbon brush changing will become problematic. In case, an AC motor is used an inverter will also be needed to convert the DC from the PV to AC which will reduce the overall efficiency because of the inverter's losses. Analysis for the optimum matching of a DC series motor driven centrifugal pump to a photovoltaic generator [52] is necessary to obtain highest wire to water efficiency.

9.5 A Hybrid System

Sometimes it is not possible to operate all the loads with photovoltaic arrays alone because of the high cost involved. For instance if the sunshine hours are small as compared to the operation time then the energy used during the sunless interval will have to be stored in batteries. The cost of the batteries, as found in Example 9.1, is exorbitant and more than the PV array itself so that either some cheaper way to store the energy has to be found or some other cheaper energy source should be located. The most common and easily available energy source is the diesel, or compressed natural gas (CNG), engine operated generator. These DG sets are widely used to augment power from the grid and also act as emergency supplies especially during power failures. The cost of the electricity produced from diesel is also comparable to that by PV without storage. To put it another way, the use of PV merely decreases the consumption of diesel thus reducing the emission of green house gases. Clearly, the two power sources have to be placed in parallel to the load such that either can be connected to it when required. This is called a *hybrid system*.

In Section 9.1 while designing a 1 kW stand-alone photovoltaic power supply, a large array was found necessary to power the load and charge the battery. It was also assumed that there was sunlight for eight hours per day for five days, a quite arbitrary assumption indeed but was satisfactory to illustrate the design procedure. If the data from Table 9.13 are taken, a more accurate picture will emerge which will also serve to illustrate the concept of a hybrid system in more detail. Although the number of solar panels will reduce in the array, thereby reducing the cost, the battery size will unfortunately remain large and consequently dictate the cost of the entire system. For an array tilted at latitude the irradiation will be the least in the month of July amounting to 4.2 kWhr/m^2/day. The maximum irradiation of 7.31 kWhr/m^2/day will fall in the month of March. One way to design the system is to calculate the array size for March and let the DG take care of the rest of the energy needed for sunless hours. If the array is to supply energy based on the data for July, its size will be largest among all other values calculated for radiation levels at other times of the year. The hybrid system is more suitable for higher latitudes, where the maximum peak sun hours may be seven hours and the minimum may be as low as one hour. For Delhi the system has two alternatives:

(i) The size of the PV array may be found for radiation received in the month of March and should operate only during daytime while the DG takes care during the rest of the time, and

(ii) The array shall supply power to the load and battery for 24 hours based on the insolation during March while the DG operates the load for the balance of time because when the irradiation is low the PV array will be insufficient.

The problem then reduces to the following form:

- Average power output = 1 kW into a DC load at a DC voltage of 108 V.
- The load is to be operated for 24 hours for all the seven days of a week.
- The insolation is 7.31 kWhr/m^2/day in the month of March in Delhi.
- Maximum depth of discharge of battery is 50%.
- Array should have a fixed tilt.
- The block Sunlight is available for five days.
- Diesel generator will supply the rest of the energy to the load.
- The battery will store the energy for the two sunless days.
- Duration of operation = 24 hours a day.

The diagram of the hybrid system is shown in Figure 9.30.

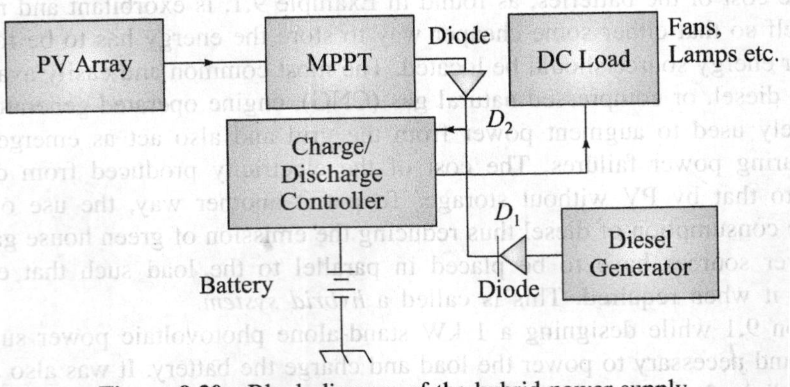

Figure 9.30 Block diagram of the hybrid power supply.

The photovoltaic array, in Figure 9.30, is connected to the charge/discharge controller for the battery through an MPPT. The diesel generator is also connected to the controller at the same point through a rectifier D_1 which not only converts the AC produced by the generator to DC but also forms an OR circuit with the other diode D_2, so that either of the two sources charges the battery at a time depending on which source produces a higher voltage. At dusk, when the PV voltage is lower the power from the generator charges the battery, and feeds the load at the same time, while during peak sun hours the PV does the same work and simultaneously disconnects the generator.

To calculate array size the total energy required the load and battery in seven days should be found. Since the power required = 1000 W

Therefore, energy needed for seven days:

$$E = 1000 \text{ W} \times 24 \text{ hr} \times 7 \text{ days} = 168{,}000 \text{ Whr}$$

For a 7.31 hour sunlit day the energy given directly to the load is:

$$1000 \text{ W} \times 7.31 \text{ hr} \times 5 \text{ days} = 36{,}550 \text{ Whr}$$

Since sunlight is available only for 5 days.

Hence, the rest of the energy must be stored in, and supplied by, the battery bank.

This energy is:

$$168{,}000 - 36{,}550 = 131{,}450 \text{ Whr}$$

It is assumed that the charge – discharge efficiency of the battery bank is 90% the energy to be supplied to the battery is:

$$131{,}450 \div 0.9 = 146{,}056 \text{ Whr}$$

A. The size of the battery bank

If the voltage of the battery bank is 108 V, as desired, then its charge capacity is:

$$146{,}056 \text{ Whr} \div 108 \text{ V} = 1352.4 \text{ Ahr}$$

Since it is stipulated that the batteries must retain 50% of the charge after supplying the required energy, their charge holding capacity must be twice this value. That is:

Total charge = 1352.4 × 2 = 2704.7 Ahr

Such high charge capacity (Ahr) cells have an output voltage of 2 V each.

To give a nominal output of 108 V a string of 54 cells, of 2 V each, must be connected in series. That is 54 cells of 2704.7 Ahr each should be attached in series to give the necessary energy. In case 2704.7 Ahr cells, which are rare, are not available then two strings of 54 cells of each 1352.4 Ahr may be connected in parallel. These must be the antimony doped deep discharge tubular type. The batteries are to be connected as shown in Figure 9.2.

Thus the energy consumed by the four diodes, is equal to 2323.4 Whr calculated in Section 9.1.1. The actual energy will be slightly more than this as the currents involved now are slightly more. However as compared to the energy needed for the load this little variation can be overlooked but this too must be supplied by the PV array. Therefore the total input energy rises to:

$$36{,}550 + 146{,}056 + 2323 = 184{,}929 \text{ Whr}$$

This energy in turn is given by the MPPT which itself has an efficiency of 90%. Hence the energy delivered at the input of the MPPT from the array is:

$$184{,}929 \div 0.9 = 205{,}477 \text{ Whr}$$

There is a 5% loss in wiring and cabling, hence the output of the array should be:

$$205{,}477 + (5/100) \times 205{,}477 = 215{,}751 \text{ Whr}$$

The array has to produce this energy in seven days. Then for one day the array will produce

$$215{,}751/7 = 30{,}822 \text{ Whr.}$$

The peak sun hours taken for this calculation is 7.31 hours. Therefore the size of the array is

$$30{,}822/7.31 = 4216.4 \text{ W}$$

An array of the above power rating will work quite well at 25°C but it is unusual to find such low temperatures in Delhi in the summers. The temperature in shade is quite often 45°C which means that the temperature under the sun will be higher. In addition, the temperature of the solar cells inside the modules will be at least 25°C above that on the surface. It appears reasonable to take a cell temperature of 70°C in month of July. The temperature difference between 70°C and 25°C, at which the panels are calibrated, is 45°C. As mentioned earlier the array loses power at the rate of 0.5% per °C. So the array, whose value has been calculated at 25°C, is reduced to a less powerful one. The loss in power is

$$(0.5/100) \times 45 \times 4216.4 = 949 \text{ W}$$

Or 22.5% of the array size has reduced. The size of the actual power of the array X can be found from the following equation

$$X\left(1-\frac{22.5}{100}\right) = 4216$$

or,

$$X = 5440 \text{ W}$$

In the month of July the peak sunshine hours reduce to 4.2, resulting in a lower energy production by (7.31 – 4.2)/7.31 = 42.5 %. This deficiency in energy will have to be supplied by the DG. Before going any further with the analysis the costs involved will be considered. The cost of the battery will remain unchanged because it will hold charge for two days due to an absence of sun or owing to the repairs needed by the DG. At ₹ 180 per watt of solar panels the cost of 5440 W of PV array is ₹ 979,200. If the array size is calculated for the month of July, the cost then escalates by 42.5% to an additional amount of ₹ 416,160; the total being ₹ 1,395,360 or $ 29,070 taking $1 = ₹ 48. The cost of the battery bank is 54 × 2704.7 Ahr × ₹ 15 = ₹ 2,190,807 = $ 45,642, assuming the cost of the battery at ₹ 15 per Ahr, which is more than twice that of the solar array.

It now becomes necessary to calculate the economics of the hybrid system. If the cost of the generator and the fuel consumed in ten years is more than that of the PV array, of a life span of ten years, then it is unsustainable. The DG will be most active in July when the PV produces the least energy. The time of the insolation is just 4.2 peak sunlight hours which leaves a gap of 7.31 – 4.2 = 3.11 hours for the DG to produce the rest of the required energy. The energy input to the charge/discharge controller is 184,929 Whr in seven days. 42.5% of this will be generated by the DG, which works out to

$$42.5/100(184{,}929) \text{ Whr} = 78{,}595 \text{ Whr}$$

The energy to be produced in one day is 78,595 Whr/7 = 11,228 Whr, which has to be produced in 3.11 hours and will be partially stored in the battery bank. Therefore the power of the DG is

$$11{,}228 \text{ Whr}/3.11 \text{ hr} = 3{,}610 \text{ W} = 3.61 \text{ kW}$$

Taking into account a 90% conversion efficiency for generator AC to DC for load operation a 4 kW generator will be just sufficient for the present purpose. The DG will operate in decreasing intervals as the winter approaches and the peak sun hour or EHFS increases. Since such a change is gradual the DG operation time varies slowly. The DG should be switched on and off depending on two primary factors:

(1) the voltage across the battery bank indicating its stored charge condition and
(2) the charging current of the PV.

To expand this reasoning further it should be noted that while the PV is engaged in supplying power to the load there is no need for the DG to do so. Also if the battery is fully charged the DG may not operate. The yearly fuel consumption of the DG shall now be calculated. The energy made in seven days is 78,595 Whr calculated earlier. In one year energy generated is

$$78{,}595 \times \frac{365}{7} \text{ Whr} = 4098 \text{ kWhr}$$

Now 1 litre of diesel oil costs ₹ 32 and weighs 0.84 kilogram (kg). Also 0.3 kg oil produces 1 kWhr. The cost for generating every kilo watt-hour or unit is

$$0.3 \times 32/0.84 = ₹\ 11.4/\text{unit}$$

Taking into account the efficiency of the DG as 90% the cost enhances to

$$11.4/0.9 = ₹\ 12.67/\text{unit}$$

The cost of the diesel oil per year

$$12.67 \times 4098 = ₹\ 51{,}921$$

If it is presumed that the price of diesel oil remains the same for the next ten years, an unlikely assumption, the cost of the oil becomes ₹ 519,210 in contrast to ₹ 416,160 for the 42.5% additional PV array required to replace the oil. The price of diesel generated electricity will escalate once the repair and maintenance cost of the generator and its initial expenditure are also included. Besides, the emission of gases from the burnt diesel oil is an environmental hazard. These considerations justify the use of PV, instead of oil, for dedicated power generation over long periods.

The total volume of the diesel oil consumed in one year by the DG is

$$4098\ \text{kWhr}/0.3 = 13{,}660\ \text{kg}$$

1 litre of oil weighs 0.84 kg; therefore 13,660 kg will have a volume of

$$V = 13{,}660/0.84 = 16{,}262\ \text{L}$$

To supply 16,262 L of oil per year a container bigger than this has to be installed near the generator. If the DG is located at a remote location either the full quantity can be transported in a single batch or it can be delivered in two consignments depending on the available space and transport.

The PV power becomes an accessory to the DG very effectively when the ratio of the peak sun hours during two seasons of the same year is 6:1 or more. In such a case the PV can be seen as lowering the oil bill and reducing the emissions.

Questions

1. Calculate the array size for the HID lamp of Section 9.2 for operation inside a factory shed from 10 a.m. to 5 p.m.
2. What should be the size of the battery if used in the above problem if the requirement is only to stabilize the voltage to prevent instability in the circuit?
3. Design a 12 W streetlight made of white Cree LEDs of 100 lumens per watt efficacy.
4. Design a constant current source with a switching regulator integrated circuit. The minimum efficiency should be 85%. Find out a suitable IC from the manufacturers' data sheets.
5. Calculate the power of the solar array and the charge capacity of a 12 V battery for the solar home light and fan system shown in Figure 9.27 when placed in Delhi. Take the number of sunless days per week as 2. Take the other parameters as given in the preceding text.

6. Calculate the cost of the system given in the problem above with the cost of the tubular battery at ₹ 9 per Ahr and the module cost at ₹ 180 per peak watt.
7. What is the area of connecting wire in Figure 9.27? Look up the wire table given in Chapter 11.
8. Calculate the array size when the array, of the example in Section 9.4, is tilted by more than 15° to the latitude of Delhi.
9. Find the amount of diesel oil saved per year, in litres, when
   (a) the maximum value of peak sun radiation in one year is 6 hours and
   (b) The minimum value of peak sun radiation is 1 hour.

References

[1] Tapan Bhattacharya, *Terrestrial Solar Photovoltaics*, Narosa Publishing House, New Delhi, 1998, p. 157.

[2] H.S. Rauschenbach, *Solar Cell Array Design Handbook*, Van Nostrand Reinhold Company, New York, 1980, p. 175.

[3] B. Bhargava and E.V.R. Sastry, Solar Lanterns: Problems and Prospects, 6th *International Photovoltaic Science and Engineering Conference (PVSEC-6)*, New Delhi, February 10–14, 1992, pp. 885–889.

[4] A.K. Mukerjee, Comparative Study of Solar Lanterns, *Energy Conversion and Management*, vol. **41**, 2000, pp. 621–624.

[5] V. Dutta, A.K. Saxena, A.K. Mukerjee and J.C. Joshi, Performance Evaluation of Photovoltaic Based Lighting System, Centre for Energy Studies, Indian Institute of Technology, Delhi, May, 20, 1993. Report submitted to the Ministry of Non-Conventional Energy Sources, Government of India.

[6] Catalogue of Central Electronics Limited (CEL), 4, industrial area, Sahibabad 201010, Ghaziabad (UP), India. www.celindia.co.in/solar_photoVolt_module.asp.

[7] OSRAM catalogue. www.osram.com/osram_com/Professionals/General_Lighting/Compact_Fluorescent_Lamps/CFL_Energy_Saver.html.

[8] M.A. Cayless and A.M. Marsden, *Lamps and Lighting*, 3rd ed., Edward Arnold Ltd., London, p. 475.

[9] Ibid., p. 183.

[10] M. George Craford, Nick Holonyak, Jr., and Frederick A. Kish, Jr., In Pursuit of the Ultimate Lamp, *Scientific American*, February 2001, p. 62.

[11] Data sheet of Cree XLamp 7090 LEDs, www.cree.com/xlamp.

[12] Kenneth R. Spring, Thomas J. Fellers and Michael W. Davidson, Molecular Expressions Microscopy Primer Physics of Light and Color—Introduction to Light Emitting Diodes. www.micro.magnet.fsu.edu/primer/lightandcolor/ledsintro.html.

[13] Evan Mills, Lawrence Berkeley National Laboratory, MS 90–4000, Berkeley, California 94720 USA. The Lumina Project, Research Memo #1*. Assessing the Performance of 5 mm White LED Light Sources for Developing-Country Applications. http://light.lbl.gov

[14] Rebecca Jones, Jianping (Tom) Du, Zachary Gentry, Ilan Gur, Evan Mills, Alternatives to Fuel-Based Lighting in Rural China, Right Light 6. Shanghai, 9–11 May 2005.

[15] G.S. Dutt, Illumination and Sustainable Development, Part I: Technology and Economics, Energy for Sustainable Development, vol. **1**(11), May 1994, p. 23.

[16] A.K. Rajvanshi, Rocket Science for Rural Development, http://nariphaltan.virtualave.net/rocketscience.pdf.

[17] A.K. Rajvanshi, Development of Improved Lanterns for Rural Areas, NimbkarAgricultural Research Institute (NARI), Phaltan–415523, Maharashtra. http://nariphaltan.virtualave.net/lantern.htm.

[18] A.K. Rajvanshi, 1987, Design and development of improved lantern for rural areas, Report prepared for the Advisory Board on Energy, New Delhi.

[19] C.S. Sinha and T.C. Kandpal, 1991, Optimal mix of technologies for rural India: the lighting sector, *International Journal of Energy Research*, vol. **15**, pp. 653–665.

[20] A.K. Mukerjee and Nivedita Dasgupta, Firing Options for CFLs Used in Solar Lanterns, *Proceedings of 3rd International Conference on Solar Radiation and Day Lighting* (SOLARIS 2007), February 7–9, 2007, New Delhi,

[21] Solar Photovoltaic Power Source for Single Channel VHF and Similar Systems, specification No. P520S93, Power Plant issue-II, September 1993. Telecommunication Engineering Centre, Khurshid Lal Bhawan, Janpath, New Delhi.

[22] David Linden (Ed.), *Handbook of Batteries*, 2nd ed., McGraw-Hill Inc., New York, 1995, pp. 25.1–25.39.

[23] D. Berndt, *Maintenance Free Batteries*, Research Studies Press, Taunton, Somerset, 1993, pp. 165–204.

[24] C.J. Yarrow, Transistor Converters for the Generation of High-Voltage Low-Current D.C. Supplies, *IEE 1959*, vol. **106**, op. cit. p. 1320.

[25] P.J. Baxandall, Transistor sine wave LC oscillators, some general considerations and new developments, *IEE 1959*, vol. **106**, p. 748.

[26] M.A. Cayless and A.M. Marsden, op. cit., p. 186.

[27] National Semiconductor, October 2006. Data sheet of LM 3402. www.national.com/JS/searchDocument.do?textfield=lm3402.

[28] Data of CAT 4201 and CAT 4240. www.catsemi.com/products/mixedsignal.html#led_drivers

[29] Nadarajah Narendran and Yimin Gu, Life of LED-Based White Light Sources, *IEEE/OSA Journal of Display Technology*, vol. **1**(1), September 2005.

[30] A.K. Mukerjee, Comparison of CFL-based and LED-based Solar Lanterns, *Energy for Sustainable Development*, vol. **11**(3), September 2007, pp. 17–25.

[31] P.N. Wood, Power MOSFET Application Notes, AN-973, International Rectifier 233 –Kansas Street, El Segundo, California 90245, USA, pp. 1–8.

[32] A.K. Mukerjee and Nivedita Dasgupta, Power Factor Controller Used as DC–DC Converter for Photovoltaic Sources, *Renewable Energy*, vol. **33**, 2008, p. 1374.

[33] A.M Tuckey and J.N. Krase, A Low Cost Inverter for Domestic Fuel Cell Applications, 33rd Annual Power Electronics Specialists Conference, 2002. *IEEE*, vol. **1**(23–27), pp. 339–346, June 2002.

[34] Data Sheet for MC 34262, MC 33262W, Semiconductor Components Industries, LLC, July 2004–Rev. 7, www.onsemi.com.

[35] Data Sheet of ML 4826–2, Power Factor Correction (PFC) and PWM Controller Combo, Fairchild Semiconductor, 2005, www.fairchildsemi.com

[36] J. Millman and S. Seely, *Electronics*, 2nd ed., McGraw–Hill Book Co. Inc., Kogakusha Company Ltd., 1951, p. 381.

[37] C.J. Yarrow, Transistor Converters for the Generation of High-Voltage Low-Current D.C. Supplies, *IEE*, vol. **106**(18), 1959, pp. 1320–1324.

[38] P.J. Baxandall, Transistor Sine-Wave L-C Oscillators, *IEE*, vol. **106**(16), 1959, pp. 748–758.

[39] A.K. Mukerjee, Comparative Study of Solar Lanterns, *Energy Conversion and Management*, vol. **41**(6), April 2000, pp. 621–624.

[40] T.–F. Wu, T.–H. Yu and H.–M. Huang, Complete Analysis and Performance-Characteristics Compromise for Self-Excited Half Bridge Parallel Resonant Electronic Ballasts, *IEEE Proceedings of Power Electronics Conference*, vol. **1**, 1994, pp. 124–131.

[41] Philips Semiconductors, Power Semiconductor Applications, Lighting, Fluorescent Lamp Control, pp. 584–587.

[42] A.I. Pressman, *Switching Power Supply Design*, 2nd ed., McGraw-Hill Book Companies Inc., International editions 1999, New York, pp. 604–606.

[43] S. Ben-Yakov and M. Gulko, Design and Performance of an Electronic Ballast for High Pressure Sodium (HPS) Lamps, *IEEE Transactions on Industrial Electronics*, vol. **44**(4), August 1997, pp. 486–491.

[44] M.A. Cayless and A.M. Marsden, op. cit. p. 229.

[45] A.K. Mukerjee and Nivedita Dasgupta, A Current-Fed Half-Bridge Oscillator for 400 Watt HID Lamp, *The First IEEE Conference on Industrial Electronics and Applications* (ICIEA 2006) 24–26 May, 2006 at Marina Mandarin Hotel, Singapore, P0121.

[46] IS: 1944 (Parts I and II)–1970, Indian Standard. Code of Practice for Lighting of Public Thoroughfares (First Revision) Fourth Reprint, April 1991. UDC 628'971'6/'7, copyright 1971. Bureau of Indian Standards, Manak Bhavan, 9 Bahadur Shah Zafar Marg, New Delhi 110002, Gr 8, February 1971.

[47] Prafulla C. Sorcar, *Energy Saving Lighting Systems*, Van Nostrand Reinhold Company, New York, 1982, p. 337.

[48] Crompton Greaves, CG-Lux, version 2.14.

[49] Roger Messenger and Jerry Ventre, *Photovoltaic Systems Engineering*, CRC Press, New York, 2000, p. 109.

[50] Pump Design Equation Formulas Calculator Horsepower. http://www.ajdesigner.com/phppump/pump_equations_water_horse_power.php.

[51] Stand-Alone Photovoltaic Systems: A Handbook of Recommended Design Practices, Report No. SAND 87–7023, Sandia National Laboratory, Albuquerque, New Mexico, USA, April 1988.

[52] P.K. Koner, J.C. Joshi, A.K. Mukerjee and K.L. Chopra, Analysis of Optimum Matching of a DC Series Motor Driven Centrifugal Pump to a Photovoltaic Generator, *Renewable Energy*, vol. **1**(5–6), 1991, pp. 683–693.

CHAPTER

10

A Grid Interactive PV System

10.1 Export of PV Power to Grid

It is evident from the design examples given in Chapter 9 that for a large photovoltaic power system, of the order of one megawatt, the space required for placing the PV array so that there is no shadowing will be very large indeed. In fact, several hectares may be needed to deploy such a system. The cost of such an enormous space will be so absurdly high that the electricity produced will be exorbitantly expensive and therefore, uncompetitive with the conventional power. In Delhi, for example, the price of land is ₹ 2 lakhs per square metre so that the cost of four thousand square metres of land itself will be rupees eighty crores or \$16.67 million assuming a conversion factor of \$1 = ₹ 48. The land value will dictate the PV economics rather than the price of the PV array itself. A sensible approach for generating competitive PV electricity will be to install the solar array at a distant place where the land is cheap and not useful for agriculture at the same time. The deserts of Rajasthan have been suggested by many scientists for economic reasons but the technical feasibility is doubtful because the sand blasting effect on the top glass surface of the PV modules, during sandstorms, may reduce their transmission with time and affect the production of electricity. Be that as it may, the point to be noted is that the solar power plant has to be located far from the place where its produce will be consumed. With this comes the question of transportation of electricity which, in normal cases, is done by a grid structure by the Power Grid Corporation of India. For PV, the picture is clear; the generated electricity has to be uploaded on to the grid and taken to the cities. For uploading the power, an inverter is a necessary accessory that not only converts the DC, made available from the PV array, to AC but also matches the requirements of grid voltage, frequency and phase thus giving it the name: *grid interactive inverter* or simply GII. The other problems that the inverter has to address are:

1. Voltage fluctuations
2. Frequency variations
3. Phase correlation

4. Real power transfer (power factor)
5. Problems with apparent power
6. Grid failure leading to a state called *Islanding*
7. Reverse power flow from the grid into the inverter.

The power grid operates at very high voltages ranging from 33 kV to 400 kV and is therefore a prohibited zone for most GIIs. A suitable scheme to upload power to the 220 V grid is presented below with certain features to overcome all the obstacles listed above.

10.2 Phase, Frequency Matching and Voltage Considerations

The power grid operates at a frequency of 50 Hz in India and some other countries while in the USA and some other countries it works at 110 V, 60 Hz. Therefore, an inverter must give an output of the same frequency and maintains a zero phase difference with a voltage level slightly higher than that of the receiving grid. The grid should be treated as an infinite sink that can receive all the power generated by an up linking inverter. If the grid fails at any time, the inverter will go into a short circuit mode. To explain the situation further, consider the diagrams of Figure 10.1, where four possibilities are presented. Figure 10.1(a) shows a state where the inverter voltage lags behind that of the grid by a phase difference of ϕ degrees. At point *A* the grid voltage is higher than that of the inverter leading to a reverse flow of current into the inverter. In such a case the transfer of power to the grid from the inverter will reduce over the entire cycle. Similarly, from the curves of Figure 10.1(b) a reverse flow

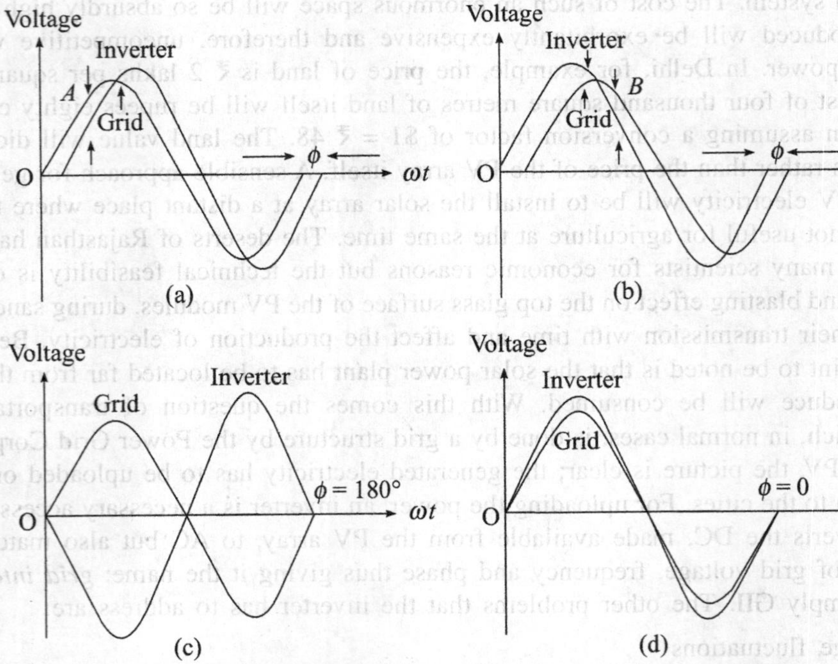

Figure 10.1 Phase relationship between inverter and grid voltages.

will also occur at point *B* when the voltage of the inverter leads that of the grid. In third case, Figure 10.1(c), the two voltages are diametrically opposite to each other in phase, which will create a short circuit condition for the inverter since the grid is considered to be an almost infinite sink. The ideal case is shown in Figure 10.1(d) where the two waveforms are exactly in phase with the inverter voltage exceeding in amplitude that of the grid. The inverter voltage must always be higher than the grid voltage for power to flow from the former to the latter. Since the voltage of the grid vary quite often, the inverter may be of a current source type or an inductor may be introduced just before the grid lines so that its reactance will offer some control over the amount of current flow into the grid. The effect of frequency variation is shown in Figure 10.2.

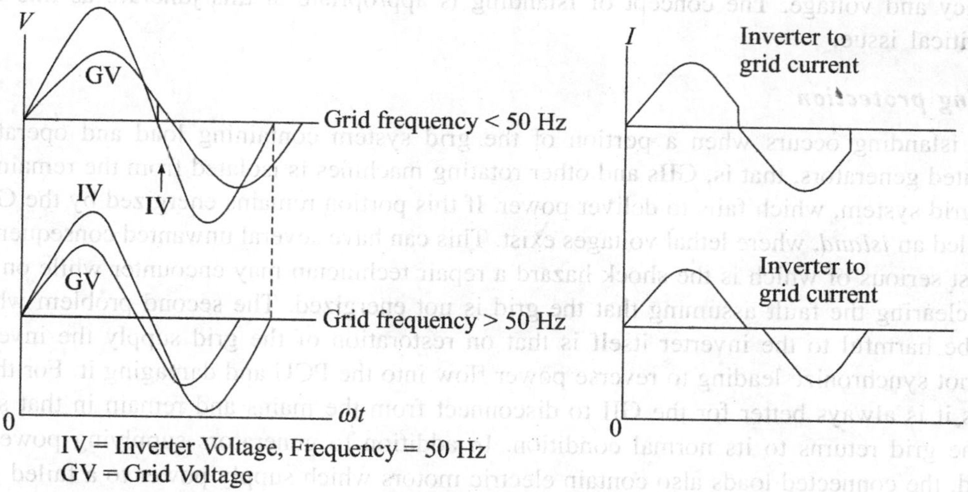

Figure 10.2 Inverter current patterns when frequency of the grid varies.

Although the inverter may maintain a stable frequency of 50 Hz, the grid frequency often varies by some amount within allowable limits unless there is a collapse, when both the voltage and the frequency of the utility drop. However, within allowable limits the transfer of current to the utility does not always follow a sinusoidal pattern. The current waveforms, when the grid frequency is either more or less than that of the inverter, are shown in Figure 10.2. It can be easily ascertained that if a mismatch of frequency is allowed to continue the transfer of power will be erratic, therefore the inverter frequency is latched on to that of the grid preferably by a phase lock loop. The amount of frequency and voltage drift allowed is restricted by the IEEE standard 929–2000 [1] which primarily defines the guidelines for avoiding an Islanding situation.

10.3 The IEEE Standard 929–2000 for Power Transfer from Inverter to Grid

The methodology for sending power from an inverter to the grid cannot be arbitrary as every manufacturer will deploy a different procedure thereby creating confusion. To avoid such a situation, the IEEE Standard 929–2000 [1] has been developed to address the technical issues associated with grid interactive PV systems. The standard fixes limits for the inverter voltage changes, its frequency variations, islanding prevention, power factor, harmonic injection, reconnection time after grid restoration from a state of failure, injection of DC into the AC system and grounding problems. The primary concern of this standard is to ensure that the inverter is disconnected from the grid when it fails for some reason even though other inverters may be pumping power into it, a state called *islanding*. Any inverter or PCU, as they are referred to in literature, certificated by UL 1741 has been tested to meet these standards. UL 1741 [2] prescribes a test procedure for utility-interactive equipment, to verify that the PCU is disconnected properly from the grid under approved conditions of grid frequency and voltage. The concept of Islanding is appropriate at this juncture as this is a very critical issue.

Islanding protection

A grid islanding occurs when a portion of the grid system containing load and operating distributed generators, that is, GIIs and other rotating machines is isolated from the remainder of the grid system, which fails to deliver power. If this portion remains energized by the GIIs, it is called an *island*, where lethal voltages exist. This can have several unwanted consequences the most serious of which is the shock hazard a repair technician may encounter while on the job of clearing the fault assuming that the grid is not energized. The second problem which might be harmful to the inverter itself is that on restoration of the grid supply the inverter might not synchronize leading to reverse power flow into the PCU and damaging it. For these reasons it is always better for the GII to disconnect from the mains and remain in that state until the grid returns to its normal condition. In addition to generators supplying power to the grid, the connected loads also contain electric motors which supply power to a failed grid because of the back electromotive force (emf) they generate when not forced to rotate by an external source. The synchronous motors are worse than the induction motors in this respect because they are better generators.

The GII is different from conventional generators. Normally the inverters are current source based instruments, which deliver a finite current into the grid even under short circuit conditions, whereas the rotating generators are constant voltage devices whose output current increases when the load increases, a condition similar to a failed grid with almost zero voltage. Therefore, the PCU should sense the utility voltage and frequency and inject current only if the voltage and frequency fall within prearranged limits. If this same circuitry is used to ensure that the current is injected in phase with the utility voltage, a high power factor for the PCU output can be obtained. The sensing circuitry usually has high input impedance and must remain connected to the grid at all times in order to monitor its voltage and frequency. Besides the IEEE 929–2000 concerning the grid interaction, the National Electrical Code (NEC) addresses areas that relate to the safety and performance of the system from the customers' viewpoint for GIIs owned by them to export excess power they produce. The restrictions imposed by these standards make a grid connected inverter far more complex than those used in isolation.

10.3.1 IEEE Standard 929–2000 Issues

Voltage changes

The American National Standards (ANSI) number ANSI C84.1 [3] recommends the voltage limits for the satisfactory performance of equipment while the IEEE standard specifies the number of cycles to disconnect it from the grid if the limits are crossed. This is given in Table 10.1. The nominal utility voltage is taken to be 240 V.

Table 10.1 Maximum voltage deviation and corresponding disconnection time in milliseconds

Maximum Voltage Deviation (percentage)	*Disconnection Time* (milliseconds)
$V < 120$ ($V < 50\%$)	100 (6 cycles)
$120 < V < 212$ ($50\% < V < 88\%$)	2000 (120 cycles)
$212 < V < 264$ ($88\% < V < 110\%$)	Normal operation
$264 < V < 330$ ($110\% < V < 137\%$)	2000 (120 cycles)
$330 < V$ ($137\% < V$)	33 (2 cycles)

The voltage levels in the original text are given for a nominal base voltage of 120 V, but these can linearly be scaled up to other voltages too along with the percentages listed in the table. The number of cycles indicates that the inverter has to trip within the specified value from the first sensing of any disturbance. More number of cycles ensures that the tripping may not be due to sudden variations in voltage like surges and the voltage returns to within the prescribed limits. Once the inverter is disconnected from the supply, it remains in that state for at least five minutes after the fault is cleared to make certain the stability of the line. The time duration is given in cycles but since the frequency in the US is 60 Hz corrections have to be made in India and other places, where it is 50 Hz. With the advent of microcontrollers and DSPs which work in the MHz region it is simple to adhere to the times specified for the inverter to stop energizing the line.

Frequency disturbances

IEEE 929–2000 requires that the PCU should shutdown within 6 cycles if the frequency of the line voltage falls below 59.3 Hz or rises above 60.5 Hz, that is, –1.16% to +0.83% of 60 Hz. Extrapolating this to a line frequency of 50 Hz it becomes 49.42 Hz to 50.42 Hz. Once again, digital sampling methods enable very accurate measurement of frequency so the shutdown algorithm becomes simple.

Normally, when the grid collapses, its voltage goes down very fast forcing the inverter to go into an overload condition so that its output voltage goes below the acceptable limit causing it to disengage from the supply line. Islanding is not possible in such a case. However, the situation is not so simple in all time. Consider a condition, where a large number of inverters are connected in parallel injecting power to the same supply line. If the grid collapses now, the inverters will keep the line energized, each taking the output of the others to be the grid, unless the connected load is large enough to overwhelm these local sources. This islanding will create a dangerous environment for the repair men to work in as the possibility of electrocution will remain a real threat. Consequently, all the GIIs should immediately and automatically be disconnected from the grid.

How quickly the grid voltage goes down to disconnection levels depends on the factors given below.

Resonance of the island motor loads to the grid frequency and their damping factors. The motors and machines like grinding wheels continue to rotate after supply failure and pump in power as long as they rotate. However, their injection frequency reduces as they slow down which might take a few tens of seconds. As mentioned earlier, the synchronous motors are worse than the induction motors for this purpose. This is the worst case load condition. The higher the Q of the circuit, the longer the circuit will oscillate at the grid frequency [4]. Hence, two algorithms were developed by the engineers at the Sandia National Laboratories and others to overcome the worst case loading condition [5], [6]. They are:

(i) The Sandia Frequency Shift (SFS) and
(ii) The Sandia Voltage Shift (SVS).

The SFS algorithm provides an accelerated frequency drift with regenerative feedback. The inverter frequency is programmed to drift away from the grid frequency unless the grid is present to lock in the inverter frequency. If the grid or the island frequency begins to move in either a higher or a lower direction, the control algorithm of the inverter frequency encourages the inverter frequency to travel even faster in the same direction. If the grid voltage is present, then at the time of zero crossing for the grid voltage, the inverter cycle is reset to grid frequency, as shown in Figure 10.2. If the grid voltage is not present to lock in the inverter current frequency, it continues to drift away from the grid frequency until it reaches the disconnection point, at which the inverter severs connection as a result of its frequency being out of the allowed range.

The SVS, slightly more complicated than the earlier one, is a similar control algorithm applied to the inverter current as voltage is sensed. The output current of the inverter is proportional to its output voltage. The feedback loop in the inverter is set so that for the inverter current increases twice for any increase in the grid voltage, that is, about a 1% grid voltage fluctuation, a 2% inverter current fluctuation occurs. These two algorithms when combined together in an inverter can overcome the worst load scenario for a load of Q less than five even when many inverters are connected to the same supply. The likelihood of an island load with a Q more than five within the utility trip frequency range is considered to be virtually zero and so any other schemes of avoiding an island is not considered. Several tests have been performed successfully to ascertain the effectiveness of this system.

Reconnection of inverter to grid after its restoration from a failure condition

The IEEE standard recommends five minutes as the reconnection time of the inverter to the grid after the latter's restoration from a state of failure. This time delay is necessary to ascertain the stability of the grid which might collapse because of several reasons starting from overloading to accidents caused by storms. Several hours may also be required to remove a major fault and to synchronize a large number of generators connected to it. Any voltage fluctuations which might result during the process of restoring the service might damage the GII. Although a steady grid is always desirable the SVS and the SFS, or similar protections, mentioned earlier will help to protect the GII.

Power factor

The power factor of the load always affects the supply line. Normally, since most of the loads consist of electric motors powering fans, coolers and air conditioning equipment for domestic users and heavy motors for industrial use the power factor of the supply is lagging. The utility therefore demands that the power factor be corrected by using leading power factor devices such as capacitors to keep a minimum power factor of 0.85 lagging below it imposes penalties. If a certain energizing unit live a GII uploads power with a power factor of 0.85 lagging, it is acceptable. The IEEE also allows the PCU to operate at a power factor greater than 0.85 leading. The power factor here is defined as the *fundamental frequency* for non-sinusoidal current delivered by the GII. Determination of this quantity should not be difficult since both the current and the voltage are monitored by the software of the inverter used to comply with the standard.

Injection of DC into the AC system

In theory the AC supply line should not contain any DC component whose presence might and will impair the function of inductors connected as load. An excess of DC will see an inductor as a short circuit and may damage it. Even if this does not happen the DC might take the inductor near its core saturation in which case the AC will face a lower reactance and will have several undesirable consequences including heating. Keeping this in mind the GII should not inject a DC current into the AC supply in excess of 0.5% of its rated AC output current under any operating conditions. Transformer coupling, though expensive, at the output of the GII is one way to keep DC from being coupled to the grid. Of course the inverter will be programmed to disconnect from the grid if this condition is ever reached.

Grounding

IEEE 929–2000 recommends that the PCU and PV system be grounded in accordance with certain codes to avoid shock hazard.

In addition to the IEEE 929–2000 standard, there are a few more among which the following two are rather important.

1. IEEE 1373: Recommended practice for field test methods and procedures for grid-connected PV systems.
2. IEEE 1374: Guide for terrestrial PV power system safety.

The International Electro-Technical Commission (IEC) has propounded its own standards listed below:

- IEC 61727 Ed. 2.0 b: Photovoltaic (PV) systems—Characteristics of the utility interface.
- IEC 62116 Ed. 1.0 b: Test procedure of islanding prevention measures for utility-interconnected photovoltaic inverters.
- IEC 61173 Ed. 1.0 b: Overvoltage protection for photovoltaic (PV) power generating systems—Guide.
- IEC 61683 Ed. 1.0 en: Photovoltaic systems—Power conditioners—Procedure for measuring efficiency.

The latter two are applicable to stand-alone power conditioners too. The first standard is relevant to the present discussion and will be taken up later.

10.4 Operation of a Grid Interactive Inverter

Figures 10.3(a) and 10.3(b) show block diagrams of a grid interactive inverter, also called a *grid connected system*, along with a sensor to disengage from the grid when a fault occurs. A 50 Hz voltage is taken from a step down transformer, or a Hall Effect based voltage sensor for better linearity, to synchronize the inverter output waveform with that of the grid to maintain a constant phase difference. A PLL LM 565 is used to lock the signal to the input from the grid. This is useful because the PLL will provide an output square waveform of the same frequency as the input yet its amplitude will not vary proportionally to the grid to reduce the possibility of regeneration. This is then passed through an amplifier tuned to the same frequency as the input, to filter out the harmonic content, and then fed to a unity gain phase shifter. The phase shifter can change the phase of the input signal in steps of 90° and also gradually alter it in between the steps. Change of phase of the input signal is necessary to align the inverter and the grid waveforms as described in Figure 10.1(d) for maximum transfer of power. To create synchronization pulses the signal from the phase shifter, attached to point '2' of the switch SW_1, is fed to the comparator LM 393A which generates a square wave. Subsequently, differentiator and clipper circuits produce positive going pulses for triggering an LM 555 based free running oscillator operating at a frequency of 200 kHz.

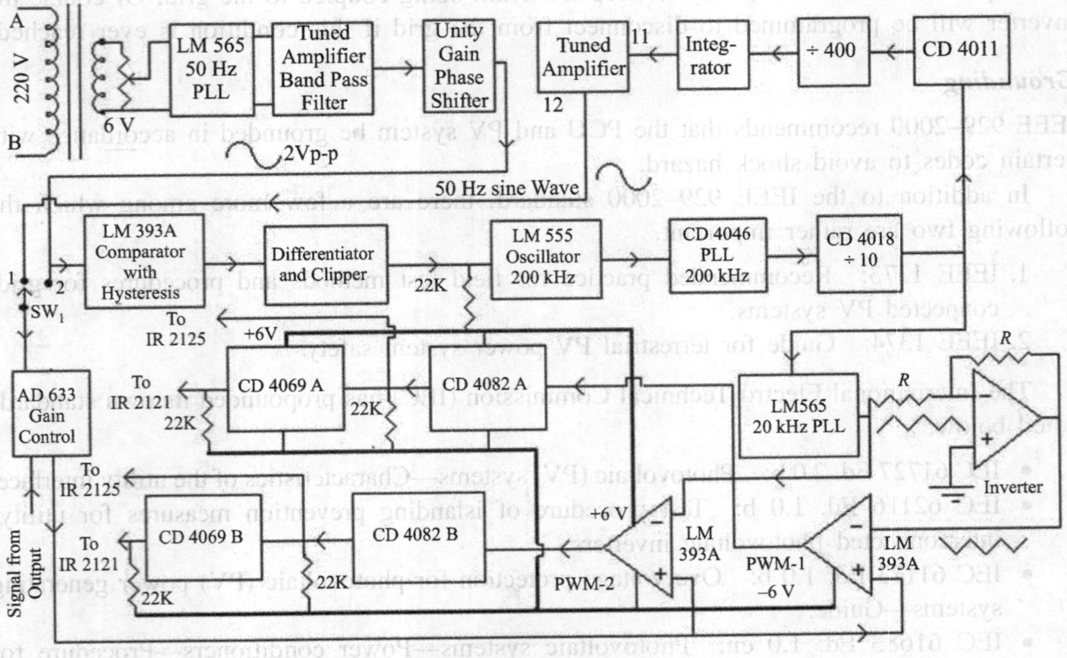

Figure 10.3(a) Block diagram of a grid interactive inverter.

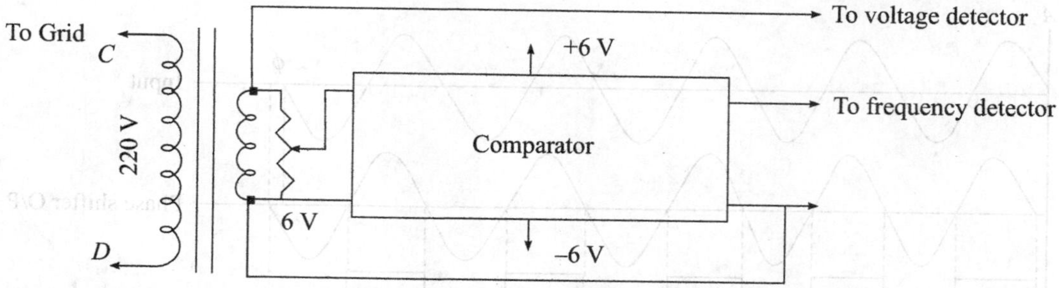

Figure 10.3(b) Grid failure sensor circuit.

Because of the disruption of the oscillator at intervals of almost 20 milliseconds, corresponding to a frequency of 50 Hz, harmonics are generated. To filter these out and keep the signal frequency intact a CD 4046 based phase lock loop, working at 200 kHz, is introduced. This device locks on to the signal produced by the oscillator and then its own output signal frequency is divided by ten by the decade counter CD 4018 to give a 20 kHz signal.

The generation of 50 Hz sinusoidal waveforms in phase with the grid is shown in Figure 10.4. It is of special interest to notice the gaps or 'blank' spaces in the oscillator output. Each pulse resets the oscillator to a zero output state by discharging the timing capacitor *C* through a transistor connected across itself. This is shown in Figure 10.5. To start the oscillations again the capacitor *C* attached to pin number 6 of the LM 555 has to charge to a certain voltage which takes some time. During this period there is no output from the oscillator. Thus, strings of pulses are initiated in synchronization with the grid signal of 50 Hz. However, the 'blank' spaces introduce harmonics which have to be filtered out first to enable frequency division. This is a very troublesome process and phase lock loops are used to ease it. A phase lock loop (PLL) integrated circuit (IC), CD 4046, locks on to the frequency of 200 kHz produced by the oscillator. If one PLL is insufficient then either another one can follow the first or a more sophisticated one may be used. The harmonics present in the pulse train interfere with the process of frequency division thus producing instability.

After division by a factor of ten, 20 kHz pulse train is produced which bifurcates further to generate a 50 Hz signal and a 20 kHz triangular signal by an LM 565 PLL. Both the signals are completely in phase with each other. These two signals then interact to form pulse width modulated signals in the two LM 393A comparators.

The two sets of PWM signals are then inverted and sent to the power stage through two sets of opto-couplers IR 2121 and IR 2125. To overcome voltage fluctuations a signal is sampled from the power output stage of the inverter and given to an analogue multiplier AD 633 along with the 50 Hz sinusoidal signal from the point 1 of the switch SW_1. When the output voltage increases, the output from the multiplier decreases to reduce the amplitude of the modulating signal given to the two modulators PWM-1 and PWM-2. Such a negative feedback stabilizes the inverter output voltage.

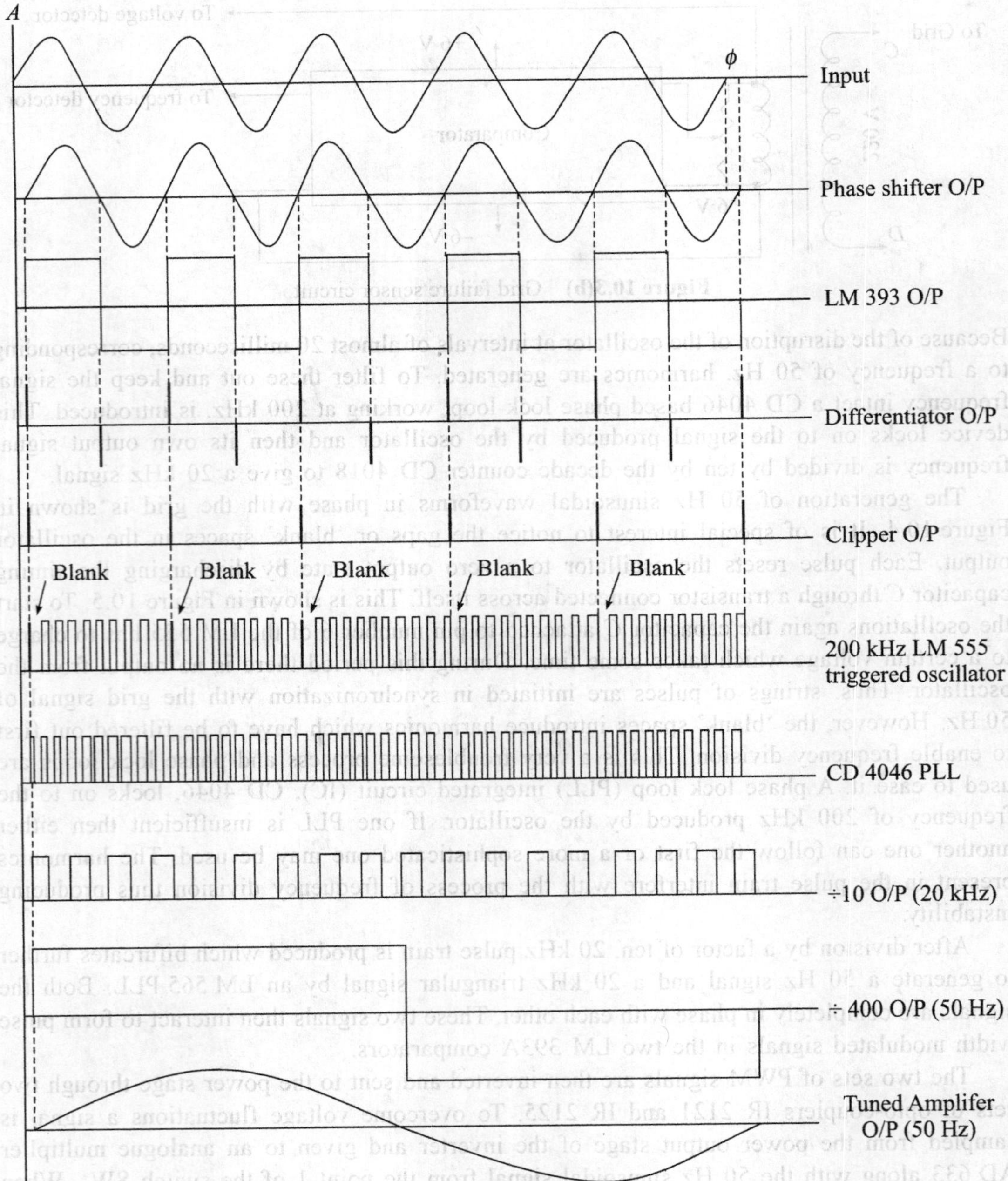

Figure 10.4 Timing sequence of the synchronizing pulses for producing sine wave.

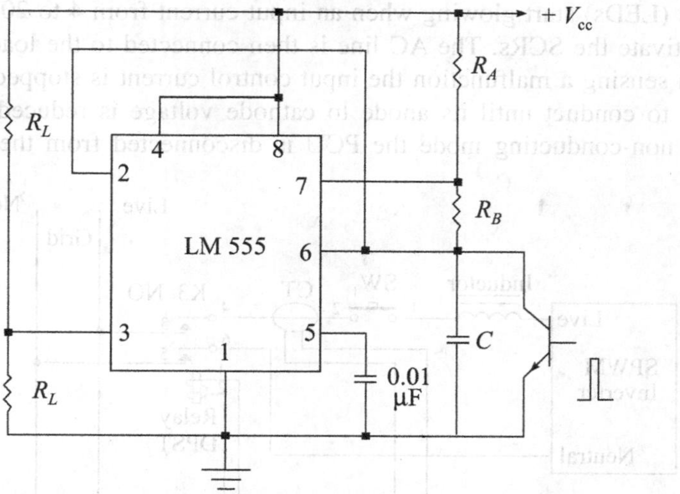

Figure 10.5 Circuit diagram of the triggered oscillator.

10.5 Protections against Islanding and Reverse Power Flow

When an inverter is connected to the grid, there is always a possibility for current to flow from the latter to the former resulting in a substantial damage to the inverter. This can happen when the grid voltage is higher than the inverter output voltage. For this reason a current source inverter is preferred since the output of the inverter, once set higher than that of the receiving grid, will continue to remain so. However, in case of an inverter malfunction its output may become lower than the voltage of the grid in which case a reverse current will flow into the inverter and seriously damage the IGBTs or the MOSFETs engaged in switching operations. To avoid any mishap the inverter must be physically disconnected from the grid. In case of DC power this isolation could be easily achieved by using a diode but in case of AC the blocking procedure is more complicated.

The circuit diagram of Figure 10.6 suggests a mechanism for disconnecting an inverter from the grid in case there is an overload or reverse current flow condition. The live output of the PCU is connected to the mechanical switch SW_1 through an inductor. SW_1 is closed once the desired voltage and phase conditions are obtained for current flowing in the proper direction. This is achieved by monitoring the waveforms on both sides of SW_1 with an oscilloscope. The inductor provides some reactance to control the current flow to the grid although it introduces a little phase change after connection. This phase difference is quite small and can be corrected with the phase shifter later. The current transformer (CT) monitors the magnitude and direction of the current and its output is fed to the circuit for reverse current and overload prevention circuit. In case either or both failures are detected the double pole, single throw (DPST) relay is thrown open for separation. Reverse power prevention flow relays from GE [7] are quite effective. Solid state relays can hasten the process of isolation in one cycle that is 20 milliseconds for 50 Hz operation [8]. This part of the circuit is essential to stop islanding according to IEEE standard 929–2000 and the IEC specifications. The functional block diagram of a solid state relay is shown in Figure 10.7. The light emitting diodes (LEDs) start glowing when an input current from 4 to 20 mA is fed to inputs 3 and 4 and activate the SCRs. The AC line is then connected to the load which in this case is the PCU. On sensing a malfunction the input control current is stopped but the conducting SCR continues to conduct until its anode to cathode voltage is reduced to zero. With both the SCRs in a non-conducting mode the PCU is disconnected from the grid.

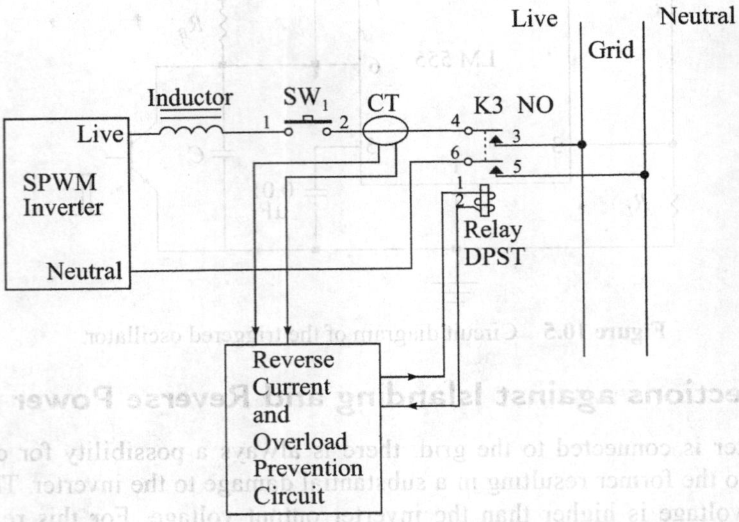

Figure 10.6 Disconnection circuit for inverter uploading power to 220 V, 50 Hz grid.

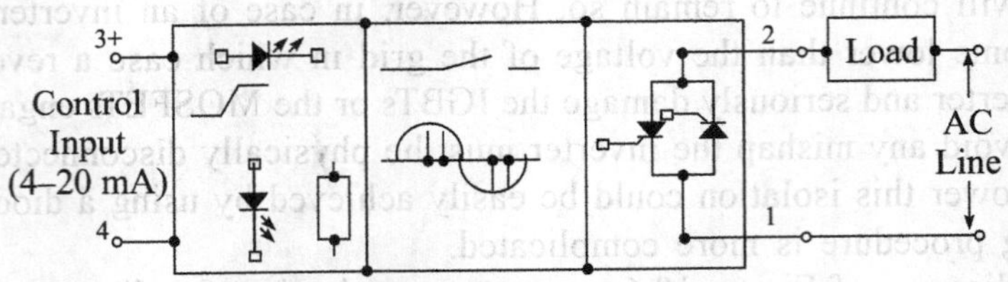

Figure 10.7 Block Diagram of a solid state relay.

Persson has described the problem of islanding in detail in his dissertation [9]. Another detailed description is given in reference [10]. A different way to detect and prevent islanding is presented in [11].

10.6 Circuit Description of a Grid Interactive Inverter

A grid interactive inverter was designed and fabricated by the authors to upload power from a solar array. To test the circuit a solar simulator was used as described in Chapter 3, Section 3.4 Figure 10.8 shows the 50 Hz filter and the phase shifter [12]. The filter in this case is needed to leave out the harmonics resident in the mains or they might interfere with the phase shifting process. The phase shifting circuit is designed for 50 Hz signals only by a certain quantity but the harmonics will change their phase by a different amount. This will affect the process of modulation and introduce the same distortion. The aim of the modulation process is to introduce as pure a 50 Hz signal to the grid as possible. Further, as shown in the block diagram of Figure 10.3(a), a phase lock loop precedes the filter to give a constant voltage square wave output. The 3rd, 5th and higher harmonics have to be eliminated to extract a distortion free mains signal. The filter, comprised of IC1 (LM 358), has two band pass amplifiers staggered slightly about 50 Hz so that they give an almost flat band pass characteristics which will not attenuate the signal in case the mains frequency deviates from its intended value. After leaving the filter the signal is buffered by high input impedance OPAMP, CA 3140, so that no further phase change occurs. The signal now enters continuous phase lead and phase lag circuits made up by IC3 (A) and 3 (B) after which another 180° can be introduced by IC3 (C) if necessary. The selected phase changed signal is then amplified and buffered by IC3 (C) and IC4. The switch WS1A enables the user to select the desired phase of the signal for modulation and eventual uploading to the grid. Figure 10.9 gives the diagram of the mains synchronization and the triggered oscillator circuits. The 50 Hz signal from switch WS1A is converted into square wave by the comparator LM 393, referenced to ground, with some built-in hysteresis to immunize it from stray noise signals. C_{14} and R_{33} differentiate the square wave while D_1 removes the negative going pulses so that a train of positive pulses, each created at the beginning of every positive going half of the square wave, is generated. The pulses are then passed through the quad 2-input NAND gates, CD 4011, and the hex inverter CD 4049 for generating pulses of proper polarities to control the 200 kHz free running oscillator comprised of LM 555 and its associated circuitry. Every time a pulse appears it is split into one negative and one positive pulse. The negative pulse rests the oscillator while the positive one switches on the transistor T_2 placed across the timing capacitor C_{16}. This action of T_2 drains away the charge on C_{16} so that the voltage across is reduced to almost zero. When the oscillations begin, after the pulses are removed, C_{16} starts to charge up leaving a 'blank' space in the oscillations as shown in Figure 10.4. The 200 kHz pulses are divided by the decade counter CD 4018B to give a pulse train frequency of 20 kHz. For greater stability it is better to introduce a phase lock loop (PLL), CD 4046, between the output of LM 555 and the input of CD 4018. The very purpose of creating 20 kHz pulses synchronous with the 50 Hz modulating signal is to avoid the formation of harmonics.

Figure 10.10 gives a selection of two phase lock loops for locking on to the 20 kHz signal and producing triangular waves subsequently for PWM generation. Whereas the PLL CD 4046 B needs an integrator IC 10A to produce triangular waves the other PLL IC 11, LM 565, produces the waveforms internally. The advantage of the latter is that in the event of any shift in frequency the amplitude of the output remains unaltered. The analogue inverters IC10B and IC12B change the phase of the signal by 180° for creating PWM signals to operate full bridge power inverters.

Two 20 kHz triangular signals are taken from pin numbers 5 and 2 of the switch TS4 and fed to the inputs of the two comparators IC13 (A) and IC13 (B) shown in Figure 10.11. The positive inputs of the comparators are connected to the 50 Hz sine signal from the pole of switch WS1A which is routed through the gain control IC AD 633 of Figure 10.14. If the output of the power bridge, Figure 10.16 (same as Figure 6.16), increases from the pre-set value the AD 633 reduces the amplitude of the 50 Hz modulating signal thus restoring the amplitude of the output. The carrier wave of 20 kHz and the modulating wave of 50 Hz interact inside the comparators to generate the PWM switching pulses. As an effective measure to turn off the inverter the PWM pulses are sent through the IC CD 4082 B, dual 4 input AND gates. The other inputs are connected to a switch TS5 which can either stop the pulses from passing to further circuits or connect to a timer, Figure 10.12, which can go up to a minute. This timer comes in handy during starting when a minute should be given to the entire circuit to stabilize. This can be further modified to give a time interval of 5 minutes required by the circuit to reconnect to grid after an islanding situation. Four PWM signals are produced, two from the outputs IC14 (A) and IC14 (B) and two by inverting these by IC15 (A) and IC15 (B). These then proceed to the inputs of opto-couplers HCPL 2211 of Figure 10.16.

Figure 10.12 shows the circuit diagram of the timer used in the instrument. The timer, LM 555 and associated components, produce oscillations of frequency 1.8 Hz that are fed to pin number 1 of the 7-stage ripple counter CD 4024 after inversion in IC19 (D). After a division by (2^7), that is 128, the pulse switches ON IC14 (CD 4082) and at the same time resets the timer IC LM 555 so that it stops oscillating. If a 5 minutes time is required a frequency dividing circuit CD 4017 may be inserted between pin number 3 of IC19 (A) and the 10 K resistor R_{110}. This circuit is a decade counter with decoded output so that a division by 5 can be selected. Any fine frequency variation can be made by trimming either or both the resistors R_{106} and R_{107}.

To produce a 50 Hz sine wave from the 20 kHz wave train the sub-block assembly of Figure 10.13 is used. The required division of 400 is done by a series of three pre-settable divide by 'N' counters in sequence of 10, 10 and 4. The resulting square wave has a high fraction of harmonics which is difficult to filter out. The input square wave is therefore first converted into a triangular one, which has a lower harmonic content, by integration in IC23 and then filtered by the band pass filter comprising IC24 (A) and IC24 (B). The resulting 50 Hz signal is synchronous with the 20 kHz input wave train and is used for generating PWM waveforms with comparators.

The output voltage amplitude of the inverter is stabilized by the circuit of Figure 10.14. The 50 Hz signal given to a load or linked to the grid is sampled and rectified by a full bridge rectifier made up of diodes D_{16} to D_{19} and given to the input of the opto-coupler 4N35. After initial filtering by the combination of resistance R_{135} and capacitor C_{67} the signal goes to the integrator IC28. The DC output from the integrator controls the gain of the analogue multiplier IC26 (AD 633) whose pin number 1 receives a 50 Hz signal from the switch WS1A. When the output of the power inverter increases for any reason, the DC gain control voltage from IC28 reduces the gain of the multiplier thus reducing its output 50 Hz voltage. This voltage reduction, in turn, reduces the modulation index of the PWM pulses, generated by this signal, that are used to operate the full bridge circuit of Figure 10.16. A lower modulation index reduces the output voltage, uploaded to the grid, thereby achieving its stabilization.

The grid interactive inverter works only when the grid is energized and should be isolated when it fails so that islanding is avoided. The inverter should therefore have two parts:

1. The main power unit that can be disconnected from the mains, and
2. An auxiliary unit which always remains connected to the mains for monitoring purposes.

It is therefore not inconceivable to have the power supply, for the control circuit of the inverter, derive its energy from the 220 V grids. This will have two benefits as follows:

1. Until the grid is restored the inverter will not function and hence accidental connection will not occur.
2. A stable voltage will be available from the grid so that simple and less expensive DC supplies can be made.

Following this reasoning the power supply of Figure 10.15 was made. It delivers power at ± 12 V and ± 6 V to the entire control circuit from Figures 10.1 to 10.14. Four standard integrated circuit voltage regulators have been used to get the required voltage levels after a 15-0-15 V (rms) step down transformer and abridge rectifier. This circuit is placed in parallel to the sensor of Figure 10.3(b).

The power output stage is shown in Figure 10.16. The switching pulses from IC15 are first electrically isolated from the power inverter by four opto-couplers type HCPL 2211and then given to the IGBT/MOSFET drivers IR 2125 and IR 2121. Whereas IR 2121 is a low side driver referenced to earth IR 2125 is a floating driver which can easily be used up to 300 V. The IGBTs shown here, IXGX 50N60BD1, are 600 V DC, 50 A devices that can be used to deliver safely a maximum power of 500 W to a load. For higher power several of these IGBTs, placed in parallel, or other high speed and more powerful IGBTs have to be used. The detailed design of this circuit is given in Chapter 6. It is interesting to note that the high frequency pulses are filtered out by the inductors L_1 and L_2 and the reflected capacitance of the capacitor C_4 placed at the output terminals of the power transformer T_1. Since high frequencies are eliminated T_1 can be a normal 50 Hz transformer instead of an expensive one. One of the outputs of T_1 is connected to the neutral wire of the grid while the other goes to the live through a 10 mH inductor [13] as indicated in Figure 10.6. The purpose of introducing a transformer between the inverter and the grid is to isolate one from the other to avoid shock hazard. There are many circuits, where the two are directly connected but for the beginner trying to gather experience in this field some precaution is of extreme importance for personal safety as well as to save the instrument from damage or destruction. However, using their own topologies some workers have proposed [14], [15] and [16] transformer less connection between the inverter and the grid.

10.7 Grounding and Other Considerations [17]

Modules connected in series are often required to produce high voltages for conversion to AC by inverters in medium to high power PV systems. In a typical 25 kW system the V_{oc} can easily be up to 600 V DC. Even for very good insulation between the positive terminal and the metal case, enclosing the solar cells, this voltage can leak to the latter. Since the ground is always situated near the array as shown in Figure 10.17, a resistance R represents this leakage path. It should be noticed that the negative terminal and the body of the inverter are connected at its input and then connected to the array earth. The dashed lines with the arrow indicate the direction of the leakage current. When the current reaches 1 A, the indicator at the switch must set off an alarm and turn off the single throw double pole isolator. One more point to be noticed here that the modules should be rated to withstand 600 V DC. *All isolating switches are to be DC rated.*

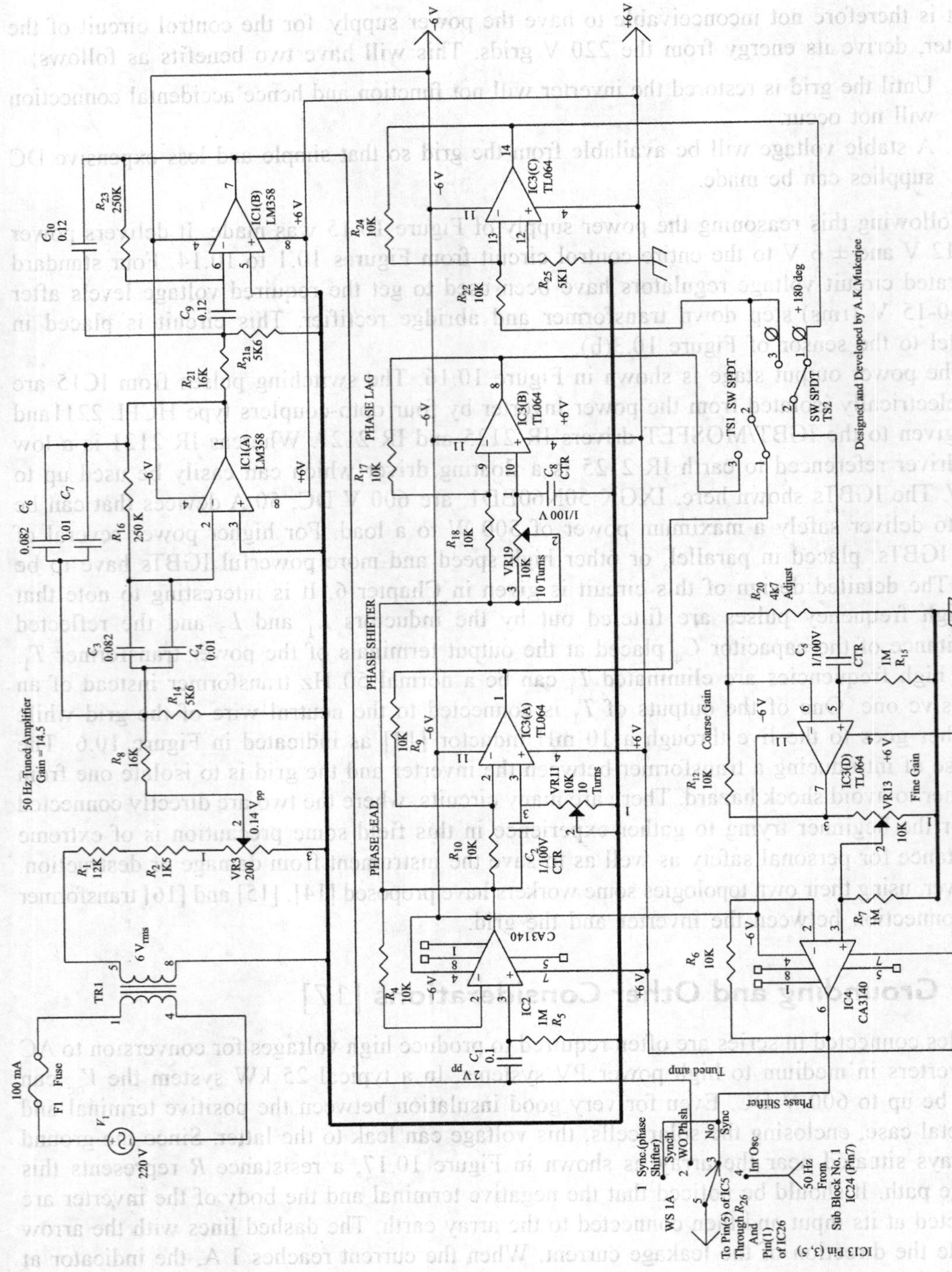

Figure 10.8 Phase shifter and tuned amplifier circuit.

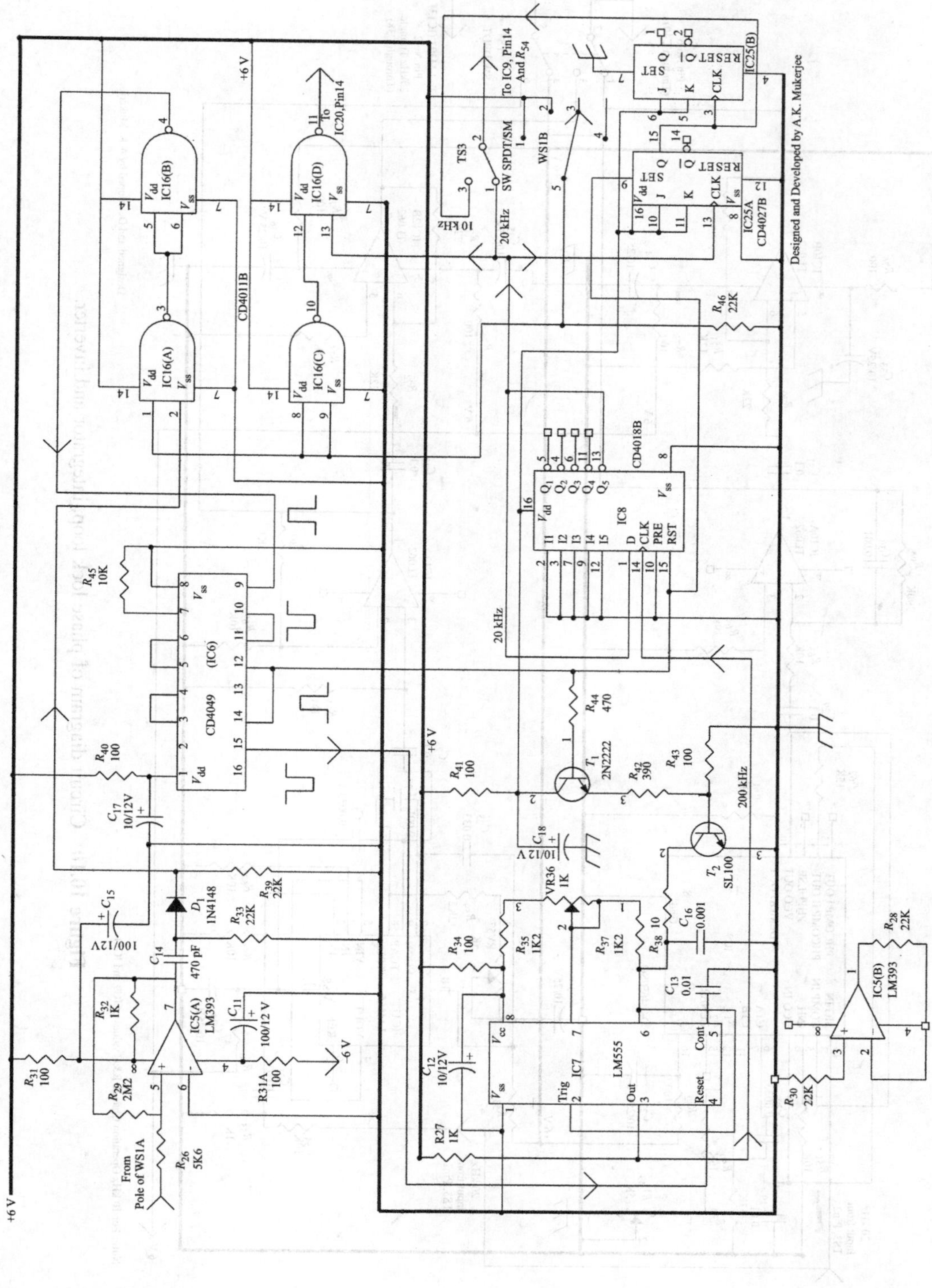

Figure 10.9 The triggered oscillator and mains synchronization circuit.

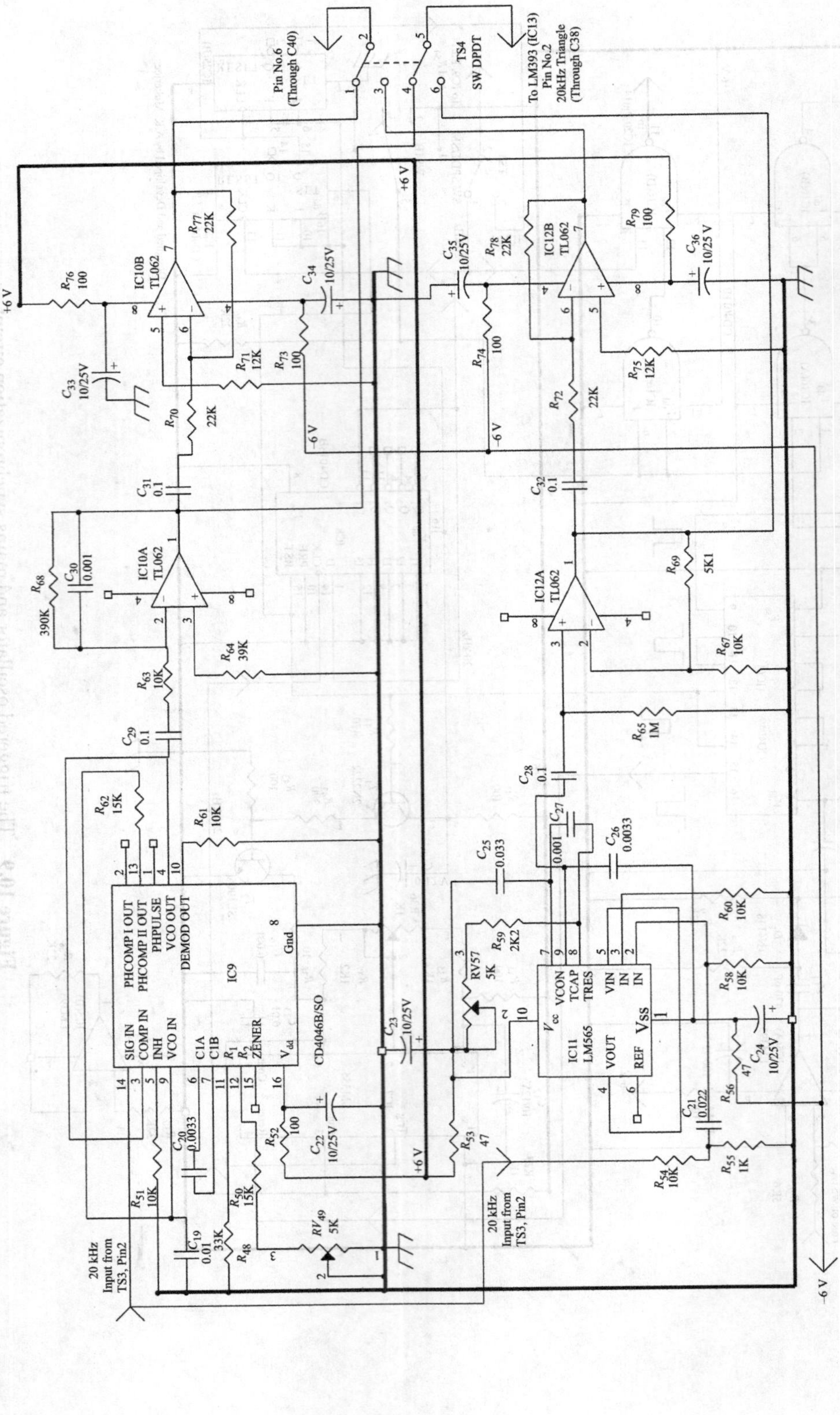

Designed and Developed by A.K. Mukerjee

Note: For 10 kHz Operation Change R&C values of IC9 and IC11

Figure 10.10 Circuit diagram of phase lock loop, integrator and inverter.

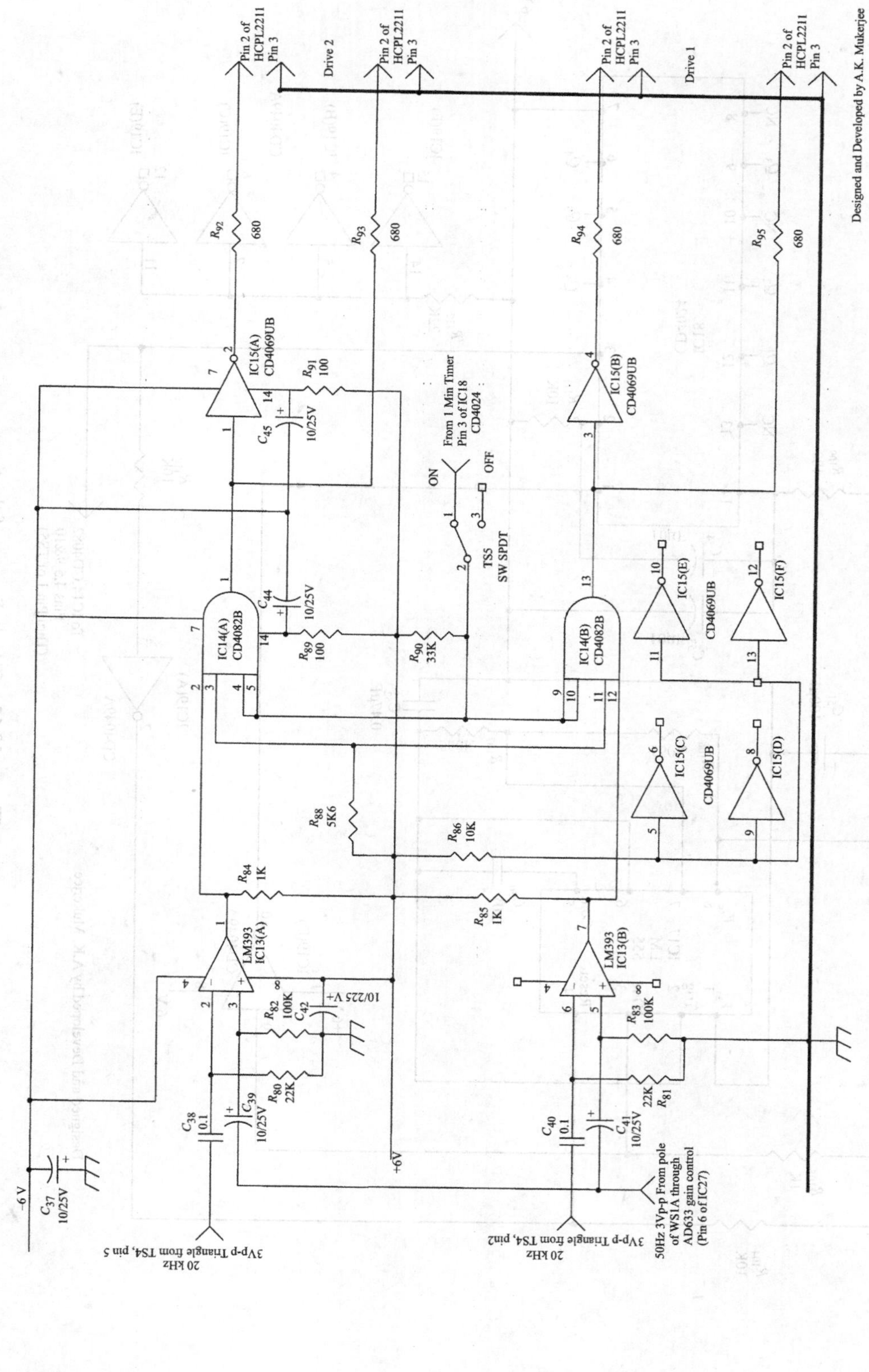

Figure 10.11 Circuit diagram of the pulse width modulator.

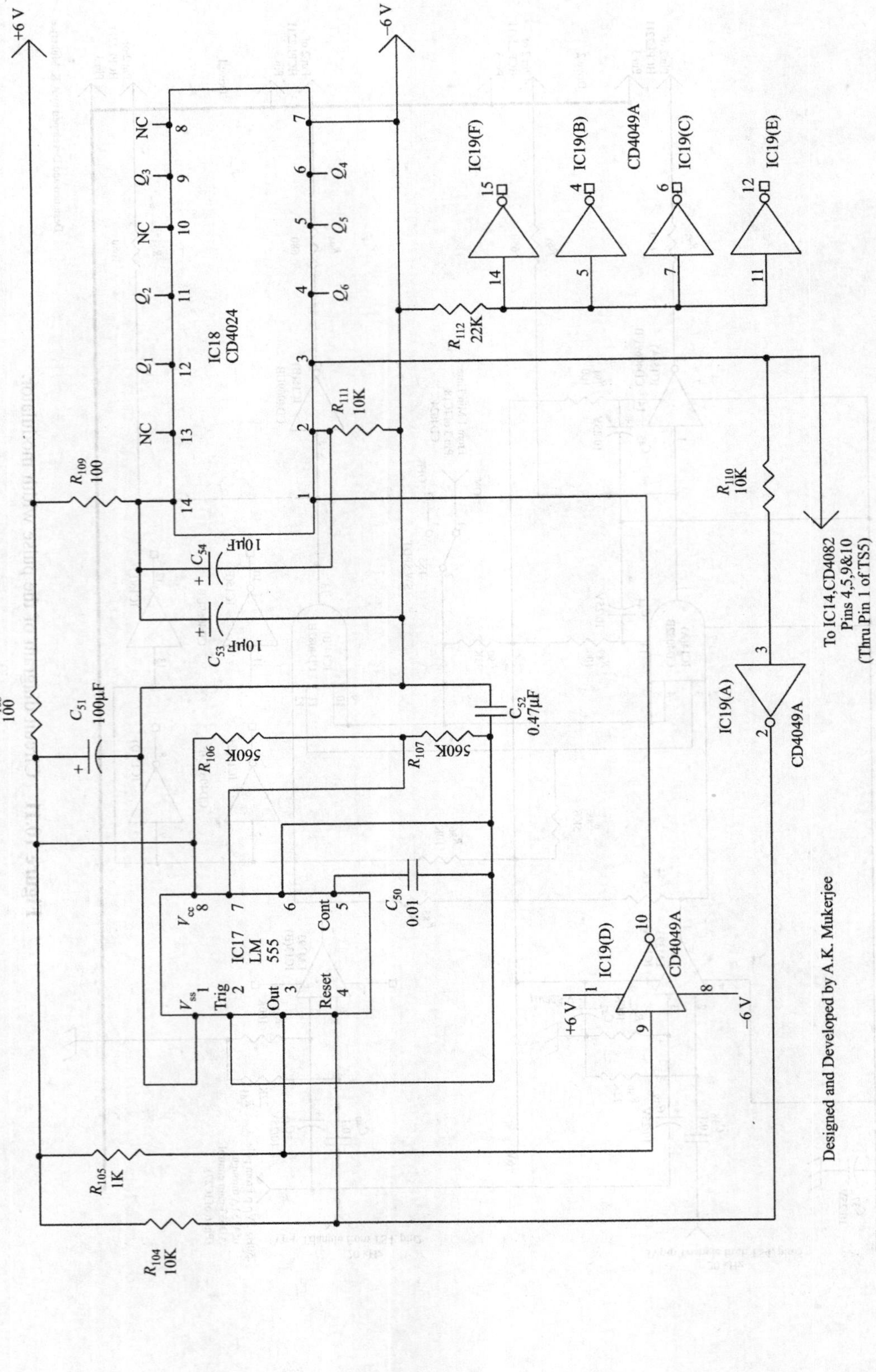

Figure 10.12 Circuit diagram of the timer.

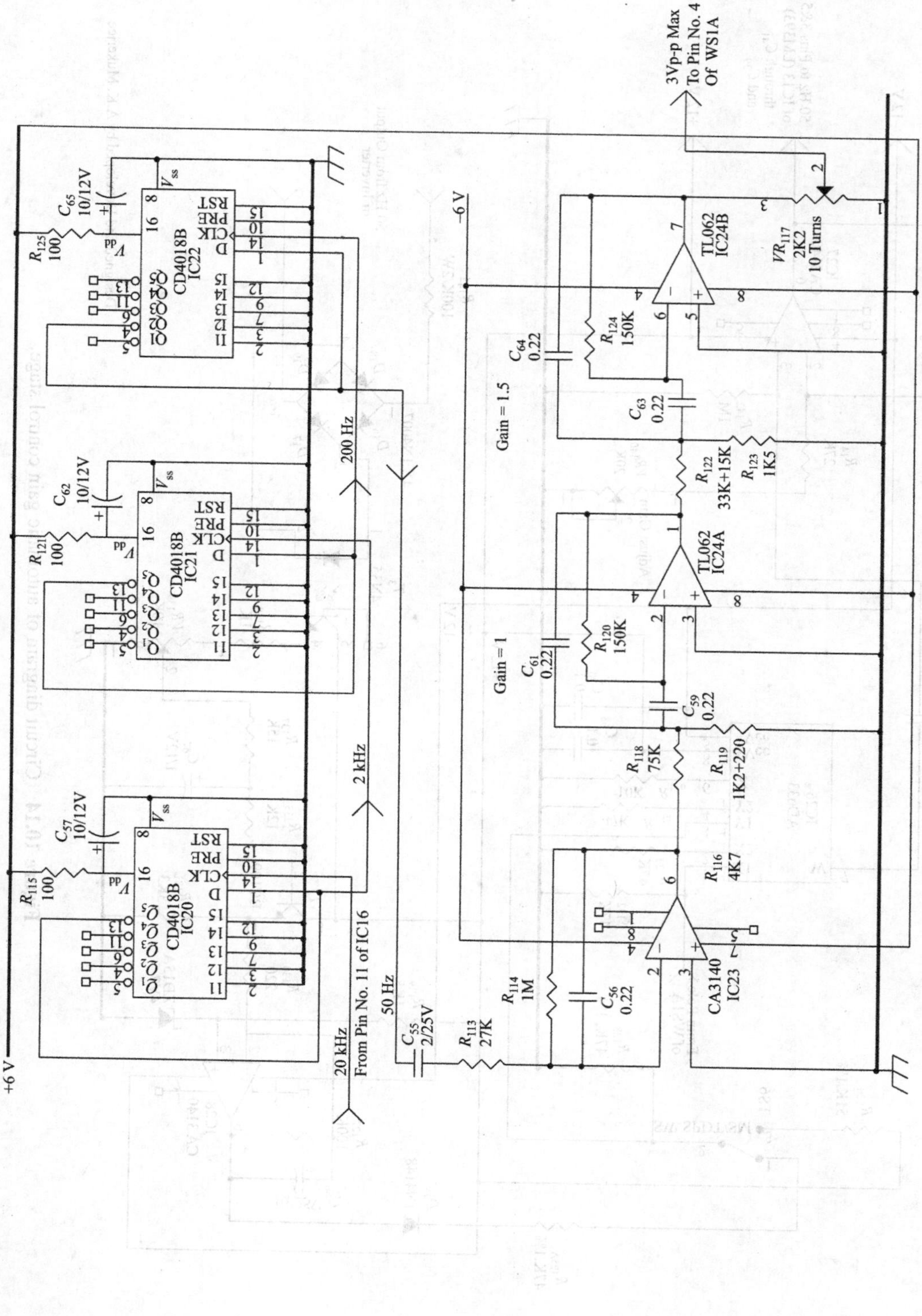

Designed and Developed by Dr. A.K. Mukerjee

Figure 10.13. Circuit diagram of divide by 400, integrator and tuned amplifier.

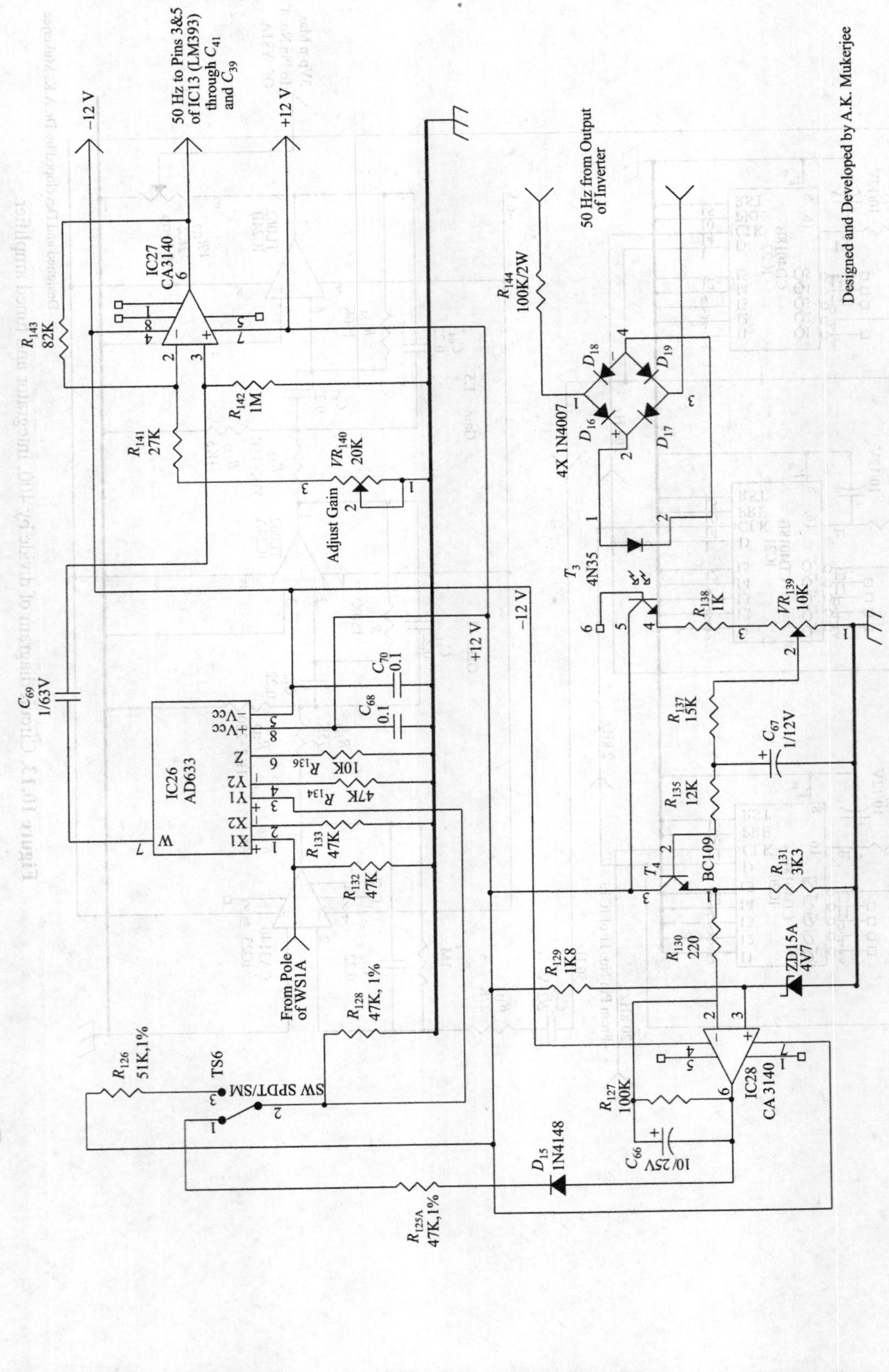

Figure 10.14 Circuit diagram of automatic gain control stage.

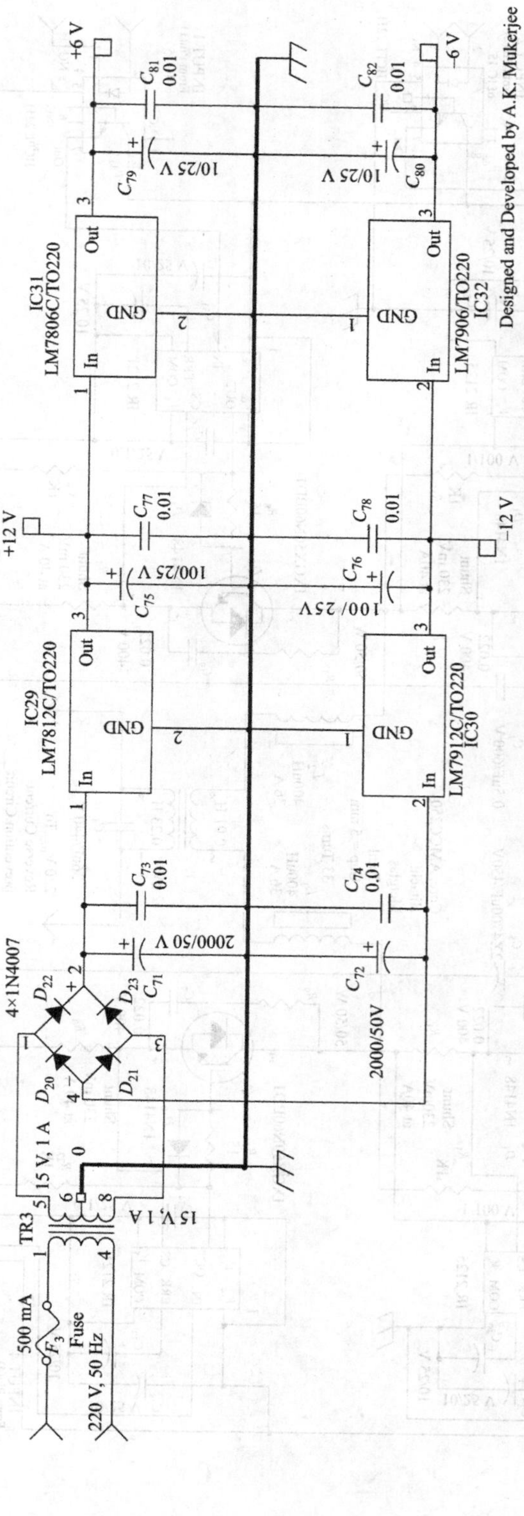

Figure 10.15 Circuit diagram of the power supply of the grid interactive inverter.

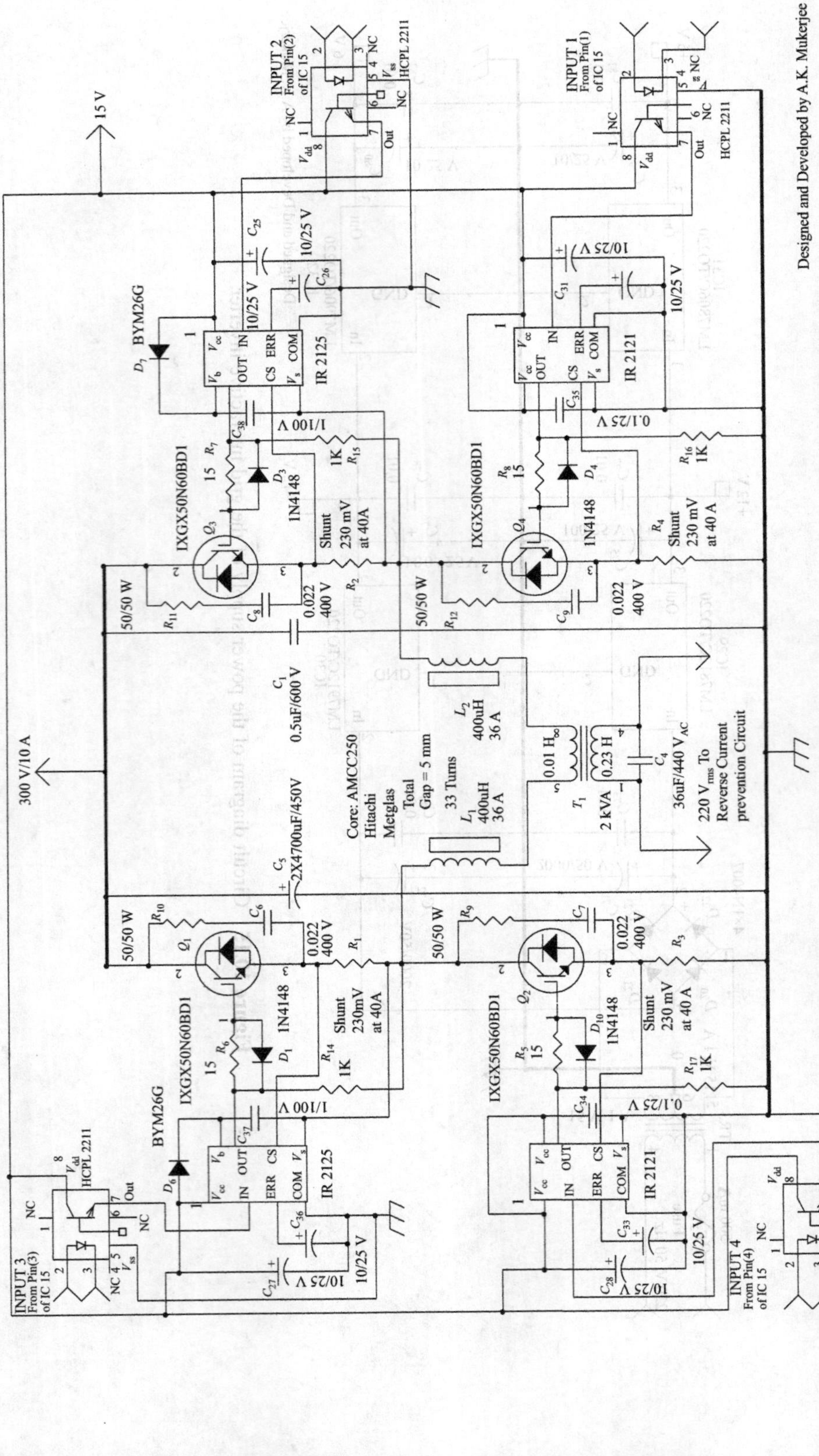

Figure 10.16 Circuit diagram of a 600 W full bridge inverter.

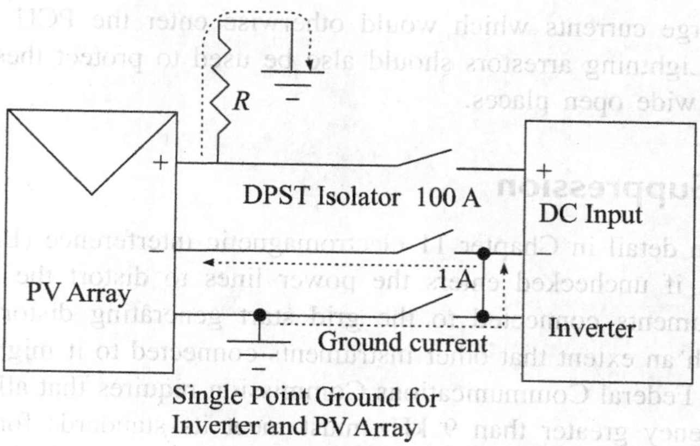

Figure 10.17 Flow of ground current and place of grounding for the entire system.

The other important question that arises is the largest size of the PCU that is to be used for high power PV units. It is abundantly clear from the above discussion that the insulation between the positive output and the ground of the array should not breakdown under the rated V_{oc}. It means therefore that the V_{oc} cannot be made arbitrarily large and must be contained within limits. Therefore, inverters have to be designed to work within these limits. Another design limit comes from the speed of the switching devices which have to be IGBTs for SPWM medium power applications. If very high power devices are used for the process of inversion, for example in a single 1 MW inverter, the switching speed of the device will be low. Apart from low speed a higher bus bar voltage will be needed to reduce the switching current which will go against the voltage considerations. Assuming a V_{oc} of 600 V_{DC} for the bus bar voltage, experimental operations suggest that the suitable size of an inverter would be between 25 and 50 kVA connected in parallel to give the required power. All the inverters have to be separately synchronized to the grid. Another problem that has to be overcome, especially in India, is the unbalanced condition of the grid. The three phases have different voltages with respect to the neutral and distortions because of unbalanced distribution of loads. Also these phases have high voltage fluctuations and frequency variations that far exceed the limits specified by the IEEE Standard 929–2000. Again it was experimentally found to be easier to tackle these problems by using three single phase inverters, one per phase against the neutral, instead of one three phase inverter.

10.8 Surge Protection

PV systems are subjected to sudden rise in grid voltages called *surge*, which may remain for a few milliseconds, originating from events such as motor starting and stopping. Protection against surge is not directly dealt with in the *NEC* [17] or in IEEE 929–2000 [1] but it is worthy of consideration. Most circuits, inverters or others, provide metal oxide varistors (MOV) for surge protection on both the DC and the AC side of the PCU. These devices act much like bilateral zener diodes, connected in opposite direction, with response times in nanoseconds. Being connected in parallel to the junction between the grid and the PCU they can conduct large currents which would otherwise enter the PCU and the PV array and destroy them. Lightning arrestors should also be used to protect these devices as they will be installed in wide open places.

10.9 EMI Suppression

As explained in detail in Chapter 11 electromagnetic interference (EMI) is generated by an instrument and if unchecked enters the power lines to distort the power signal. If many powerful instruments connected to the grid start generating distortions the grid may be polluted to such an extent that other instruments connected to it might malfunction. For this reason, the US Federal Communications Commission requires that all electronic devices that run at a frequency greater than 9 kHz must meet its standards for EMI. For this reason suitable filters must be placed between the PV array and the inverter and also between the inverter and the grid. Since there are two ways the EMI can infiltrate the grid, first by conduction and second by radiation, the inverter should be placed inside a box made of an electromagnetic material such as steel to contain EMI emission. Of course the first will be taken care of by the filter. The EMI is generated inside the inverter by the high frequency switching of high currents which pass through inductors and wires.

10.10 Wire Sizes and Voltage Ratings

The *National Electrical Code* stipulates that the overall voltage drop in a PV system can be less than 5% meaning thereby that the power loss should be restricted to within the same percentage. A PV system designer, however, should limit the voltage drop within 2% since the extra cost of wire is nearly always less than the cost of additional PV modules needed to supply the power lost in wiring.

If section 690.9(B) of the NEC is followed then fuses in the wiring circuit can be omitted according to section 690.9(A). According to 690.8(B), the required ampacity (current carrying capacity in amperes) of circuit conductors should not be less than 125% of the maximum current they will carry. The maximum current is defined as 125% of the rated *short-circuit current* for modules in series. The overall size of the conducting wires is therefore 1.56 times the module short circuit current. Still, if the voltage drop across the conductor exceeds 2%, then the area of the wire should be increased.

There are two definitions for the voltages involved: *system voltage* and *maximum system voltage* or *rated voltage*. It is important to distinguish one from the other. The *maximum, or rated, system voltage* is the voltage that is available under open circuit conditions or V_{oc} of the array while the *system voltage* refers to that value at which the system operates. Undoubtedly, it is the voltage that corresponds to the maximum power point (V_M) in well designed systems. If, for example, three modules are connected in series with each having V_{oc} = 22 V, the *maximum system voltage, or rated voltage*, is 66 V while the *system voltage* for the same combination reduces to almost 46 V. Corrections for temperature are also given in the NEC section 690.7. It is well known that the voltage of silicon solar cells decrease by 2.3 mV per degree centigrade rise in temperature—a fact that is of utmost importance in array design at high temperatures in the tropics. Similarly, when the temperature decreases at higher latitudes and altitudes the voltages of the cells, and therefore those of the arrays, increase. This correction to the V_{oc} of the arrays must be accounted for while determining their *rated voltage*.

10.11 AC Modules

A large GI system might not suit a domestic requirement either by size or cost. The average household plans for a gradual increase in expenditure on equipment rather that one time outlay. Of course loans can be arranged from banks but the cost increases once the interest payments are taken into account. Further a PV rooftop project may not get priority considering its initial cost. To popularize PV power generation the cost of the system has to be progressive. The concept developed so far is that the household that installs the PV power supply should be able to export the power to the grid when not consuming it. The power so exported to the grid will in a way bought by the utility and its cost will therefore be subtracted from the utility bill. This requires a bi-directional energy measurement meter that keeps track of the grid power consumed by the consumer and should be capable of subtracting the PV power sent to the grid. For instance, if a household consumes x units of power from the grid and sends y units of PV power when its members have gone out to work will be charged for $(x - y)$ units of electricity. This way millions of houses will send power to the grid to be used eventually in factories, shops and other places. This will reduce the dependence on fossil fuels and reduce the production of green house gases. This system is prevalent in the USA, Germany and some other countries. Figure 10.18 shows such an arrangement. PV power is generated on the rooftop of the house and synchronized with the grid before being fed to the load in the house through the switch SW. The power from the PV supply may not be enough for the load so the load also draws power from the grid which the meter records as x units. When the load is switched off during the day, the PV supplies the generated energy to the grid which the meter records as $- y$ since the power flow now is in the reverse direction. The total energy in kWhr is therefore $(x - y)$. When not exporting, its power the PV reduces that drawn from the grid and is effective in lowering the electricity bill.

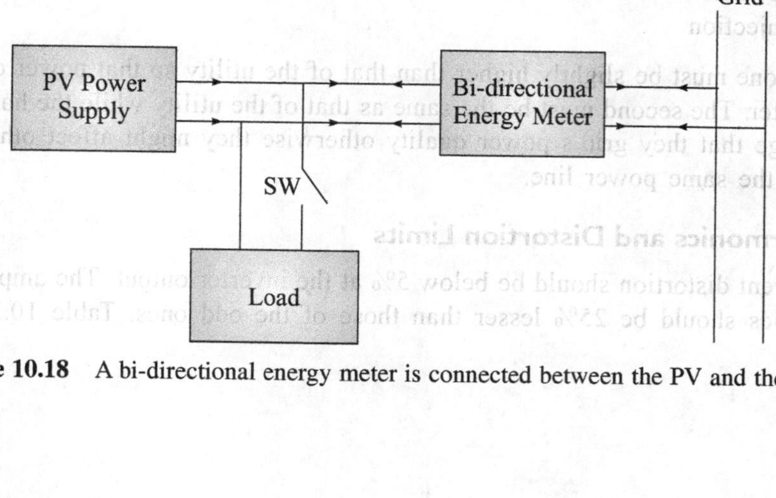

Figure 10.18 A bi-directional energy meter is connected between the PV and the grid.

To support the gradual payment policy some manufacturers have introduced the modular approach whereby the consumer buys as many modules as he can afford at a time. Each module has its own MPPT and grid interactive inverter and link up to the grid independent of the others. If many modules are purchased over the years they can be placed parallel to each other. The positive points are:

- The electronics can be made compact as it can be operated at higher frequencies because the module voltage is low. This enables faster MOSFETs and smaller inductors and transformers.
- The cost of individual module is lower so that the number of modules purchased can suit the household budget.
- The number of modules may be selected to fit in the available space.

The only disadvantage is:

- The cost for the same size power supply may be slightly higher because each module has the entire electronics fitted to it but use of application specific integrated circuits (ASICs) [18] have reduced the difference.

Each module qualifies under different standards like IEEE 929 and UL 1741 requirements in the USA. The difference between a normal DC module and the one is enormous since this one gives an AC output at the required grid voltage and frequency and can give an uncomfortable shock if its terminals are inadvertently touched. The DC module just gives a low voltage DC without all the electronics.

10.12 Requirements of IEC 61727

The *utility interface* is defined as the point of interconnection between the utility distribution and the grid interactive inverter. While connecting to the interface the inverter must comply with the requirements of the utility. These are measured at the interface and are listed below:

- Voltage
- Frequency
- Harmonics
- Power factor
- DC injection

The first one must be slightly higher than that of the utility so that power does not flow into the inverter. The second must be the same as that of the utility while the harmonics must not be so large that they grid's power quality otherwise they might affect other equipment connected to the same power line.

10.12.1 Harmonics and Distortion Limits

The total current distortion should be below 5% at the inverter output. The amplitudes of the even harmonics should be 25% lesser than those of the odd ones. Table 10.2 gives these limits.

Table 10.2 Distortion limits of current harmonics

Number of the harmonics	*Maximum distortion* (%)
3rd, 5th, 7th and 9th	Below 4%
2nd, 4th, 6th and 8th	Below 1%
11th,13th and 15th	Below 2%
10th, 12th, 14th up to 32nd	Below 0.5%
17th, 19th and 21st	Below 1.5%
23rd to 33rd	Below 0.6%

10.12.2 Power Factor

The output of the photovoltaic driven inverter should have a power factor of 0.9 lagging when its power exceeds 50% of its rated capacity.

10.12.3 Islanding

The mechanism of islanding has been discussed earlier and it is agreed by all concerned that the PV system should disconnect itself from the utility in case the grid fails for some reason. This must be done for personnel protection and equipment safety irrespective of whether other generators or loads are connected to the utility. The PV system must stop supplying power to the utility within 2 seconds of the grid's failure.

10.12.4 Under or Over Voltage and Frequency

Normally, the inverter follows the utility voltage and frequency up to a certain limit but when these factors get abnormally beyond specified limits the power conditioner must take corrective measures to save personnel and connected equipment by disconnecting from the grid. These voltage limits along with disconnection times are listed in Table 10.3. However, a frequency deviation of ±1 Hz is allowed beyond which the inverter shall isolate itself from the utility in 0.2 second. This time is allocated to the inverter so that it does not respond to transients, or short time disturbances, and thereby disrupt smooth operation particularly in noisy grids.

Table 10.3 Trip time for abnormal voltages

Utility voltage	*Disconnection time* (seconds)
$V \geq 135\%$ of $V_{nominal}$	0.05
135% of $V_{nominal} \geq V \geq 110\%$ of $V_{nominal}$	2.0
110% of $V_{nominal} \geq V \geq 85\%$ of $V_{nominal}$	Continuous operation
85% of $V_{nominal} \geq V \geq 50\%$ of $V_{nominal}$	2.0
$V < 50\%$ of $V_{nominal}$	0.1

† V is the actual voltage of the utility at the point of connection.

†† $V_{nominal}$ is the normal utility voltage. In India it is 240 V.

From Table 10.3 it can be seen that the PV inverter should operate continuously if the utility voltage is between 85% and 110% of the nominal voltage. If the nominal grid voltage in Delhi is 240 V, then the grid connected inverter here should work smoothly between 85% of 240 V and 110% of 240 V. This works out to 204 V and 264 V. The inverter must cease to work within 0.05 second when the utility voltage equals 135% of 240 V, that is, 324 V.

10.12.5 Recovery Time after Disconnection

Once the inverter has tripped because of some reason cited above it is usually given enough time to sample the grid for quite a while before reconnecting to the grid. This time changes from place to place depending upon local conditions. It varies from 20 seconds to 5 minutes in which the utility voltage and frequency attain their nominal values and stabilize.

10.12.6 Earthing, Short Circuit Protection and Isolation

IEC 60364-7-712 gives the guidelines for earthing or grounding the utility interface equipment, which is the inverter. The same document also provides details of short circuit protection and a method of isolation and switching.

An exhaustive discussion on the subject of connection between the PV system and the grid is given in [19].

Questions

1. For a 1 kW grid interactive system operating from V_M = 96 V, calculate the area of the conductor used for interconnections according to the NEC. Use the wire table given in Chapter 11.
2. Draw a block diagram of the system of Question 1 and show the earth points. Also specify the fuse ratings.
3. Choose a solid state relay for the Question 1 to disconnect the system from the grid in the event of a reverse power flow situation. What will be the maximum time taken for cut-off of the relay?
4. If the system of Question 1 works for six hours a day and the house consumes five kWhr in twenty four hours, what will be the reduction in the electricity bill per month? Assume the cost of electricity to be ₹ 5.00 per unit (kWhr).
5. Design a 2 kW residential grid interactive inverter using PV AC modules. Draw the interconnections with the grid and the load. Compare this with the connections of Question 1. The two block diagrams should be compared to find the differences.
6. Select a microcontroller to generate a single phase AC wave form. Trigger the generation from a pulse to synchronize the signal with that of the grid.
7. A circuit is given in the text to stabilize the output voltage of the inverter using an AD 633 analogue multiplier. Can this process be realized using an OPAMP?
8. Why should so many IEEE and IEC standards be used for designing the grid interactive PCUs?
9. Can the PCU be disconnected from the grid without the Sandia Voltage Shift and the Sandia Frequency Shift methods?
10. Why should the negative of the PV array and that of the inverter be first connected together and then attached to Earth? What will happen if the Earth is connected at both the PV array end and the inverter negative separately?

11. What are the differences in the specifications of the IEC 61727 and the IEEE standard 929–2000? Make a comparative chart.

References

[1] IEEE Standard 929–2000 (Revision of IEEE Std. 929–1988), "IEEE Recommended Practice for Utility Interface of Photovoltaic (PV) Systems", IEEE Standards Coordinating Committee 21, Photovoltaics, 2000.

[2] UL Subject 1741, 'Standard for Power Conditioning Units for use in Residential Photovoltaic Power Systems', Underwriters Laboratories Inc., 1997.

[3] ANSI C84. 1–1995, Electric Power Systems and Equipment—Voltage Ratings (60 Hertz), American National Standards Institute, New York, 1995.

[4] Roger A. Messenger and Jerry Ventre, *Photovoltaic Systems Engineering*, Second edition, Chapter 8, CRC Press, Boca Raton, London, New York, Washington, D.C., 2005.

[5] Kern, G.A., Bonn, R.H., Ginn, J., and Gonzalez, S., Results of Sandia National Laboratories grid-tied inverter testing, *Procedings 2nd World Conf and Exhibition on PV Solar Energy Conversion*, Vienna, Austria, July 1998.

[6] Kern, G.A., "New Techniques for Islanding Protection," Interconnecting Small PV Systems to Florida's Utility Grid, A Workshop for Florida Utilities, Cocoa, FL, October 22, 1998.

[7] http://pm.geindustrial.com/FAQ/Documents/GGP/GER-2309A.pdf

[8] Satronix India Limited, www.satronixindia.com/pdf/analog_proportional_control_ssr.pdf.

[9] Daniel Persson, "Islanding detection in power electronic converter based distributed generation", Industriell Elektroteknik och Automation, Lunds Tekniska Högskola, Box 118, SE-221 00 LUND, Sweden.

[10] Z. Ye, R. Walling, L. Garces, R. Zhou, L. Li, and T. Wang, 'Study and Development of Anti-Islanding Control for Grid-Connected Inverters' General Electric Global Research Center, Niskayuna, New York, http://www.osti.gov/bridge.

[11] Byunggyu Yu a, Mikihiko Matsui b, Youngseok Jung a, Gwonjong Yu a, 'Modeling and design of phase shift anti-islanding method using non-detection zone', *Solar Energy*, vol. **81**, 2007, pp. 1333–1339.

[12] David F. Stout and Milton Kaufman, *Handbook of Operational Amplifier Circuit Design*, McGraw-Hill Book Company Inc., US, 1976.

[13] Chemmangot V. Nayar, Mochamad Ashari, and W.W.L. Keerthipala, 'A Grid-Interactive Photovoltaic Uninterruptible Power Supply System Using Battery Storage and a Back Up Diesel Generator', *IEEE Transactions on Energy Conversion*, vol. **15**(3), September 2000, p. 348.

[14] Joy Mazumdar, Issa Batarseh, Nasser Kutkut and Osman Demirci, 'High Frequency Low Cost DC-AC Inverter Design with Fuel Cell Source for Home Applications', 0-7803-7420-7/02/$17.00 © 2002 IEEE, p. 789.

[15] Haimin Tao, Jorge L. Duarte, Marcel A.M. Hendrix, "Control of Grid-Interactive Inverters as Used in Small Distributed Generators", 0197-2618/07/$25.00 © 2007 IEEE, p. 1574.

[16] K. Masoud and G. Ledwich, "Grid Connection without Mains Frequency Transformers", *JEEA*, Australia, vol. **19**(1), June, 1999, pp. 31–36.

[17] John Wiles, "Photovoltaic Power Systems and the 2005 National Electrical Code: Suggested Practices", ELECTRONIC VERSION 1.9, Dated November 26, 2008.

[18] Radhe Das, Stan Krauthamer and Alec Bulawka, 'Smart Integrated Power Circuit Technologies for Photovoltaic System Power Conditioning Applications', p. 1564. http://ieeexplore.ieee.org/stamp/stamp.jsp?arnumber=00656654.

[19] Remus Ieodorescu, Marco Liserre and Pedro Rodriguez, *Grid Converters for Photovoltaic and Wind Power Systems*, Wiley, U.K. 2011.

CHAPTER 11

Design of Transformers, Inductors and Line Filters

11.1 Magnetic Field in Coils

When a direct current (DC) is passed through an air-core coil of wire, a magnetic field is generated as shown in Figure 11.1.

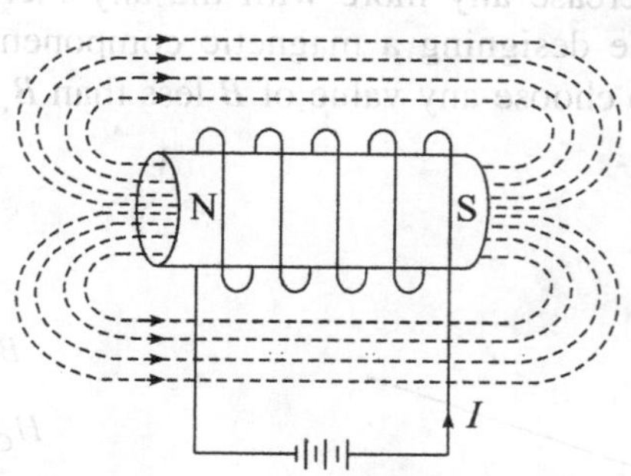

Figure 11.1 Magnetic lines of force produced by a DC source.

However, if the DC source is replaced by an alternating current generator, the poles of the magnet change places with the direction of the current and the picture looks like the one shown in Figure 11.2 with no fixed polarity on either side of the coil. Now if the field B, the flux density, is plotted against the magnetizing force H the resulting curve is linear, Figure 11.3, showing a lack of core saturation. This is the greatest advantage of the air core for which it is extensively used in radio frequency applications. However, the shortcoming of the air-core is that it is unable to produce a high flux density required for inductors and transformers for power applications as the size of the component will be unacceptably large. To achieve this objective a magnetic material of high permeability μ is introduced around which the wire is wound to carry current. The relation between the flux density B and the magnetizing force being:

$$B = \mu \times H \tag{11.1}$$

For vacuum, $\mu = \mu_o$ is called permeability of free space and is same for air or any non-magnetic material.

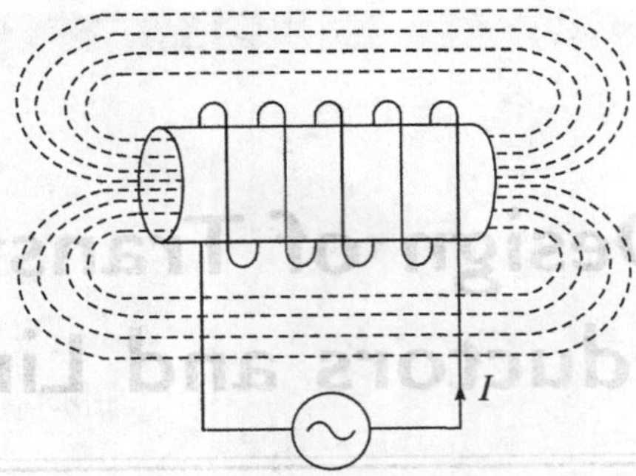

Figure 11.2 Magnetic lines of force produced by an AC source.

When a magnetic core is introduced, its own characteristics come into picture as can be observed from Figure 11.4. In this the *B–H* curve is plotted over an entire cycle for a neutral magnetic material. Starting from the origin *O* the field *H* is gradually increased and the corresponding value of *B* is plotted on the *Y*-axis. The flux density *B* increases linearly up to a certain value beyond which it changes its slope towards the horizontal axis and at the point B_s becomes parallel to the *X*-axis. This is called the saturation value of the flux density *B* after which it does not increase any more with the any increment in *H*. Care should be taken never to reach B_s while designing a magnetic component. In fact, a designer should exercise his own discretion to choose any value of *B* less than B_s and call it B_m, the maximum value of the flux density.

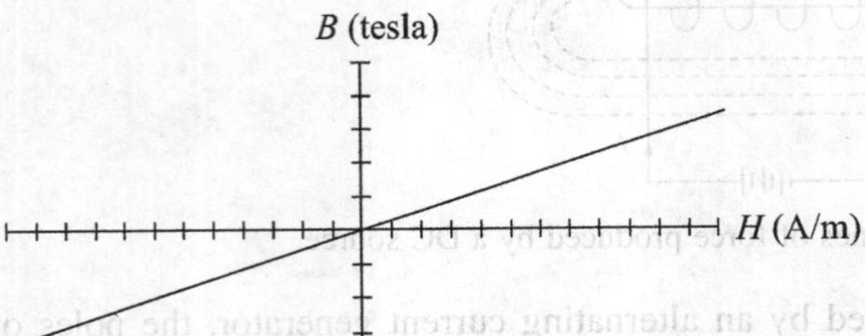

Figure 11.3 The relationship between *B* and *H* ($B = \mu_o H$) produced by an AC source.

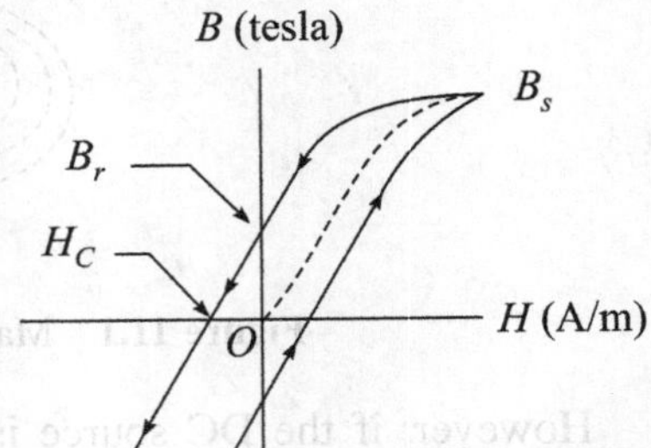

Figure 11.4 A typical hysteresis loop of a magnetic material.

The field *H* is reversed at B_s but the flux density now does not follow the dashed line it took from *O* to B_s but takes a different path and intersects the *Y*-axis at B_r when *H* is zero as the core still remains magnetized. B_r is called the *remnant flux*. If the magnetizing force *H* continues in the reverse direction to reduce B_r to zero, it cuts the *X*-axis at H_c, which is called the *coercive force*. For soft magnetic materials, used in transformers and inductors, H_c is small whereas for permanent magnets, it is large. On further increasing the value of $-H$

the saturation flux density on the reverse side, $-B_s$, is reached. Thus, a symmetrical hysteresis loop is generated that has a finite area which is proportional to the loss in the core.

When subjected to an alternating magnetizing force as encountered transformers, this curve is traversed many times per second depending on the frequency of the field. Thus, the losses of a transformer depend on the frequency. For 50 Hz operation a silicon steel core will suffice to contain the losses within reasonable limits, but if this material is subjected to higher frequencies in the range of kHz the losses will be so excessive that the component will not be usable. The area of the hysteresis loop also increases with the increase in frequency. In addition, *eddy currents* are set up inside the core which can acquire large values because of the low resistivity, ρ, of steel. It is necessary for high frequency operation that the hysteresis curve of the core material is narrow, that is the area of the loop is small, and its resistivity large. Ferrites fit such requirement but only for small power applications because of their low values of flux density B_m which is generally in the range of 300 to 400 millitesla with $\rho = 100\ \Omega$cm. For higher power applications amorphous materials, commercially known as *Metglas*, can be used as their B_m is approximately 1.2 T.

11.1.1 Transformers

If two coils of wire are wound on the came core and current is passed through one of them, a current is induced in the second one according to Faraday's laws. Michael Faraday had annunciated two laws given below:

The first law states that when the magnetic flux linked with a wire changes an electromotive force (emf) is introduced in it.

The second law states that the magnitude of the induced emf is proportional to the rate of change of the flux.

Suppose a coil of wire of N turns is placed in a magnetic field. If the flux changes from ϕ_1 Wb to ϕ_2 Wb in time t seconds, then the initial and final flux linkages are $N\phi_1$ and $N\phi_2$. Therefore the induced emf is:

$$e = \frac{N\phi_2 - N\phi_1}{t}\ \text{Wb/s or V} \tag{11.2}$$

or,

$$e = \frac{N(\phi_2 - \phi_1)}{t}\ \text{V} = \frac{d}{dt}(N\phi) = -N\frac{d\phi}{dt}\ \text{V} \tag{11.3}$$

The negative sign is put in Equation (11.3) to indicate that the induced emf sets up a current in such a direction that the magnetic field produced by it opposes the very cause that produced it. This effect is called *Lenz's law.*

A simple air core transformer is shown in Figure 11.5. An alternating current (AC) is passed through the first coil, labelled as the primary winding, which generates a magnetic field as described above. Some of the lines of force, called *mutual flux*, cut the secondary winding coils and generate a current there. However, because of the separation between the two some of the flux is not coupled to the secondary and does not add its contribution to the induced current. This is called the *leakage flux.* There are losses in a transformer due to hysteresis and eddy currents and the leakage flux also has its share. Apart from the core losses there are the copper losses which are due to the heat generated by the primary and the secondary currents flowing through the respective copper wires. If the two currents are I_1 and I_2, in amperes, and the resistances of the wires are R_1 and R_2, in ohms, the copper loss is given by:

$$\text{Loss} = (I^2R) = (I_1^2R_1 + I_2^2R_2)\ \text{W} \tag{11.4}$$

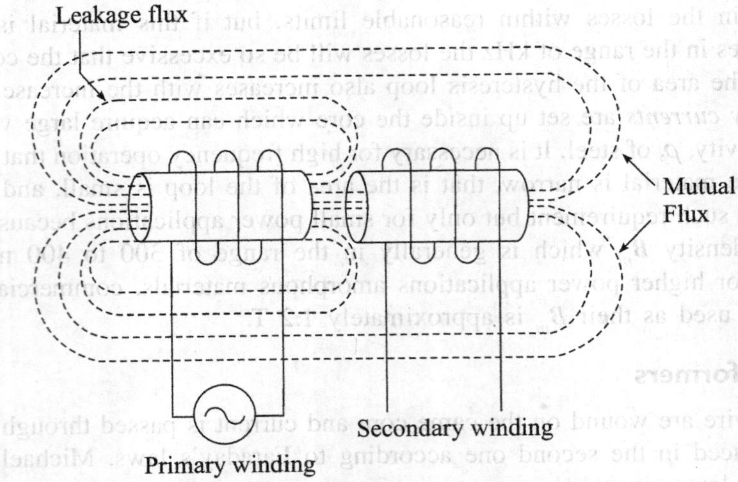

Figure 11.5 A simple air-core transformer.

To keep the temperature of the device low the area of the copper should be large enough to carry the current. According to the British system the current carrying capacity of copper is 1200 A/inch^2 which works out to 186 A/cm^2. The current that can rupture the wire, that is called its fusing current, is forty times this value. Based on these properties the equivalent circuit of the transformer is given in Figure 11.6.

P and *S* in Figure 11.6 denote the primary and secondary windings of an ideal transformer with no losses or leakage. However, a practical device has losses as mentioned above which have to be shown in appropriate places. R_1 and L_1 represent the resistance of the primary winding and the inductance due to leakage flux which carry the input current I_1. This current is then bifurcated with I_o going through *R* and *L*. *R* represents the hysteresis and eddy current losses, carrying current I_{o1}. The inductive reactor *L* is such that it takes a reactive current equal to the magnetizing current I_{o2} of the actual transformer. The rest of the current,

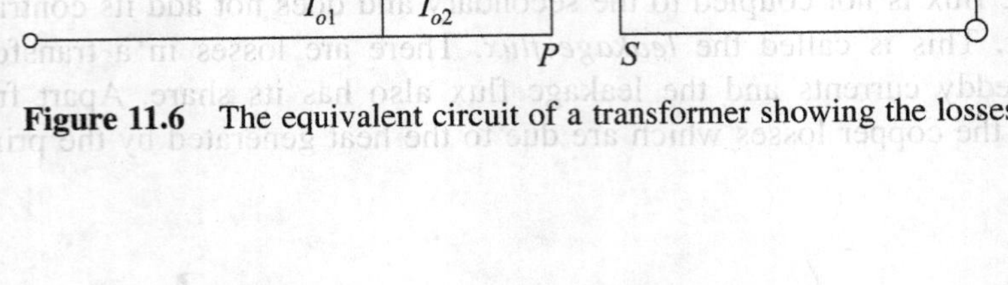

Figure 11.6 The equivalent circuit of a transformer showing the losses.

$I_1 - I_o$, goes to P. Similarly, the output current I_2 passes through L_2, leakage inductance, and the secondary winding resistance R_2 to give an output voltage V_2. When connected to a load V_2 will reduce because of the voltage drop equal to $I_2 \times R_2$, with consequent production of heat equivalent to $I_2^2 \times R_2$.

11.1.2 Design of Transformers

Transformers generally encountered in photovoltaic work operate at high frequencies. The normal ones for 50 Hz operation are available in the market but those needed for high frequency work have to be mostly fabricated. For this purpose, a working knowledge of the design work with available components is extremely necessary. With the design procedure given below transformers of any frequency can be designed. There are two ways to calculate the values of a transformer of given specifications.

First method

The area of the core window W and the core area A_c of a transformer core are shown in the Figure 11.7. A_c is the product of E and F, that is:

$$A_c = E \times F \tag{11.5}$$

The window area is given by:

$$W_a = [(B - E)/2]\, D \tag{11.6}$$

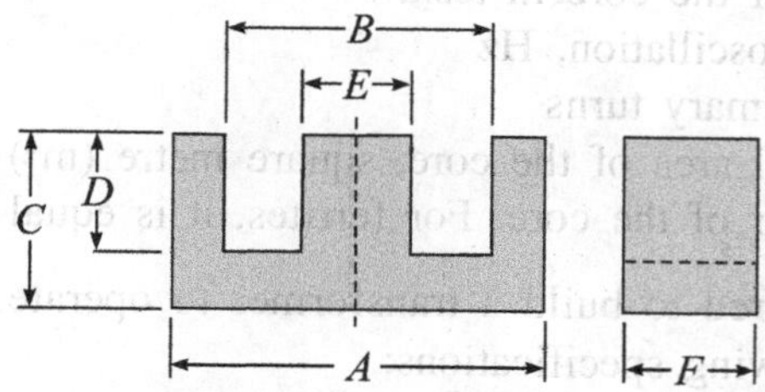

Figure 11.7 The dimensions of an E-core for calculating the area product A_p.

A necessary step in this procedure is to estimate the ratio of the total wire area to the core window area W. This ratio is the K factor, which can be written as [1]:

$$K = \frac{XN_1A_{w1} + YN_2A_{w2}}{W_a} \tag{11.7}$$

where

K = ratio of the total wire area to core window area W, also called the *window utilization factor*
A_{w1}, A_{w2} = area of a single strand of wire in m²
N_1 = number of turns of the primary winding
N_2 = number of turns of the secondary winding
W_a = area of the core window in m²
X = number of primary windings. For centre-fed primary $X = 2$
Y = number of secondary windings

If the turns ratio is T, that is $N_2 = TN_1$, then N_2 can be replaced in Equation (11.7) to give

$$K = \frac{XN_1A_{w1} + YTN_1A_{w2}}{W_a} \tag{11.8}$$

If the number of layers or windings of the primary and secondary is one, then putting $X = Y = 1$ in Equation (11.8) we get Equation (11.9):

$$K = (N_1 A_{w1} + T N_1 A_{w2})/W_a = N_1 (A_{w1} + TA_{w2})/W_a \tag{11.9}$$

The K factor is taken as 0.2 to assure ease in winding the transformer with its bobbin taking up some space. Separating the known and unknown quantities, Equation (11.8) results in:

$$\frac{N_1}{W_a} = \frac{K}{A_{w1} + TA_{w2}} \tag{11.10}$$

The basic transformer expression is given by:

$$V = K_f \times B \times f \times N_1 \times S \times A_c \tag{11.11}$$

where

K_f = 4 for square wave and 4.44 for sine wave.
V = root mean square (rms) voltage impressed across the core primary winding N_1 in volts
B = flux density of the core in tesla
f = frequency of oscillation, Hz
N_1 = number of primary turns
A_c = cross sectional area of the core, square metre (m^2)
S = stacking factor of the core. For ferrites, it is equal to 1.

Example 11.1: It is required to build a transformer to operate a high intensity discharge (HID) lamp with the following specifications:

1. Frequency of operation $f = 60 \times 10^3$ Hz
2. RMS voltage across the primary $V = 266.6$ V
3. Primary winding current (rms) $I_p = 6.7$ A
4. Lamp current (rms) flowing through the secondary winding $I_s = 4$ A
5. Window utilization factor $K = 0.2$
6. Maximum flux density $B = 0.2$ T
7. The transformer has a turns ratio $T = N_2/N_1 = 1.68$.

Solution: Since the only unknowns in the basic transformer expression of Equation (11.11) are the number of primary turns N_1 and the area of the core A_c, Equations (11.9) and (11.11) can be solved simultaneously for the product of W and A_c:

$$W_a A_c = V (A_{w1} + 1.68A_{w2})/4.44KBf \tag{11.12}$$

The nearest gauge of wire which can carry the primary current of 6.7 A (rms) is no. 13 SWG (standard wire gauge) with a capacity to carry 7.978 A (rms). Therefore:

A_{w1} = area of the primary winding wire of no. 13 SWG = 4.289 mm^2 = 4.289 × 10^{-6} m^2. Hence two strands of Litz wire of 4 A (rms) were used to increase the primary current. As the operating frequency increases the current tends to flow near the surface of a conductor because of "skin effect" thus reducing the effective area of the wire. In such cases a large number of very thin wires are used instead of one thick wire. Such a bunch of insulated (enameled) wires, each in the range of 42 to 48 SWG, and put together in a single outer cotton or polyester insulating sleeve is called *Litz wire*. The depth of penetration of the current into copper is given by:

$$d = \frac{75}{\sqrt{f}} \text{ mm} \tag{11.13}$$

where f is the frequency in hertz (Hz). The penetration depth in different metals is shown in Figure 11.8. Another way to wind a high frequency transformer is to use thin copper strips of necessary thickness, 0.1 mm is sufficient in most cases, and of sufficient width to provide the desired area of cross section to carry the required current.

A_{w2} = area of the secondary winding wire = 2.627 mm^2 = 2.627 × 10^{-6} m^2 is required to carry 4.884 A of secondary current.

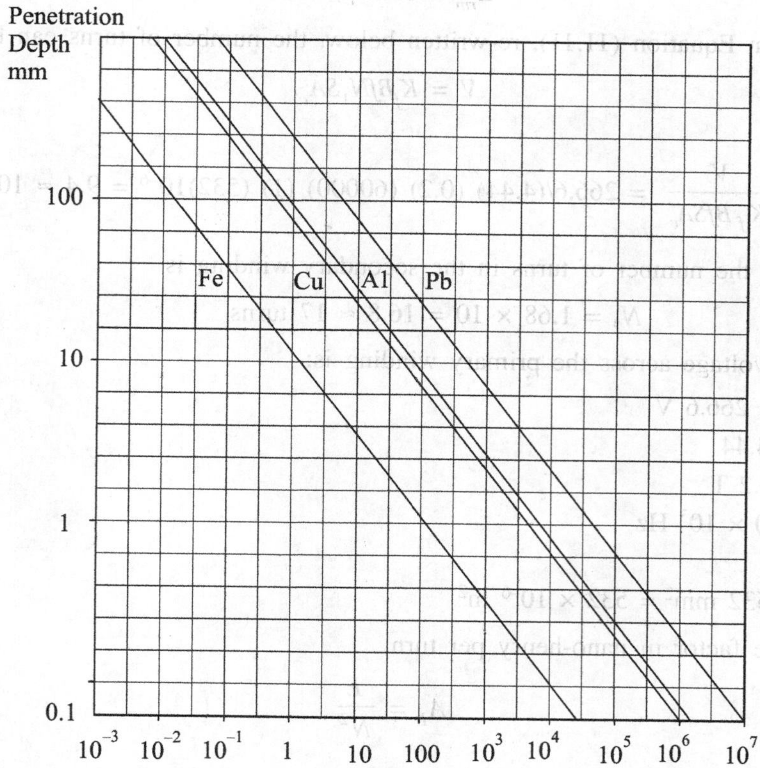

Figure 11.8 The variation of depth of penetration of signals in different metals with frequency.

Therefore,

$$W_a A_c = 267(4.289 + 1.68 \times 2.627)10^{-6}/(4.44)\ (0.2)\ (0.2)\ (60)10^3$$
$$= 218 \times 10^{-9}\ \text{m}^4 = 218000\ \text{mm}^4$$

From the data sheet of COSMO, given in Table 11.1 (see the Appendix to this chapter), for the single ferrite core EE 6527:

Area of the core $A_c = 532\ \text{mm}^2$, and

Window area of the core $W = 268.6\ \text{mm}^2$

To find the window area required for the windings of the transformer, 218000 mm^4 is divided by the core area of EE 6527. Thus,

$$W_a = 218000/532 = 409.9\ \text{mm}^2 \approx 410\ \text{mm}^2$$

Since the window area of this core is only 268.6 mm^2, which is less than that needed, two cores are added in an EE configuration to double the area to 537.2 mm^2.

To determine the inductance factor A_L of the core, to avoid saturation and the number of turns for the primary and the secondary windings it is noted that the primary inductance value calculated is:

$$L_{rm} = 63.9\ \mu\text{H}$$

Again from Equation (11.11), re-written below, the number of turns can be found.

$$V = K_f B f N_1 S A_c,$$

hence

$$N_1 = \frac{V}{K_f B f S A_c} = 266.6/(4.44)\ (0.2)\ (60000)\ (1)\ (532)10^{-6} = 9.4 \approx 10\ \text{turns}$$

Therefore, the number of turns in the secondary winding is

$$N_2 = 1.68 \times 10 = 16.8 \approx 17\ \text{turns}$$

Here the voltage across the primary winding is:

1. $V_{rms} = 266.6$ V
2. $K_f = 4.44$
3. $B = 0.2$ T
4. $f = 60 \times 10^3$ Hz
5. $S = 1$
6. $A_c = 532\ \text{mm}^2 = 532 \times 10^{-6}\ \text{m}^2$

Inductance factor in nano-henry per turn,

$$A_L = \frac{L}{N^2} \tag{11.14}$$

where

L = inductance in nanohenrys
N = number of turns

Therefore, using the number of primary turns $N_1 = 10$

$$A_L = 63900/(10)^2 = 639 \approx 640 \text{ nH/turn}$$

The maximum voltage swing permissible at the gate of the 2SK 1512, used in the oscillator is ± 20 V. To calculate W_3 and W_4 windings to drive the MOSFETs the drive should therefore be restricted to ± 15 V or 10.6 V (rms) in terms of AC. Since the voltage across the primary of the transformer is 266.6 V (rms) the turns-ratio is:

$$T_{P-W_3,W_4} = 266.6/10.6 = 25.2$$

Thus, the number of W_3 and W_4 turns is = 10/25.2 = 0.397 ≈ 0.4. One turn was provided for each of the two windings. This will give a drive signal of 25 V (rms) at the gates of the MOSFETs that make the oscillator. The gate of a MOSFET can withstand a maximum voltage of ± 20 V. In order to provide safety to the transistors two 15 V zener diodes with a 100 Ω resistor can be placed at each transistor's gate.

The final composition of the transformer is given below:

1. Core = COSMO ferrite EE 6527, $A_L = 640$
2. Primary winding = Two strands of 4.884 A of Litz wire
3. Number of primary turns $N_1 = 10$
4. Secondary winding = Single strand of 4.884 A of Litz wire
5. Number of secondary turns $N_2 = 17$
6. Number of turn for the drive windings W_3 and $W_4 = 1$

The power consumed by W_3 and W_4 is negligible, as compared to the 400 W needed for the lamp, because of the high input impedance of the MOSFETs which is mainly capacitive. This will not put any appreciable burden on the transformer.

Second method

A second method [2] of calculating the transformer details is as follows:

Since the power transformer has to be designed to accommodate the primary power P_{in} and the secondary winding power P_o then the apparent power:

$$P_t = (P_{in} + P_o) \text{ W} = V_{in} \times I_{in} \cos\phi + V_{out} \times I_{out} \cos\phi \quad (11.15)$$

P_{in} can be expressed as $P_{in} = P_o/\eta$ W, where η is the transformer efficiency. When a low power factor cos ϕ exists, the total *VI* capacity of the transformer is taken for calculations which consist of the sum of the primary $V \times I$ and the secondary $V \times I$. Hence the transformer windings will be sized according to the secondary current and the dependent primary current. The primary turns can be expressed using Faraday's law:

$$N_p = \frac{V_p}{K_f B A_c f} \text{ turns} \quad (11.16)$$

where

N_p = number of primary turns
A_c = area of the core in m²
V_p = rms voltage for sine wave across the primary winding

K_f = waveform coefficient = 4 for square wave and 4.44 for sine wave
B = magnetic flux density in tesla
f = frequency in Hz

The winding area is fully utilized when

$$KW_a = N_p A_{wp} + N_s A_{ws} \tag{11.17}$$

where

K = window utilization factor = ratio of the total wire area A_w to core window area W_a
W_a = core window area in m²
N_p = number of primary turns
A_{wp} = wire area of primary in m²
N_s = number of secondary turns
A_{ws} = wire area of secondary winding in m²

By definition

$$A_w = \frac{I}{J} \tag{11.18}$$

where

I is the rms current
J is the current density in A/m²

Rearranging Equation (11.17), we get

$$K\,W_a = N_p \frac{I_p}{J} + N_s \frac{I_s}{J} \tag{11.19}$$

From Equations (11.16) and (11.19), we get

$$K\,W_a = \left(\frac{V_p}{K_f B A_c f}\right)\left(\frac{I_p}{J}\right) + \left(\frac{V_s}{K_f B A_c f}\right)\left(\frac{I_s}{J}\right) \tag{11.20}$$

Rearranging Equation (11.20), we obtained

$$W_a A_c = \frac{V_p I_p + V_s I_s}{K\,K_f BfJ} \tag{11.21}$$

The current density J = 1200 A/inch² in the British FPS system, this is equal to 186 A/cm² = 186 × 10⁴ A/m².

In the present case, that is the design of the transformer for the HID lamp:

$$f = 60 \times 10^3 \text{ Hz}$$

$$V_p \text{ (rms)} = 266.6 \text{ V}$$

$$K = 0.2$$

$$B = 0.2 \text{ T}$$

$$J = 186 \times 10^4 \text{ A/m}^2$$
$$K_f = 4.44$$
$$I_p = 6.68 \text{ A}$$
$$I_s = 4 \text{ A}$$
$$V_s \text{ (rms)} = 448 \text{ V}$$

Turns ratio of the secondary and primary $N_s/N_p = 1.68$

Substituting these values in Equation (11.21), we get

$$W_a A_c = \frac{266.6 \times 6.68/0.9 + 448 \times 4}{0.2 \times 4.44 \times 0.2 \times 60000 \times 186 \times 10^4}$$

$$= 3770.76 \times 10^{-4}/1982016 = 0.0019 \times 10^{-4} \text{ m}^4$$

$$= 190000 \text{ mm}^4$$

The factor of 0.9 in the numerator above is the transformer efficiency.

$$A_c = 532 \text{ mm}^2 \text{ for EE 6527}$$

Therefore,

$$W_a = 190000/532 = 357.14 \text{ mm}^2$$

This is larger than that listed for this core, which is 268.6 mm^2. Therefore, two cores forming an EE structure has been chosen.

To calculate the number of turns in the primary winding, the values are substituted in Equation (11.16), that is

$$N_p = \frac{V_p}{K_f B A_c f} \text{ turns}$$

$$N_p = \frac{267}{4.44 \times 0.2 \times 60000 \times 532 \times 10^{-6}} = 9.42 \text{ turns}$$

Therefore, $N_s = 1.68\ N_p = 1.68 \times 9.42 = 15.82 \approx 16$ turns

here the voltage across the primary winding is:

1. V (rms) $= 266.6$ V ≈ 267 V
2. $K_f = 4.44$
3. $B = 0.2$ T
4. $f = 60 \times 10^3$ Hz
5. $A_c = 532$ mm $^2 = 532 \times 10^{-6}$ m^2

Inductance factor in nanohenry per turn,

$$A_L = L/N^2 \tag{11.22}$$

where

L = inductance in nanohenrys = 63900 nH (63.9 μH)
N = number of turns

Therefore,

$$A_L = 63900/(9.4)^2 = 723.1$$

A detailed calculation involving the regulation, temperature rise and losses in the transformers is given in reference [2]. The SWG copper wire data are given in Table 11.2 (see the Appendix to this chapter) and the data for the ferrite EE core are tabulated in Table 11.1.

Example 11.2: It is required to design a transformer for a solar lantern with the following specifications:

1. RMS voltage across the primary winding = 26.6 V
2. Operating frequency = 30,000 Hz
3. Primary current (rms) = 1.8 A
4. Secondary voltage (rms) = 250 V
5. Secondary current (rms) = 0.16 A.

Solution: We take

1. Window utilization factor $K = 0.4$
2. Waveform coefficient $K_f = 4.44$
3. Flux density $B = 0.3$ T
4. Current density $J = 186 \times 10^4$ A/m²
5. Transformer efficiency $\eta = 80\% = 0.8$

From Equation (11.21) the area product

$$A_p = W_a \times A_c = \left(\frac{\dfrac{250\times 0.16}{0.8} + 250\times 0.16}{0.4\times 4.44\times 0.3\times 30{,}000\times 186\times 10^4} \right) \text{m}^4$$

or

$$W_a \times A_c = 3.00 \times 10^{-9} \text{ m}^4 = 3000 \text{ mm}^4$$

The result has to be converted to mm⁴ since the data given in the manufacturers' catalogue are in this unit. From Table 11.3 (see the Appendix to this chapter) it is found that the pot core 26 × 16 has the area product A_p = 5015 mm⁴. Its A_c = 94 mm² and W_a = 53.35 mm².The number of primary turns is calculated from Equation (11.16). Substituting the values in this expression:

$$N_p = \frac{V_p}{K_f B\, A_c f} \text{ turns} = 26.65/4.44 \times 0.3 \times 94 \times 10^{-6} \times 30{,}000 = 7.1 \approx 8 \text{ turns}$$

Since the transformer will be used in a push-pull oscillator, it will have a centre tap in the primary for DC feed. Hence the two sides of the centre tap should be symmetrical and closely coupled. The total primary turns are divided in equal halves of four turns each which are then bifilar wound. The number of secondary turns will be:

$$N_s = 8 \times 250/26.6 = 75 \text{ turns}$$

For feedback it will suffice to provide one turn for each transistor base.

To calculate the A_L value of the core, the inductance of the primary winding must be calculated. Since a current of 1.8 A_{rms} must pass through the coil when 26.6 V_{rms} is impressed across it the reactance

$$X_L = 26.6/1.8$$

Therefore,

$$L = \frac{26.6}{1.8 \times 2\pi \times 30,000} = 78\ \mu\text{H} = 78000\ \text{nH}$$

and $A_L = 78000/8^2 = 1219$. However, the standard A_L value available closest to this is 1000 nH/Turn. The turns may now be re-calculated with this value of A_L. The diagram of the transformer is shown in Figure 11.9(a).

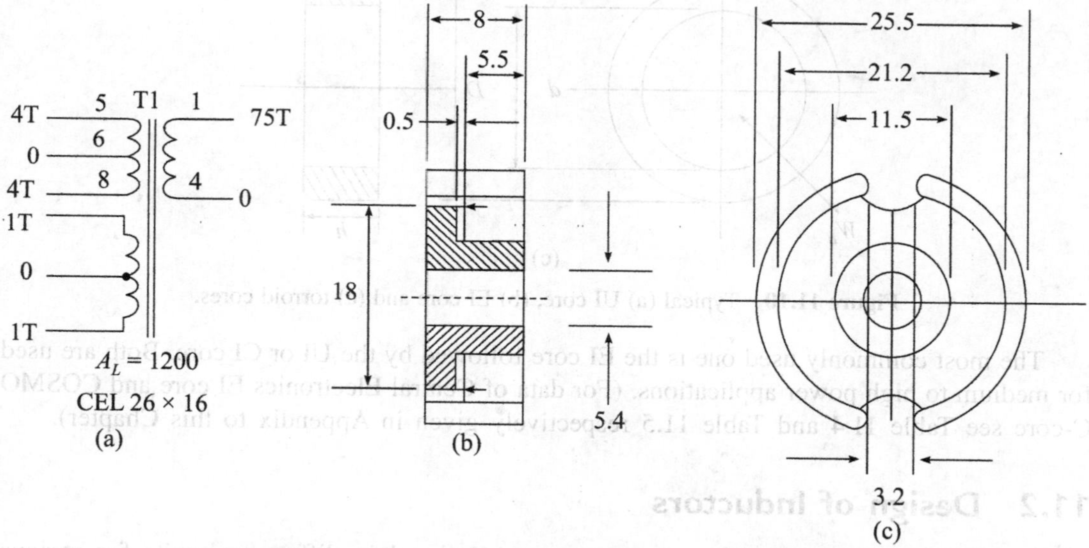

Figure 11.9 The diagram of a 26 × 16 Pot core (all the dimensions are in mm).

Figure 11.9 shows the diagram of a pot core used in the calculations. Only one half of the core is shown here whereas another similar half is attached to it either by a clamp or a non magnetic screw-nut made of brass or nylon. The air gap that determines the A_L value is made by filing off the central stem. Cores of some other geometry are also shown in Figure 11.10.

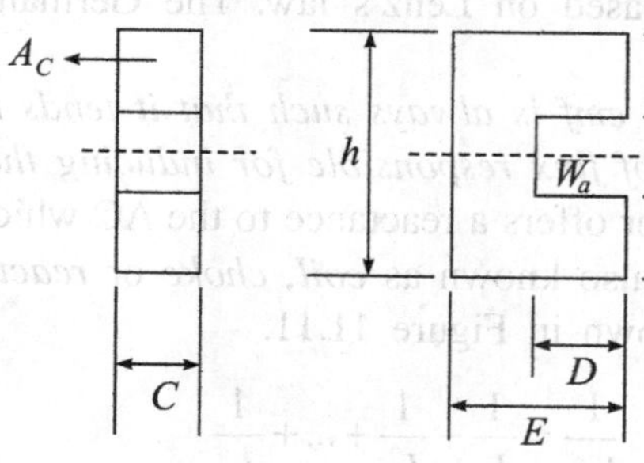

(a)

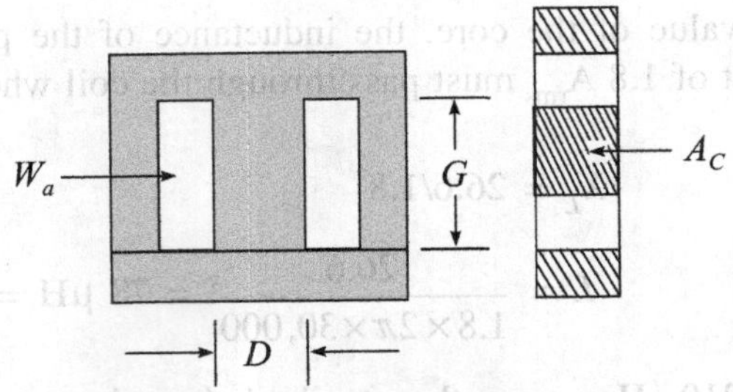

(b)

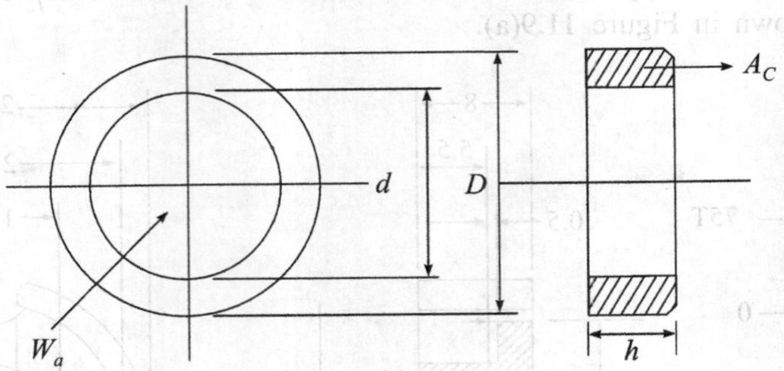

(c)

Figure 11.10 Typical (a) UI core, (b) EI core and (c) torroid cores.

The most commonly used one is the EI core followed by the UI or CI core. Both are used for medium to high power applications. (For data of Central Electronics EI core and COSMO C-core see Table 11.4 and Table 11.5 respectively given in Appendix to this Chapter).

11.2 Design of Inductors

Inductors are required to offer reactance to an AC signal in different circuits for various reasons. In current fed oscillators they limit the current to the oscillator which contains both AC and DC currents whereas in fluorescent lamps they control the AC current to the lamp so that it is not damaged. In this case it acts as a stabilizer or ballast and is commonly known as *choke*. Thus, the inductor in its many capacities has to perform with pure AC or a mixture of AC and DC currents.

Operation of inductors is based on Lenz's law. The German physicist Heinrich Lenz expressed this rule thus:

The direction of an induced emf is always such that it tends to set up a current which opposes the motion or change of flux responsible for inducing that emf.

As a consequence an inductor offers a reactance to the AC which induces a flux opposing the AC current. An inductor is also known as *coil*, *choke* or *reactor*. These when placed in parallel add up inversely as shown in Figure 11.11.

$$\frac{1}{L_{eq}} = \frac{1}{L_1} + \frac{1}{L_2} + \ldots + \frac{1}{L_n}$$

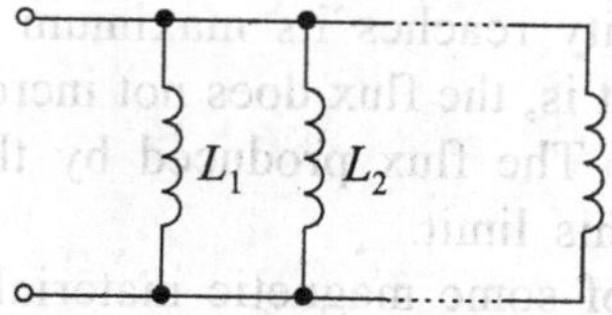

Figure 11.11 Inductors placed in parallel.

However, they add up when placed in series as shown in Figure 11.12.

$$L_{eq} = L_1 + L_2 + \ldots + L_n$$

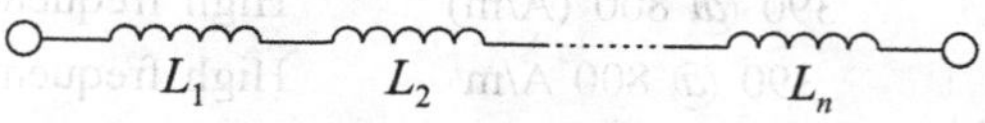

Figure 11.12 Inductors placed in series.

The energy E stored in an inductor is given in Equation (11.23), where L is the inductance in henry and I is the current in amperes.

$$E = \frac{1}{2} LI^2 \tag{11.23}$$

The reactance of the inductor is given by $X_L = -j\omega L$, where $j = \sqrt{-1}$ and $\omega = 2\pi f$. Therefore,

$$|X_L| = 2\pi f L \tag{11.24}$$

where f is the frequency of the signal.

The design of an inductor depends on the four factors listed as follows:

1. Required inductance
2. Alternating current (AC)
3. Direct current (DC)
4. Power loss and therefore rise in its temperature.

It is necessary that the core on which the coil is wound should not go into saturation otherwise the value of the inductance will drastically reduce. Figure 11.13 shows that when

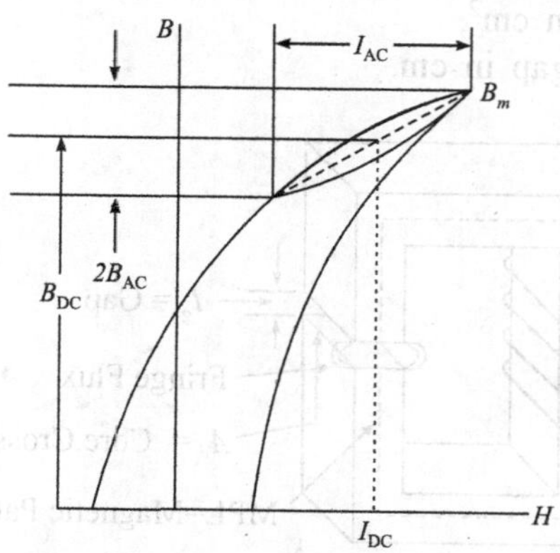

Figure 11.13 Variation of flux density with current showing core saturation.

$I_{DC} + I_{AC}$ increases the flux density reaches its maximum value of B_m beyond which the B–H curve becomes horizontal, that is, the flux does not increase with an increase in current. This is known as core saturation. The flux produced by the sum of the DC and the AC currents should stay well within this limit.

The maximum flux densities of some magnetic materials are given in Table 11.6.

Table 11.6 Maximum flux densities of commonly used magnetic materials

Material	*Max. flux density* B_m *in millitesla* (mT)	*Comments*
CEL HP3B Ferrite[a]	390 @ 800 (A/m)	High frequency, Low to medium power
CEL HP4 Ferrite[a]	390 @ 800 A/m	High frequency, Low to medium power
CEL HP5 Ferrite[a]	390 @ 800 A/m	High frequency, Low to medium power
Metglas[b] ***MICROLITE®***	1560	High frequency, Low to high power
Silicon Steel[c]	1200	Low frequency, Low to high power

(a) Professional Ferrites Catalogue 94, Central Electronics Limited, 4, Industrial area, Sahibabad – 201010, Ghaziabad, UP, India.
(b) www.metglas.com/downloads/microlite.pdf
(c) R. Lee, *Electronic Transformers and Circuits*, 2nd edition, John Wiley & Sons, New York, 1958.

The actual values taken for calculations are usually less than those shown to lower the losses and contain excessive rise in temperature.

To avoid core saturation an air gap is introduced in the core, as shown in Figure 11.14, which effectively tilts the hysteresis loop so that the core goes into saturation at a higher magnetizing force caused by a larger current through the windings. This becomes necessary particularly in square loop materials like ferrites. The inductance of a coil wound on a gapped or cut core is given by:

$$L = \frac{0.4\pi N^2 A_c \times 10^{-8}}{l_g} \text{ H} \tag{11.25}$$

where

N = number of turns
A_c = area of the core in cm^2,
l_g = length of the air gap in cm.

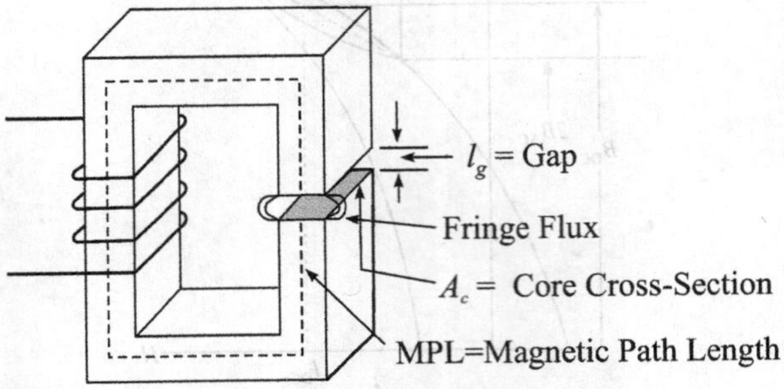

Figure 11.14 The parameters of a cut core are shown.

These parameters are illustrated in Figure 11.14. Some magnetic flux bulges out at the air gap and is called *fringe flux* because there is no magnetic material there to conduct it. While mounting the core care must be taken to use a non magnetic clamp otherwise the gap will be short circuited. After having decided on the value of the inductance needed, the energy in the inductor is calculated from Equation (11.23), that is

$$\text{Energy} = \frac{LI^2}{2} \text{ W s}$$

Then, the area product of the core is determined from:

$$A_p = \left(\frac{2(\text{Energy}) \times 10^4}{BKJ}\right) \text{cm}^4 \tag{11.26}$$

where

B = flux density (T)
K = window utilization factor (0.2 to 0.4)
J = current density (A/cm²)

The number of turns of wire required for the inductance is given by [2]

$$N = \frac{W_{a\text{eff}} \times S_2}{A_W} \approx \frac{W_a \times S_2}{A_W} \text{ turns} \tag{11.27}$$

Substituting the values of N, A_c (= A_p/W_a) and L in Equation (11.25) the air-gap length l_g can be found. However, cores are sold in the market by their A_L value which is the inductance factor of the core and is given in nanohenrys per turn [3], [4] by the following expression:

$$A_L = L/N^2$$

Example 11.3: It is required to make an inductor for a maximum power point tracker of the following specifications:

1. Frequency of operation = 10 kHz
2. Inductance = 450 μH
3. DC current carrying capacity = 4 A
4. AC current = 20% of DC current = 0.8 A (peak to peak)

Because a high frequency of 10 kHz is being filtered a ferrite core is preferred.

Solution: The current can be determined from Figure 11.13 so that core saturation does not occur.

$$I = I_o + I_{AC}/2$$
$$= 4 + 0.8/2 \text{ A} = 4.4 \text{ A}$$

Therefore, from Equation (11.23),

$$\text{Energy} = \frac{LI^2}{2} \text{ W s}$$

$$= \frac{450 \times 10^{-6} \times 4.4^2}{2}$$

$$\text{Energy} = 0.0044 \text{ Ws}$$

Calculating the area product from Equation (11.26) with B = 0.3 T, K = 0.3 and J = 186 A/cm^2

$$A_p = \left(\frac{2(\text{Energy}) \times 10^4}{BKJ}\right) \text{cm}^4$$

$$= (2 \times 0.0044 \times 10^4)/(0.3 \times 0.3 \times 186) = 5.26 \text{ cm}^4$$

From Table 11.1 (Appendix), the ferrite core EE 5521 is selected. This core has the following parameters:

A_p = 13.24 cm^4, W_a = 3.74 cm^2, A_c = 3.54 cm^2. From this data the number of turns of wire that can be accommodated in the core can be found using Equation (11.27)

$$N = \frac{W_{a\text{eff}} \times S_2}{A_W} \approx \frac{W_a \times S_2}{A_W} \text{ turns} = 3.74 \times 0.3/0.02627 = 42.7 \approx 43 \text{ turns}$$

A_W Area of wire in cm^2.

Now, the inductance factor can be calculated from the relation

$$A_L = L/N^2 = 450 \times 10^{-6}/43^2 = 243 \text{ nH/turn.}$$

A standard value for the core is

$$A_L = 250 \text{ nH/turn.}$$

Core EE4220 will also work but it will heat up.

Example 11.4: A coupled inductor required for a current-fed oscillator has to be designed such that there are two windings of 250 μH each wound on the same core so that they are coupled with each other as shown in Figure 11.15.

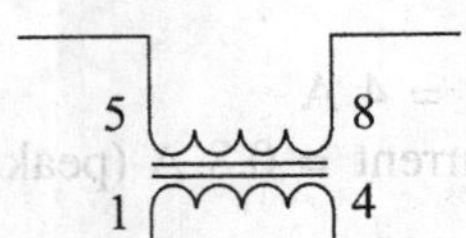

Figure 11.15 The coupled inductor for a current fed oscillator.

Specifications of the inductor:

Frequency of operation = DC + 60 kHz

$$L = 250 \text{ μH}$$

$$I_{DC} = 1.125 \text{ A}$$

$$V_{DC} = 85 \text{ V (min) and } 200 \text{ V (max)}$$

Solution: Take $B = 0.2$ T

Current through the inductor = I_{DC} + 20% of I_{DC} = 1.125 + 20% of 1.125 = 1.35 A [1]

From Equation (11.23),

$$\text{Energy} = \frac{LI^2}{2} = (250 \times 10^{-6}) \times (1.35)^2/2 = 227.8 \times 10^{-6} \text{ Ws or J}$$

Area Product, $A_p = W_a \times A_c$

From Equation (11.26), putting $J = 186$ A/cm² we get,

$$A_p = \frac{2 \times \text{Energy} \times 10^4}{BKJ} = \frac{2 \times 227.8 \times 10^{-6} \times 10^4}{0.2 \times 0.4 \times 186} = 0.306 \text{ cm}^4$$

However, there are two windings of 250 μH each, therefore the area product is twice the above value, that is, $A_p = 0.612$ cm⁴. Since the frequency of operation is 60 kHz the core material must either be ferrite or amorphous material commercially known as *metglas*. Ferrite has been chosen being a cheaper option for low power.

The nearest core with area product higher than this is EE 4012 with $A_p = 1.29$ cm⁴ as obtained from Table 11.1 (Appendix).

Hence, the number of turns N which the core can accommodate is given by Equation (11.27)

$$N = \frac{W_{a\text{eff}} \times S_2}{A_W} \approx \frac{W_a \times S_2}{A_W} = \frac{0.84 \times 0.4}{0.0081} = 41.5 \text{ turns} \approx 42 \text{ turns}$$

From EE4012 ferrite core data, window area $W_a = 0.84$ cm²; $S_2 = K = 0.4$; A_W = wire area = 0.0081 cm² for 1.5 A wire from Table 11.2 (Appendix).

Thus, the number of turns which the core can accommodate, $N = 42$ turns.

Since two windings of 250 μH each have to be wound on the same core on a split bobbin each coil should have 21 turns.

Hence, the inductance factor of the core

$$A_L = L/N^2 = 250 \times 10^{-6}/21^2 = 566 \text{ nH/turn}$$

The standard core available had an inductance factor of $A_L = 360$ nH/turn. So the number of turns works out to $N = 26$. This means that a thinner wire has to be used for accommodating the turns. The next lower current capacity wire is of size 20 SWG which takes 1.22 amperes. That means the heating will be slightly more due to copper losses. Ideally a copper ribbon of 0.1 mm or some other suitable thickness should be used for winding such inductors. EE3512 is also suitable.

11.2.1 Inductance of a Coil or Solenoid [5]

Single layer air core coils

Consider a long thin solenoid consisting of a single layer of strip wound such that the distance between each turn is negligibly small. Such an arrangement approximates an ideal "current sheet" in which the current density is uniform over the entire surface of the winding. The inductance of this coil is given by:

$$L=\frac{4\pi N^2 A}{l}\times 10^{-9}\ \text{H} \tag{11.28}$$

where

A = the area of the coil in cm^2
N = the number of turns of wire
l = mean path length of the magnetic circuit (length of coil) in cm

If the cross section is circular, with a radius r cm, the area will be (πr^2), and the inductance is then

$$L=\frac{4\pi^2 r^2 N^2}{l}\times 10^{-9}\ \text{H} \tag{11.29}$$

As the coil is made shorter the field inside is no longer uniform and the inductance becomes a function of the ratio of length to radius. Equation (11.29) must then be multiplied by a correction factor K_n, the *Nagaoka constant*, whose value is [6]

$$K_n=\frac{1}{1+0.9\left(\frac{r}{l}\right)-0.02\left(\frac{r}{l}\right)^2} \tag{11.30}$$

Multilayer coils

A typical multilayer solenoid is shown in Figure 11.16

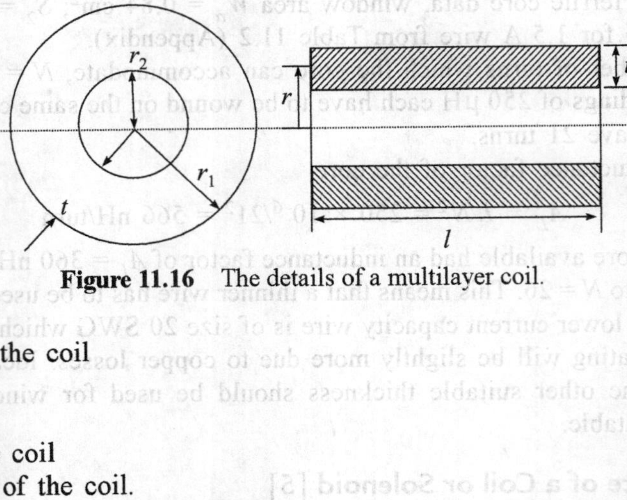

Figure 11.16 The details of a multilayer coil.

where

t = thickness of the coil
r_1 = outer radius
r_2 = inner radius
l = length of the coil
r = mean radius of the coil.

The Nagaoka correction factor for this case, when the coil has considerable winding thickness, is given by

$$K_n=\frac{1}{1+0.9\left(\frac{r}{l}\right)+0.32\left(\frac{t}{r}\right)+0.84\left(\frac{t}{l}\right)} \tag{11.31}$$

here, r is the mean radius of the coil = $(r_1 + r_2)/2$. These correction factors tend to become a bit tedious when followed in actual practice. Wheeler [7] therefore gave certain approximations for the design of solenoids with single and multiple layers of windings are as follows:

Wheeler's formulas [7]

(*a*) For a single layer air core solenoid

$$L = \frac{r^2 \times N^2}{9r + 10l} \tag{11.32}$$

where

L = inductance in microhenrys
r = coil radius in inches
l = coil length in inches
N = number of turns

The accuracy of this formula is better than 1% if the length of the coil l is more than 0.8r.

(*b*) For a multilayer air core solenoid

$$L = \frac{31.6 \times r_2^{\,2} \times N^2}{6r_2 + 9l + 10(r_1 - r_2)} \tag{11.33}$$

where

L = inductance in microhenrys
r_2 = inside coil radius in metres
r_1 = outside coil radius in metres
l = coil length in metres
N = total number of turns

An alternate expression for the above mentioned configuration is

$$L = \frac{0.8 \times r^2 \times N^2}{6r + 9l + 10t} \tag{11.34}$$

L = inductance in microhenrys
r_2 = inside coil radius in inches
r_1 = outside coil radius in inches
l = coil length in inches
N = number of turns total
r = average winding radius in inches = $(r_1 + r_2)/2$
t = thickness of the coil, that is, difference between r_2 and r_1 in *inches* = $r_1 - r_2$

All these parameters are explained in Figure 11.16.

When the dimensions are converted into centimetres, Equation (11.35) is obtained.

$$L = \frac{0.314 \times r^2 \times N^2}{6r + 9l + 10t} \tag{11.35}$$

To simplify matters further, the geometry of the coil should be fixed to some convenient value so as to fit the coil in a suitable place on the circuit board. For example, if the length $l = 4r$ so that the requirement of $l \geq 0.8\ r$ is fulfilled, then the Equation (11.35) reduces to Equation (11.36)

$$L = \frac{0.157 r^2 N^2}{31r - 10r_2} \tag{11.36}$$

Example 11.5: Find the number of turns required for a 30 μH, 2 cm long coil capable of carrying 2 A current and has r_2 = 0.3 cm.

Solution: Since $l = 4r$ with l = 2 cm then r = 0.5 cm. Substituting in Equation (11.36), we get

$$L = \frac{0.157(0.5)^2 N^2}{31 \times 0.5 - 10 \times 0.3} = 0.0393\ N^2/12.5 = 30\ \mu\text{H}$$

or $\quad N \approx 98$ turns

The window area available for winding the coil is

$$2(r_1 - r_2) = 2 \times (0.7 - 0.3) = 0.8\ \text{cm}^2$$

If the window utilization factor K_u = 0.4, then the available area for winding is

$$0.8 \times 0.4 = 0.32\ \text{cm}^2$$

The wire that fits in this area must have an area of

$$0.32/98 = 0.00326\ \text{cm}^2 = 0.326\ \text{mm}^2$$

This corresponds to the wire gauge of 23 SWG which unfortunately can carry a current of only 0.81 A which is certainly not applicable for operation at 2 A. An SWG 20 wire with a current density of 280 A/cm² must be used. However, to accommodate 98 turns of this wire a larger coil former is needed. Hence, a second iteration is required on the above guidelines. Thus, for $l = 4r$ dimension, if r = 0.65 cm, then a coil of length 2.6 cm will require only 88 turns as calculated below.

$$L = \frac{0.157(0.65)^2 N^2}{(31 \times 0.65) - (10 \times 0.3)} = 30\ \mu\text{H}$$

or $\quad N = 88$ turns

The window area of this coil:

$$W_a = 2.6\ (r_1 - r_2).$$

Now, $\quad (r_1 + r_2)/2 = r$

or $\quad r_1 = 0.65\ (2 - 0.3) = 1.0$ cm

Therefore, $\quad r_1 - r_2 = 0.7$

Hence

$$W_a = 2.6 \times 0.7 = 1.82\ \text{cm}^2$$

If the window utilization factor $K_u = 0.4$, then the area available for winding is

$$1.82 \times 0.4 = 0.728 \text{ cm}^2.$$

Therefore, the area available per turn of wire is

$$0.728/88 \text{ cm}^2 = 0.82 \text{ mm}^2.$$

This is more than the area of a 20 SWG wire which is 0.6567 mm². In fact, a 19 SWG wire with an area of 0.8107 mm² and current capacity of 2.25 A can be accommodated. Some other coil geometries can be tried like $l = 3r$. For this condition the expression becomes

$$L = \frac{0.314r^2N^2}{53r - 20r_2} \tag{11.37}$$

If $l = 2$ cm, then

$$r = l/3 = 2/3 = 0.67 \text{ cm}$$

$$r_2 = 0.3 \text{ cm}$$

Therefore,

$$L = \frac{0.314 \times 0.667^2 N^2}{53 \times 0.667 - 20 \times 0.3} = 30 \ \mu\text{H}$$

or

$$N = 79 \text{ turns}$$

Again, the window area

$$W_a = 2(r_1 - r_2) = 2(1.034 - 0.3) = 1.468 \text{ cm}^2$$

For a window utilization factor $K_u = 0.4$, the effective winding area available equal to $1.468 \times 0.4 = 0.5872$ cm². Therefore, the area available to each winding is 0.743 mm². This coil can comfortably hold 79 turns of 20 SWG wire. The above calculations illustrate the dependence of the inductance value on the size and geometry of the coils. To make a coil of the desired value, a few iterations are needed as the wire current rating is not included in Wheeler's approximations although he has simplified the design procedure drastically. In 1950, Wheeler [8] published another paper where he gave a chart to find the inductance of a coil of several dimensions. Lundin has also given a procedure for calculating inductance values [9]. Many software programmes are available on the Internet to perform the necessary calculations for multilayer coils [10], 12] and [13]. A programme is also available for anyone interested in storing it [11].

Example 11.6: This example illustrates the calculation of a multilayer inductor using the Nagaoka correction. The value calculated with this method will be compared with the one calculated from Wheeler's formula.

Solution: A coil was constructed with the following parameters:

1. Length of the coil = 1.8 cm
2. Outer diameter of the coil = 1.8 cm
3. Inner diameter of the coil = 0.6 cm
4. The number of turns of copper wire in the coil = 85.

Therefore,

1. The mean radius $r = (r_1 + r_2)/2 = 0.6$ cm. This will be taken to calculate the area.
2. $l = 1.8$ cm
3. Thickness of the coil $t = r_1 - r_2 = 0.6$ cm
4. $N = 85$

The dimensions of the coil are shown in Figure 11.17.

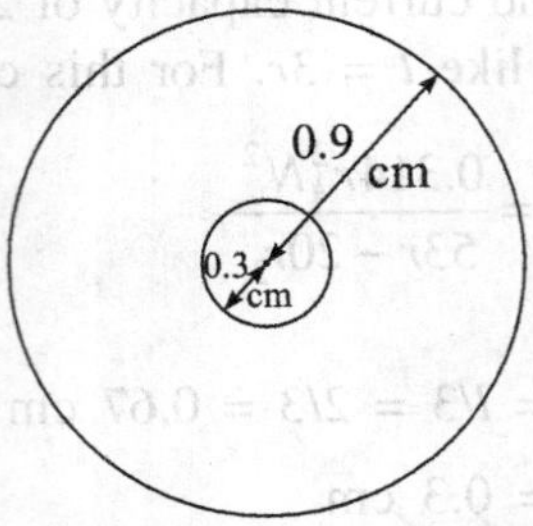

Figure 11.17 Cross section of the coil.

Therefore from Equation (11.29)

$$L = \frac{4\pi^2 \times (0.6)^2 \times 85^2}{1.8} \times 10^{-9} \text{ H} = 57 \ \mu\text{H} \qquad \text{(i)}$$

To calculate the Nagaoka constant to take into account the finite length and thickness of the coil, proper values are inserted in Equation (11.31) giving the expression

$$K_n = \frac{1}{1 + 0.9\left(\frac{0.6}{1.8}\right) + 0.32\left(\frac{0.6}{0.6}\right) + 0.84\left(\frac{0.6}{1.8}\right)} = 0.526 \qquad \text{(ii)}$$

Multiplying (i) and (ii) the value of the inductance comes to

$$L \text{ (corrected)} = 57 \times 0.526 = 30 \ \mu\text{H}.$$

Thus, the value of the inductor reduces to almost half because of comparable dimensions of the coil. The inductance of the coil was actually measured by an LCR meter which yielded a close value of 30.9 μH giving an error of 3%.

This value can now be compared with that found using Wheeler's approximate formula given in Equation (11.34) and included in Table 11.7. For calculations the first step is to convert all the dimensions of the coil to inches. Thus, the following results are obtained:

1. Length of coil $l = 1.8$ cm = 1.8/2.54 = 0.709 inch
2. Average radius of the coil $r = 0.6$ cm = 0.6/2.54 = 0.236 inch
3. Thickness of the coil $t = 0.6$ cm = 0.6/2.54 = 0.236 inch

Hence

$$L = \frac{0.8 \times r^2 \times N^2}{6r + 9l + 10t} = \frac{0.8 \times (0.236)^2 \times 85^2}{(6 \times 0.236) + (9 \times 0.709) + (10 \times 0.236)} = 31.69 \ \mu\text{H}$$

This value is different from the one calculated earlier. The percentage deviation is

$$\{(31.69 - 30)/30\} \times 100 = 5.65\%$$

This error is acceptable in most cases since the percentage deviation in values of components ranges from 5% to 10%.

The magnetizing force H

To find the magnetizing force H we consider Equation (11.29).

$$L = \frac{\mu_0 N^2 KA}{l} = \frac{\mu_0 NAK}{I}\left(\frac{NI}{l}\right) = \frac{\mu_0 NA}{I}(KH)$$

where $H = NI/l$ and K is the Nagaoka constant for a multilayer coil. It can be interpreted that the value of H is now effectively modified by K that is the field produced by a current I passing through the coil of this specification produces a magnetizing force multiplied by K which is less than unity. Thus, the modified force

$$H' = KH = 0.526H \tag{11.38}$$

Now, the field produced by the coil is

$$H = \frac{NI}{l} = \frac{85 \times 2}{0.018} = 9444 \text{ AT/m}$$

or

$$H' = KH = 0.526H = 0.526 \times 9444 = 4968 \text{ AT/m}$$

This can be verified experimentally from the following calculations.

Inductance of the coil

$$L = N\phi/I \tag{11.39}$$

where the notations have their usual meaning, ϕ is the magnetic flux in weber. Substituting the values

$$L = 30.9 \times 10^{-6} = 85\phi/2$$

or

$$\phi = 0.727 \times 10^{-6} \text{ Wb}$$

Since the average area of the coil is 0.6 cm its area is $\pi \times (0.006)^2$ m^2. Therefore, flux density

$$B = \frac{\phi}{A} = \frac{0.727 \times 10^{-6}}{\pi \times (0.006)^2}$$

$$= 0.00643 \text{ T or } 64.3 \text{ gauss.}$$

Also permeability of free space

$$\mu_0 = B/H$$

or

$$H = 0.00643/4\pi \times 10^{-7}$$

$$= 5117 \text{ AT/m}$$

Since the error is 3% it can be concluded that the effective magnetizing force is the product of the calculated value of H, and the Nagaoka constant for the coil.

For solenoids with ferrite cores there does not seem to be a theoretical calculation giving an exact value of the inductance although a quite complicated method is given in reference [14]. The easiest way is to insert a ferrite bead of about 0.5 cm in diameter inside the coil former and measure the value. Experimentally, it has been found that such a process increases the inductance of an air core coil by a factor of approximately five.

Example 11.7: The inductor to keep the current constant in the push-pull inverter used in a solar lantern is usually a coil of wire, a solenoid, with a ferrite bead inside it. This not only keeps the cost low but also ensures that the core does not saturate even when a DC current of approximately 2 A flows through it. The required value of the inductor is 150 μH. The insertion of the ferrite bead as mentioned above increases its inductance by a factor of 5. Hence an air core inductor of 30 μH has to be made with enough space to house the bead. This will require a coil former with a 0.5 cm hole running cylindrically through the middle. It can be noticed that throughout the above examples the same geometry and coil value have been used for calculations. Example 11.6 gives the rest of the calculations.

No discussion involving ferrite materials is complete without referring to the exhaustive work by E.C. Snelling [15], [16] which deals with almost every aspect starting with their preparation to their use in transformers. Unfortunately, his work does not cover the use of ferrites in the calculation of inductance of multilayer coils. Another comprehensive work on inductor design is authored by F.W. Grover [17] while F.E. Terman's handbook [18] is necessary to calculate inductance of chokes of different shapes and sizes. Table 11.7 gives some expressions for calculating inductance values for different geometries and configurations which may be used for design purposes.

Table 11.7 Formulae for calculating inductance values of different geometries

Construction	*Formula*	*Dimensions*
Cylindrical coil	$L = \frac{\mu_0 K N^2 A}{l}$	L = inductance in henrys (H) μ_0 = permeability of free space = $4\pi \times 10^{-7}$ H/m K = Nagaoka coefficient N = number of turns A = area of cross-section of the coil in square metres (m^2) l = length of coil in metres (m)
Straight wire conductor	$L = l\left(\ln\frac{4l}{d} - 1\right)200\times10^{-9}$	L = inductance (H) l = length of conductor (m) d = diameter of conductor (m)
	$L = 5.08 \times l\left(\ln\frac{4l}{d} - 1\right)$	L = inductance (nH) l = length of conductor (in) d = diameter of conductor (in)
Short air-core, single layer cylindrical coil	$L = \left(\frac{r^2N^2}{9r + 10l}\right)$ Wheeler's Formula[1]	L = inductance (μH) r = outer radius of coil (in) l = length of coil (in) N = number of turns

(Contd.)

Table 11.7 Formulae for calculating inductance values of different geometries (*Contd.*)

Multilayer air-core coil	$L = \left(\dfrac{0.8r^2 N^2}{6r + 9l + 10t} \right)$ Wheeler's Formula	L = inductance (μH) r = mean radius of coil (in) l = physical length of coil winding (in) N = number of turns t = thickness of coil (outer radius minus inner radius) (in)
Flat spiral air-core coil	$L = \dfrac{r^2 N^2}{(2r + 2.8t) \times 10^5}$	L = inductance (H) r = mean radius of coil (m) N = number of turns t = thickness of coil (outer radius minus inner radius) (m)
	$L = \dfrac{r^2 N^2}{8r + 11t}$	L = inductance (μH) r = mean radius of coil (in) N = number of turns t = thickness of coil (outer radius minus inner radius) (in)
Torroidal core (circular cross-section)	$L = \mu_0 \mu_r \dfrac{N^2 r^2}{D}$	L = inductance (H) μ_0 = permeability of free space = $4\pi \times 10^{-7}$ H/m μ_r = relative permeability of core material N = number of turns r = radius of coil winding (m) D = overall diameter of torroid (m)

[1]This formula is given in reference [7] and is accurate to within one per cent for $l > 0.8r$, that is, if the coil is not too short.

Table 11.8 gives some magnetic properties of ferromagnetic materials which come in handy while designing both inductors and transformers. The vast variation of initial and final values of relative permeability makes their use difficult in any design work. Also different sources give different units that can create confusion and lead to erroneous results. Since it is normal practice now to use the SI system of units all the values must be converted to this system.

Table 11.8 Some magnetic properties of ferromagnetic materials

Material	*Treatment*	*Initial Relative Permeability*	*Maximum Relative Permeability*	*Coercive Force* (oersteds)	*Remnant Flux Density* (gauss)
Iron, 99.8% pure	Annealed	150	5000	1.0	13,000
Iron, 99.95% pure	Annealed in hydrogen	10,000	200,000	0.05	13,000

(*Contd.*)

Table 11.8 Some magnetic properties of ferromagnetic materials (*Contd.*)

Material	*Treatment*	*Initial Relative Permeability*	*Maximum Relative Permeability*	*Coercive Force* (oersteds)	*Remnant Flux Density* (gauss)
78 Permalloy	Annealed, quenched	8,000	100,000	0.05	7,000
Superpermalloy	Annealed in hydrogen, controlled cooling	100,000	1,000,000	0.002	7,000
Cobalt, 99% pure	Annealed	70	250	10	5,000
Nickel, 99% pure	Annealed	110	600	0.7	4,000
Steel, 0.9% C	Quenched	50	100	70	10,300
Steel, 30% Co	Quenched	...	...	240	9,500
Alnico 5	Cooled in magnetic field	4	...	575	12,500
Silmanal	Baked	...	...	6,000	550
Iron, fine powder	Pressed	...	...	470	6,000
Mumetal (75% nickel, 2% chromium, 5% copper, 18% iron)			20,000	...	...
Ferrite U60			8	...	...
Ferrite M33			750	...	...
Ferrite N41			3,000	...	...
Ferrite T38			10,000	...	...
Silicon GO steel			40,000	...	...

In this table the remnant flux density is the retained *magnetic field B*, and the SI unit for *B* is the tesla (T). 1 Tesla = 10,000 gauss. The coercive force is the applied reverse *magnetic field strength H* required to force the net magnetic field back to zero after magnetization. The SI unit for *H* is A/m, and 1 A/m = 0.01257 oersteds.

11.3 Electromagnetic Interference (EMI) Filter Design

Line filters prevent excessive noise from being conducted from any electronic equipment, particularly involving power switching, to the AC line so that the latter is not contaminated. The purpose of the AC line is to carry a pure 50 Hz, 220 V power to all the equipment working in a certain area. If some of these generate large quantities of electromagnetic interference (EMI) or radio frequency interference (RFI) signals the other instruments working on the same line will be affected and therefore may perform erroneously. Figure 11.18 shows the use of a common mode filter between the AC lines and a noisy power supply. The direction of common mode noise, which occurs simultaneously on both the lines referred to earth common, is from the load and into the filter, where the noise common to both lines becomes sufficiently attenuated. The resulting common mode output of the filter onto the AC line through the impedance matching circuitry, is then low.

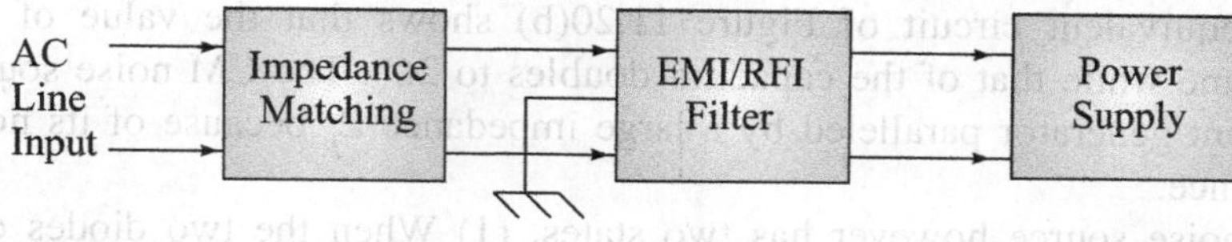

Figure 11.18 Block diagram showing the location of the EMI/RFI filter.

The common mode filter design is essentially the design of two identical filters one for each of the two polarity lines. The inductors are wound on a single, usually a torroid core, while the capacitors are placed from the lines to the common ground. A wound inductor is shown in Figure 11.19. The current that flows from (*A*) to (*B*) through L_1 returns from (*B*) to (*A*) through L_2 after passing through the load connected at the terminals of (*B*). It is clear from the direction of flow of these two currents that the net magnetic flux which is coupled between the two inductors is zero because they cancel the flux created by each other. It is also noteworthy that for power line frequency the inductors are essentially shorted whereas they offer appreciable reactance to higher frequencies generally encountered.

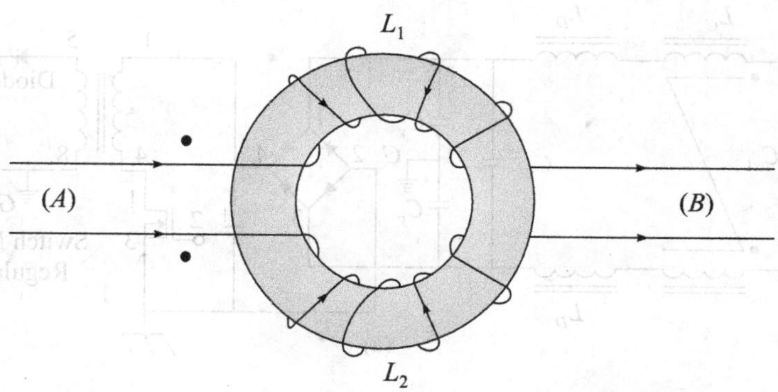

Figure 11.19 The common mode inductor.

Any inductance encountered by the differential signal is then the result of imperfect coupling of the two chokes; they perform as independent components with their leakage inductances responding to the differential signal: the leakage inductances attenuate the differential signal. However, when identical signals with respect to the ground are present at the two inductors L_1 and L_2, called a common mode signal, they each contribute a net, non-zero flux in the core they share. The inductors therefore act as separate components but the mutual inductance between them is responsible for attenuating this common signal.

Figure 11.20(a) shows a switch mode power supply connected to an EMI filter which reduces the noise generated in the equipment before passing it on to the power line. There are clearly two paths for the noise to travel to the line. First there is a common earth connection between the output of the switching regulator marked *G* and point *G* at the junction of the two capacitors C_y of the filter. With respect to this common point noise is introduced at the live and neutral wires *L* and *N*. This is called the *common mode* (CM) noise. The other noise is introduced in the circuit from the two ends of the bridge rectifier marked 1 and 3. This is called *differential mode* (DM) noise. A detailed analysis is given in reference [19].

The CM equivalent circuit of Figure 11.20(b) shows that the value of the inductor remains the same while that of the capacitor doubles to $2C_y$. The CM noise source is shown here as a current generator paralleled by a large impedance Z_p because of its normally large output impedance.

The DM noise source however has two states. (1) When the two diodes of the bridge rectifier conduct their impedance connecting the filter to the power supply is small but there comes a second state (2) when all the four diodes are off at the cross over of the AC power voltage. At this point the input impedance of the power supply is large and therefore the noise generator too has large output impedance because it has to pass through the diodes which are non-conducting. For the first case the noise generator is represented by a voltage source which has low output impedance and for the second case it is represented by a current source which of course has inherently large output impedance. In one cycle of the power line frequency the diodes switch states (ON/OFF) twice, therefore this noise generator changes its nature from voltage to current source at twice the line frequency. This is shown in Figure 11.20(c).

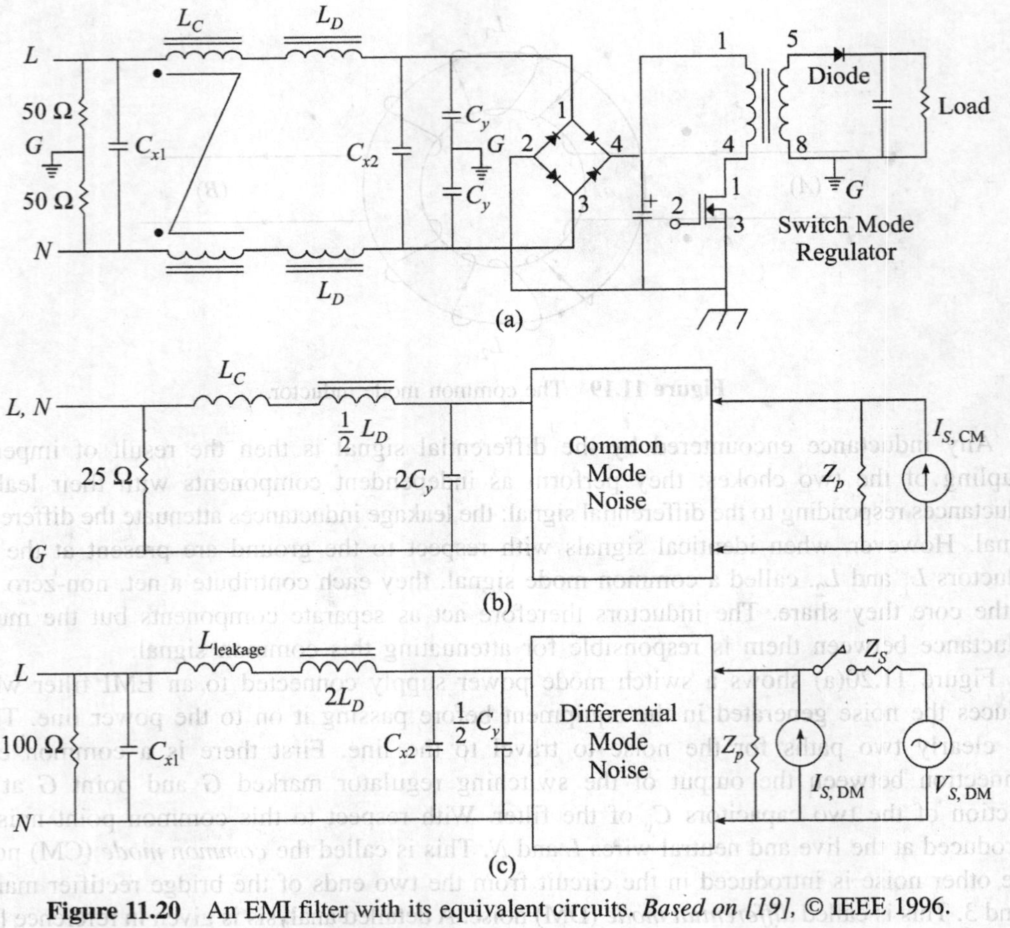

Figure 11.20 An EMI filter with its equivalent circuits. *Based on [19],* © IEEE 1996.

Calculation of the filter components requires certain assumptions and reduction of the noise sources to acceptable circuit equivalents. The following analysis will help in this direction.

- *Common Mode Filter*

From Figure 11.20(b), the CM noise equivalent circuit can be represented by Figure 11.21(a). If the reactance of the capacitor to the noise frequency, $2C_y = 1/\omega\,(2C_y)$ is far less than the output impedance Z_p of the current source and the reactance of the inductor $\omega\,(L_C + 1/2\,L_D)$ is much greater than 25 Ω then Figure 11.21(a) reduces to Figure 11.21(b). *The reciprocity theorem* can convert this current source representation to a voltage source equivalent circuit shown in Figure 11.21(c). *This theorem states that the open circuit voltage ratio is equal to the reverse ratio of the short circuit current.*

From Figure 11.21 the transfer function for the noise voltage V_N of the CM filter is given by:

$$\frac{V_N\ \text{(without filter)}}{V_N\ \text{(with filter)}} = V_{S,\text{CM}}/V_{O,\text{CM}} \approx \frac{I_{S,\text{CM}}\ 25\ \Omega}{I_{O,\text{CM}}\ 25\ \Omega} \tag{11.40}$$

At the resonance frequency $f_{R,\text{CM}}$, the reactance of inductor L_{CM} is equal to that of the capacitor C_{CM} that is:

$$\omega L_{\text{CM}} = \frac{1}{\omega C_{\text{CM}}}$$

Therefore it follows that:

$$f_{R,\,\text{CM}} = \frac{1}{2\pi\sqrt{L_{\text{CM}}.C_{\text{CM}}}} = \frac{1}{2\pi\sqrt{\left(L_C + \frac{1}{2}L_D\right)2C_y}} \approx \frac{1}{2\pi\sqrt{L_C.2C_y}} \tag{11.41}$$

If $L_C >> \frac{1}{2}L_D$

Since it is a second order low pass filter the attenuation is 40 dB per decade. The filter attenuation is determined by L_{CM} (= $L_C + L_D/2$) and C_M (=$2C_y$) and is independent of source impedances. Figure 11.21(d) shows the plot of CM noise attenuation versus frequency. Normally, $L_C >> ½\,L_D$ and the corner frequency $f_{R,\text{CM}}$ is mainly determined by L_C and C_y values. In practical filters all the assumptions made above are achievable.

- *Differential Mode*

Figure 11.20(c) showed the equivalent circuit of the DM filter and also explained that the alternation between high and low impedances of the source depended on the conducting states of the rectifiers. Figure 11.22 shows the two noise equivalent circuits leading to the derivation of the DM filter attenuation.

When the rectifier diodes are off, a current source model for the noise source is used and the equivalent circuit is shown in Figure 11.22(a). When diodes are on, the low-Z model shown in Figure 11.22(d) is used. When the impedance conditions are met, as they really are in practice, Figure 11.22(a) reduces to a more simplified form of Figure 11.22(b) and then is converted to a voltage source based circuit of Figure 11.22(c) by using the *reciprocity theorem* as discussed earlier. The assumptions are:

$$\frac{1}{\omega C_{x1}} \gg 100\,\Omega, \quad \frac{1}{\omega C_{x2}} \ll Z_P \text{ and } \omega L_{\text{DM}} \gg 100\,\Omega$$

If again it is assumed that

$$C_{x1} = C_{x2} \equiv C_{\text{DM}}$$

The corner frequency becomes

$$f_{R,\text{DM}} = \frac{1}{2\pi\sqrt{L_{\text{DM}}C_{\text{DM}}}} = \frac{1}{2\pi\sqrt{(2L_D + L_{\text{leakage}})C_{\text{DM}}}} \tag{11.42}$$

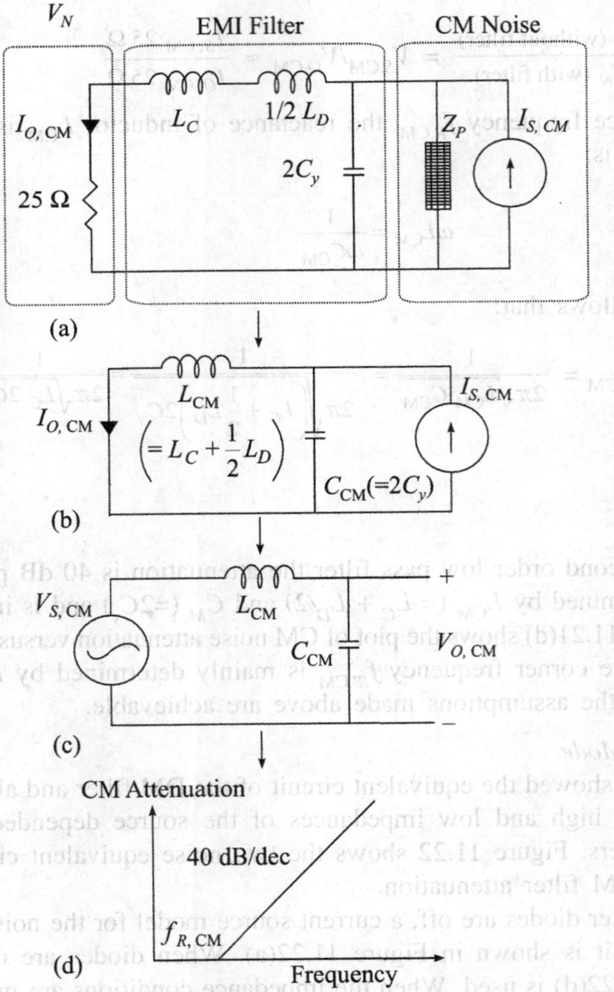

Figure 11.21 Common mode noise filter and its equivalent circuit. *Based on [19]*, © IEEE 1996.

Therefore at this frequency the noise voltage is reduced by 6 dB and it continues to reduce by 40 dB per decade, or ten times, of frequency increase. This can also be expressed as 12 dB per octave or doubling of frequency.

The same results will be obtained if the voltage source equivalent circuit of Figure 11.22(d) is considered as the starting point. With the assumptions below the circuits of Figures 11.22(e) and 11.22(f) are generated.

(i) If $\dfrac{1}{\omega C_{x2}} \gg Z_S$ Figure 11.22(e) is formed from Figure 11.22(d)

(ii) If $\dfrac{1}{\omega C_{x1}} \ll 100\ \Omega$ Figure 11.22(f) follows from Figure 11.22(e).

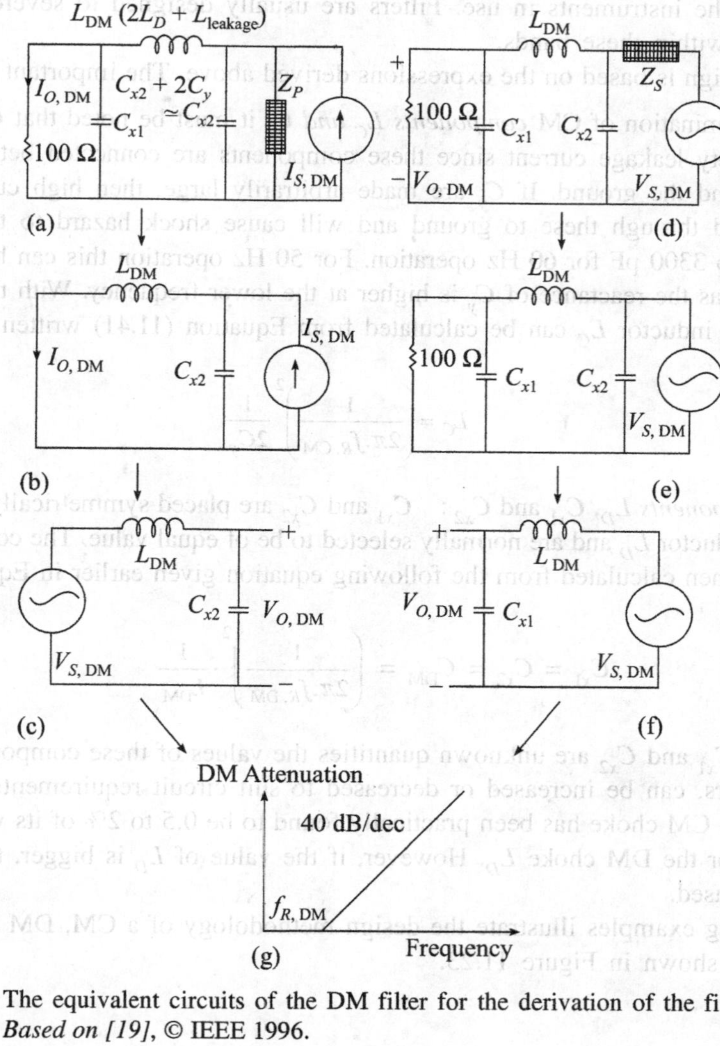

Figure 11.22 The equivalent circuits of the DM filter for the derivation of the filter characteristics. *Based on [19]*, © IEEE 1996.

It is to be noted from a comparison of Figures 11.22(c) and 11.22(f) that both the capacitors C_{x1} and C_{x2} are needed to reduce the two types of DM noise. However, the CM noise reduction method, as shown in Figure 11.21, requires only one series combination of two C_y capacitors to be placed *near the noise generator* to sufficiently reduce the unwanted signals.

11.3.1 Design of Filters

Conducted noise limits for EMI are given below for both VDE and FCC standards:

1. Verband der Elektrotechnik (VDE) limit: 10 kHz to 30 MHz
2. Federal Communications Commission (FCC) limit: 450 kHz to 30 MHz

Noise in these frequency ranges for the two standards is not supposed to be injected into the mains from the instruments in use. Filters are usually designed to severely curtail the noise amplitude within these bands.

The filter design is based on the expressions derived above. The important equations are:

(a) For determination of CM *components* L_C *and* C_y it must be noted that C_y is restricted by a safety leakage current since these components are connected between the live, neutral and the ground. If C_y are made arbitrarily large, then high currents will be conducted through these to ground and will cause shock hazard to the user. C_y is limited to 3300 pF for 60 Hz operation. For 50 Hz operation this can be enhanced to 4700 pF as the reactance of C_y is higher at the lower frequency. With the value of C_y fixed the inductor L_C can be calculated from Equation (11.41) written below as:

$$L_C = \left(\frac{1}{2\pi . f_{R,\text{CM}}}\right)^2 \frac{1}{2C_y} \tag{11.43}$$

(b) DM *components* L_D, C_{x1} and C_{x2}: C_{x1} and C_{x2} are placed symmetrically on both sides of the inductor L_D and are normally selected to be of equal value. The corner frequency $f_{R,\text{DM}}$ is then calculated from the following equation given earlier in Equation (11.42).

$$C_{x1} = C_{x2} = C_{\text{DM}} = \left(\frac{1}{2\pi . f_{R,\text{DM}}}\right)^2 \cdot \frac{1}{L_{\text{DM}}} \tag{11.44}$$

Since L_{DM}, C_{x1} and C_{x2} are unknown quantities the values of these components, inductor and the capacitors, can be increased or decreased to suit circuit requirements. The leakage inductance of the CM choke has been practically found to be 0.5 to 2% of its value. This can be easily used for the DM choke L_D. However, if the value of L_D is bigger, then additional chokes may be used.

The following examples illustrate the design methodology of a CM, DM filter based on Figure 11.20(a), shown in Figure 11.23.

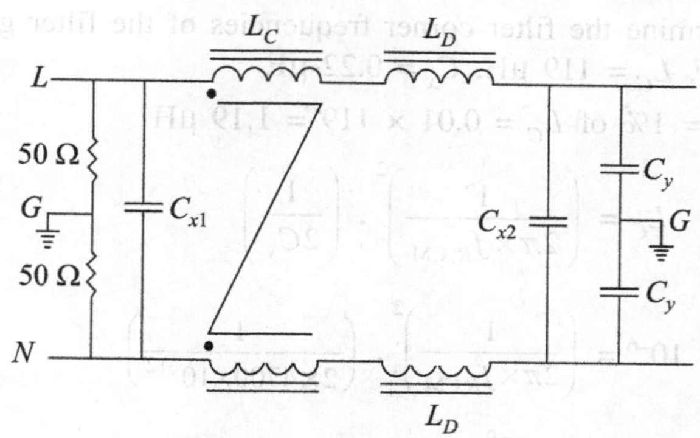

Figure 11.23 A typical EMI filter reproduced here for the design examples.

Example 11.8: Design a filter for corner frequencies $f_{R,\text{CM}}$ =10 kHz and $f_{R,\text{DM}}$ =150 kHz. Assume C_y = 4700 pF and L_{DM} = 1% of L_C.

Solution: First the value of L_C is calculated from the following equation.

$$L_C = \left(\frac{1}{2\pi \times f_{R,\text{CM}}}\right)^2 \cdot \left(\frac{1}{2 \times C_y}\right)$$

$$L_C = \left(\frac{1}{2\pi \times 10 \times 10^3}\right)^2 \cdot \left(\frac{1}{2 \times 4700 \times 10^{-12}}\right)$$

$$= 2.69 \times 10^{-2} = 26.9 \times 10^{-3} \sim 27 \text{ mH}$$

Therefore, L_{DM} = 0.01 × 27 = 0.27 mH. Now the C_x capacitors are determined from the equation:

$$C_x = \left(\frac{1}{2\pi \times f_{R,\text{DM}}}\right)^2 \cdot \left(\frac{1}{L_{\text{DM}}}\right)$$

$$= \left(\frac{1}{2\pi \times 150 \times 10^3}\right)^2 \cdot \left(\frac{1}{0.27 \times 10^{-3}}\right)$$

or $$C_x = 4.16 \times 10^{-9} = 4.16 \text{ nF}$$

These values are shown in Figure 11.24.

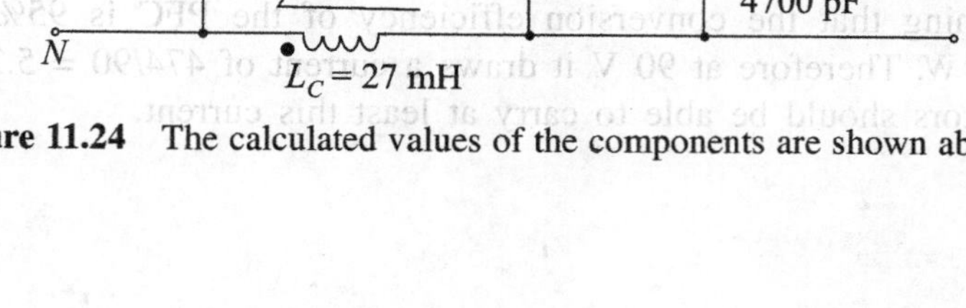

Figure 11.24 The calculated values of the components are shown above.

Example 11.9: Determine the filter corner frequencies of the filter given in Figure 11.25. Assume C_y = 4700 pF, L_C = 119 µH, C_x = 0.22 µF.

Solution: Take L_{DM} = 1% of L_C = 0.01 × 119 = 1.19 µH

$$L_C = \left(\frac{1}{2\pi \times f_{R,CM}}\right)^2 \cdot \left(\frac{1}{2C_y}\right)$$

or

$$119 \times 10^{-6} = \left(\frac{1}{2\pi \times f_{R,CM}}\right)^2 \cdot \left(\frac{1}{2 \times 4700 \times 10^{-12}}\right)$$

Therefore, $2\pi\, f_{R,CM} = \dfrac{1\times 10^9}{1057.63} = 0.94 \times 10^6$

Hence CM corner frequency $f_{R,CM} = \dfrac{0.94}{2\pi} \times 10^6 = 0.1496 \times 10^6 = 149.6$ kHz. To determine $f_{R,DM}$ use the following expression

$$C_x = \left(\frac{1}{2\pi \times f_{R,DM}}\right)^2 \cdot \left(\frac{1}{L_{DM}}\right)$$

$$2\pi\, f_{R,CM} = 1.95 \times 10^6$$

$$f_{R,DM} = \left(\frac{1.95}{2\pi}\right) \times 10^6 = 0.311 \times 10^6 = 311 \text{ kHz}$$

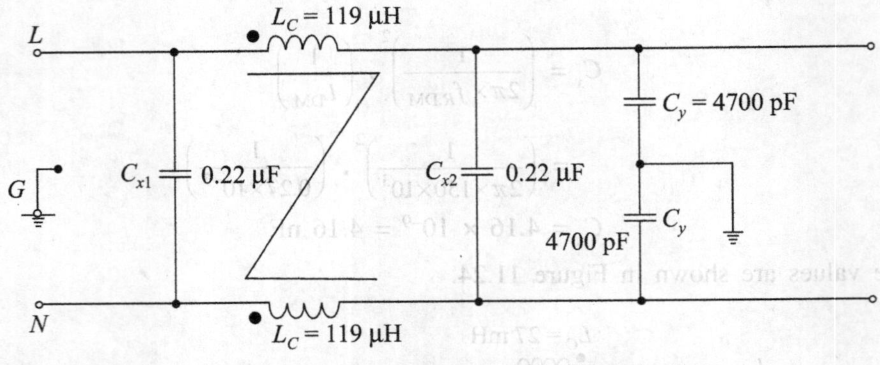

Figure 11.25 Component values for Example 11.9.

Example 11.10: It is desired to make an EMI filter for a power factor controller MC 34262 operating at a switching frequency of 25 kHz. The PFC input varies from 90 V to 270 V and has an output of 450 W.

Solution: Assuming that the conversion efficiency of the PFC is 95% its input power = 450/0.95 = 474 W. Therefore at 90 V it draws a current of 474/90 = 5.26 A. This means that all the inductors should be able to carry at least this current.

If the switching frequency is 25 kHz, then the filter cut off frequencies $f_{R,CM}$ and $f_{R,DM}$ should be at least half this value to provide an attenuation of 12 dB. Thus:

$$f_{R,DM} = 12500 \text{ Hz}$$

$$f_{R,CM} = 12500 \text{ Hz}$$

Assume

$$C_y = 4700 \text{ pF}$$

Therefore,

$$L_C = \left(\frac{1}{2\pi\times 12500}\right)^2 \frac{1}{2\times 4700\times 10^{-12}} = 17 \times 10^{-3} \text{ H from Equation (11.43),}$$

Since this is quite a large inductance its leakage inductance can be used as L_D. Let the leakage inductance be 1% of this value of L_{DM}. Therefore

$$L_{DM} = 170 \times 10^{-6} \text{ H.}$$

Substituting this value of L_{DM} in Equation (11.44)

$$C_x = C_{x1} = C_{x2} = \left(\frac{1}{2\pi\times f_{R,DM}}\right)^2 \times \frac{1}{L_{DM}} = \left(\frac{1}{2\pi\times 12500}\right)^2 \times \frac{1}{170\times 10^{-6}} = 9.52 \times 10^{-7} \approx 1 \text{ μF}$$

C_{x1} and C_{x2} reduce the noise when the rectifiers are ON and OFF respectively. Thus if two cut-off frequencies are required then the two capacitors can be of different values. In this case C_{x2} will be retained at 1 μF for an $f_{R,DM}$ of 12500 Hz for the condition when the rectifiers are OFF. If, however, $f_{R,DM}$ is chosen to be 40 kHz when the rectifiers are ON, then a new value of C_{x1} may be calculated as follows:

$$f_{R,DM1} = \frac{1}{2\pi\sqrt{L_{DM}\times C_{x1}}}$$

$$\text{or } C_{x1} = \left(\frac{1}{2\pi\times f_{R,DM1}}\right)^2 \times \left(\frac{1}{L_{DM}}\right) = \left(\frac{1}{2\pi\times 40000}\right)^2 \left(\frac{1}{170\times 10^{-6}}\right) = 0.09312 \times 10^{-6} \text{ F} \approx 0.1 \text{ μF}$$

These values are shown in Figure 11.26.

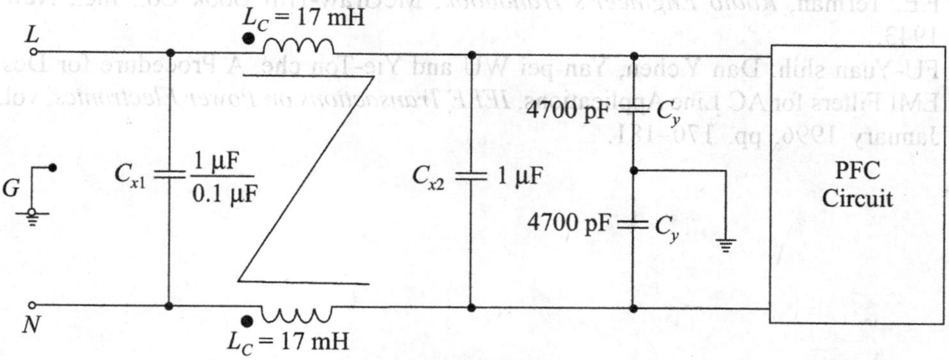

Figure 11. 26 An EMI filter for power factor controller.

References

[1] Engineering staff of Texas Instruments Incorporated, *Transistor Circuit Design*, McGraw-Hill Book Company, Inc., New York, Kogakusha Company, Ltd., Tokyo, 1963, pp. 433–446.

[2] Colonel Wm. T. McLyman, *Transformer and Inductor Design Handbook*, 2nd ed., Marcel Dekker, Inc., New York, 1988, pp. 95–144.

[3] Central Electronics Limited, Ferrites Soft-Magnetic Material Data Book 1993–94.

[4] COSMO Ferrites Limited Catalogue, URL: www.cosmoferrites.com.

[5] V.G. Welsby, *The Theory and Design of Inductance Coils*, 2nd ed., 1960, Macdonald, London.

[6] H. Nagaoka, The Inductance Coefficients of Solenoids, *J. Coll. Sci.*, vol. **27**, pp. 18–33, 1909.

[7] Harold A. Wheeler, Simple Inductance Formulas for Radio Coils, *Proceedings of the I.R.E.*, October 1928, pp. 1398–1400.

[8] H. A. Wheeler, Inductance Chart for Solenoid Coils, *Proceedings of the I.R.E.*, vol. **38**, 1950, p. 1398.

[9] Richard Lundin, A Handbook of formula for the inductance of a single–Layer Circular Coil. *Proceedings of the IEEE*, vol. **73**(9), September 1985, pp. 1428–1429.

[10] www.circuits.dk/calculator_multi_layer_aircore.htm.

[11] Program available at: www.moshier.net/coildoc.html.

[12] www.pronine.ca/multind.htm.

[13] http://www.coolcircuit.com/tools/multi_layer_coil_calculator/.

[14] Manfred Albach, Alexander Stadler, and Matthias Spang, The influence of a ferrite characteristics on the inductance of coil with rod cores. *IEEE transactions on magnetics*, vol. **43**(6), June 2007, p. 2618.

[15] E.C. Snelling, *Soft Ferrites Properties and Applications*, London, Iliffe Books Ltd., 1969.

[16] E.C. Snelling and A.D. Giles, *Ferrites for Inductors and Transformers*, Research Studies Press, John Wiley & Sons Inc., 1983.

[17] Frederick W. Grover, *Inductance Calculation Methods: Working Formulas and Tables*, Dover Publications Inc., New York, 1946.

[18] F.E. Terman, *Radio Engineer's Handbook*, McGraw-Hill Book Co., Inc., New York, 1943.

[19] FU-Yuan shih, Dan Y.chen, Yan-pei WU and Yie-Ton che, A Procedure for Designing EMI Filters for AC Line Applications, *IEEE Transactions on Power Electronics*, vol. **II**(1), January 1996, pp. 170–181.

APPENDIX

Table 11.1 COSMO Ferrite EE-core Data

CORE	A_p (cm^4)	W_a (cm^2)	A_c (cm^2)	*A* (cm)	*B* (cm)	*C* (cm)	*D* (cm)	*E* (cm)	*F* (cm)	*G* (cm)	*MPL* (cm)	*MLT* (cm)
EE 6527	28.56	5.37	5.32	6.6	4.42	3.55	2.22	2.0	2.7	3.25	14.7	7.35
EE 5525	15.78	3.74	4.22	5.5	3.75	2.78	1.85	1.72	2.5	2.78	12.0	6.0
EE 6513	14.25	5.36	2.66	6.6	4.4	3.25	2.2	2.0	1.37	3.25	14.7	7.35
EE 5521	13.24	3.74	3.54	5.5	3.75	2.78	1.85	1.72	2.1	2.78	12.0	6.0
EE 4220	6.144	2.56	2.40	4.2	2.95	2.12	1.48	1.2	2.0	2.12	9.7	5.0
EE4215	4.633	2.56	1.81	4.2	2.95	2.12	1.48	1.2	1.5	2.12	9.7	5.0
EE4716	4.554	1.92	2.36	4.715	3.16	1.97	1.2	1.56	1.56	1.97	8.94	4.5
EE4112	2.487	1.704	1.46	4.07	2.85	1.64	1.05	1.2	1.2	1.64	7.75	3.81
EE4012	1.29	0.84	1.53	4.1	2.85	1.74	1.02	1.2	1.2	1.74	7.9	4.0
EE3512A	0.9263	0.81	1.13	3.43	2.6	1.41	0.98	0.93	1.2	1.41	6.9	3.5
EE3512	0.79	0.66	1.2	3.52	2.5	1.46	0.9	1.03	1.2	1.46	6.73	3.4
EE3213	0.75	0.66	1.13	3.19	2.27	1.4	0.96	0.89	1.2	1.4	6.64	3.32
EE3007	0.3579	0.59	0.60	3.0	1.95	1.95	0.97	0.72	0.73	1.95	6.7	3.35
EE2811	0.3378	0.39	0.84	2.82	1.92	1.92	0.66	0.71	1.07	1.92	5.07	2.53
EE2532	0.3213	0.81	0.42	2.53	1.93	1.93	1.27	0.65	0.7	1.93	7.35	3.075
EE2511	0.30	0.43	0.84	2.5	1.75	1.28	0.87	0.75	1.1	1.28	5.75	2.875
EE2507	0.23	0.43	0.52	2.5	1.75	1.28	0.87	0.75	0.75	1.28	5.75	2.875
EE2011S	0.1871	0.28	0.65	2.0	1.41	0.98	0.70	0.59	1.1	0.98	4.49	2.245
EE2506	17.29	0.43	0.40	2.54	1.96	0.95	0.65	0.63	0.75	0.95	4.8	2.4
EE2506M	0.1629	0.43	0.38	2.54	1.96	0.978	0.678	0.63	0.625	0.978	4.9	2.4
EEL1905	0.1209	0.51	0.23	1.9	1.40	1.3	1.1	0.48	0.48	1.3	6.17	3.08
EE2005S	0.0961	0.28	0.33	2.0	1.41	1.01	0.7	0.59	0.59	1.01	4.49	2.24
EE2005K	0.0742	0.23	0.31	2.0	1.28	1.02	0.63	0.52	0.53	1.02	4.3	2.15
EE1905S	0.0622	0.27	0.22	1.9	1.45	0.79	0.56	0.47	0.52	0.79	3.93	1.96
EE1905	0.0605	0.24	0.25	1.9	1.4	0.79	0.55	0.52	0.52	0.79	3.8	1.9
EE1605	0.0378	0.18	0.20	1.6	1.1	0.82	0.57	0.47	0.47	0.82	3.76	1.88
EE1306	0.0277	0.12	0.22	1.26	0.9	0.65	0.45	0.37	0.63	0.65	3.96	1.48
EE1304	0.0160	0.12	0.13	1.26	0.9	0.65	0.45	0.37	0.37	0.65	2.96	1.48
EE1011	0.0134	0.11	0.12	1.0	0.7	0.55	0.415	0.24	0.50	0.55	2.54	1.27

Table 11.2 SWG Copper wire Data

Gauge no. SWG	*Nominal Copper diameter*		*Nominal Cross Section*			*Overall Diameter mm*	*Nominal resistance per unit length* Ω/m	*Current carrying capacity* A
	Mil	mm	circular mil	Inch²	mm²			
50	1.0	0.02504	1.0	0.0000007854	0.0005067	0.03556	33.43	0.001
49	1.2	0.03048	1.440	0.000001131	0.0007297	0.04064	23.22	0.0013
48	1.6	0.04064	2.56	0.000002011	0.001297	0.05080	13.06	0.0024
47	2.0	0.0508	4.00	0.000003142	0.002027	0.06069	8.357	0.0037
46	2.4	0.06096	5.760	0.000004524	0.002919	0.07366	5.804	0.0054
45	2.8	0.07112	7.84	0.0000006158	0.003973	0.08382	4.264	0.0074
44	3.2	0.01828	10.24	0.000008043	0.005189	0.09652	3.265	0.0096
43	3.6	0.09144	12.96	0.00001018	0.006567	0.1067	2.580	0.0121
42	4.0	0.1016	16.0	0.00001257	0.008107	0.1194	2.089	0.0151
41	4.4	0.118	19.36	0.00001521	0.009810	0.1321	1.727	0.0182
40	4.8	0.1219	23.04	0.00001810	0.01167	0.1422	1.451	0.0217
39	5.2	0.1321	27.04	0.00002124	0.01370	0.1549	1.236	0.0254
38	6.0	0.1524	36.0	0.00002827	0.01824	0.1753	0.9286	0.0340
37	6.8	0.1727	46.24	0.00003632	0.02343	0.2007	0.7230	0.0436
36	7.6	0.1930	57.76	0.00004536	0.02927	0.2210	0.5787	0.0545
35	8.4	0.2134	70.56	0.00005542	0.03575	0.2438	0.4738	0.0665
34	9.2	0.2337	84.64	0.00006648	0.04289	0.2642	0.3950	0.0798
33	10	0.2540	100	0.00007854	0.05067	0.2870	0.3343	0.0942
32	10.8	0.2743	116.6	0.00009161	0.05910	0.3099	0.2866	0.1099
31	11.6	0.2946	1346	0.0001057	0.06818	0.3302	0.2485	0.1268
30	12.4	0.3150	153.8	0.0001208	0.07791	0.3531	0.2174	0.1450
29	13.6	0.3454	185	0.0001453	0.09372	0.3861	0.1807	0.1744
28	14.8	0.3759	219	0.0001720	0.1110	0.4166	0.1526	0.2064
27	16.4	0.4166	269	0.0002112	0.1363	0.4623	0.1243	0.2537
26	18	0.4572	324	0.0002545	0.1642	0.5055	0.1032	0.3054
25	20	0.5080	400	0.0003142	0.2027	0.5588	0.08358	0.3770
24	22	0.5588	484	0.0003801	0.2452	0.6096	0.06907	0.4561
23	24	0.6096	576	0.0004524	0.2919	0.6629	0.05804	0.5429
22	28	0.7112	784	0.0006158	0.3937	0.7696	0.04264	0.7390
21	32	0.8128	1024	0.0008042	0.5189	0.8738	0.03265	0.9650
20	36	0.9144	1296	0.001018	0.6567	0.9804	0.02580	1.2215
19	40	1.016	1600	0.001257	0.8107	1.085	0.02089	1.508
18	48	1.219	2304	0.001810	1.167	1.293	0.01451	2.1715
17	56	1.422	3136	0.002463	1.589	1.501	0.01066	2.956
16	64	1.626	4096	0.003217	2.075	1.707	0.008162	3.860
15	72	1.829	5184	0.004072	2.627	1.915	0.006449	4.884
14	80	2.032	6400	0.005027	3.243	2.123	0.005223	6.032
13	92	2.337	8464	0.006648	4.289	2.433	0.003950	7.978
12	104	2.642	10820	0.008495	5.481	2.743	0.003091	10.194
11	116	2.95	13456	0.010568	6.835	3.056	0.00285	12.682
10	128	3.25	16384	0.012868	8.296	3.366	0.002040	15.442

Table 11.3 Data of CEL pot cores

Pot Core	*Area Product* A_p (mm^4)	*Window Area* W_a (mm^2)	*Effective core area* A_c (mm^2)	*Core Factor* l_c/A_c (mm^{-1})	*Effective Length* l_c (mm)	*Effective Volume* V_e (mm^3)	*Approximate Weight* grams/set
9 × 5	65.448	6.48	10.10	1.24	12.50	126.00	1.30
11 × 7	153.252	9.46	16.2	0.956	15.5	251.0	1.8
14 × 8	392.0	15.68	25.0	0.789	20.0	500.0	3.2
18 × 11	1145.52	26.64	43.0	0.597	26.0	1120.0	6.0
22 × 13	2478.94	39.1	63.4	0.497	31.5	2000.0	13.0
26 × 16	5014.9	53.35	94.0	0.4	37.5	3520.0	20.0
30 × 19	12184.25	74.75	163.0	0.33	45.0	6120.0	36.0
36 × 22	20496.94	101.47	202.0	0.264	53.2	10700.0	54.0
42 × 29	48146.525	181.685	265.0	0.259	68.6	18200.0	104.0
47 × 28	56316.0	180.5	312.0	0.232	72.5	22600.0	125.0

Table 11.4 Central Electronics EI-core Data

Core Number	A_p (cm^4)	W_a (cm^2)	A_c (cm^2)	*A* (cm)	*B* (cm)	*C* (cm)	*D* (cm)	*E* (cm)	*F* (cm)	*G* (cm)	*MPL* (cm)	*MLT* (cm)
EI65 × 32 × 13	7.14	2.68	2.66	6.65	4.42	3.25	2.22	2.0	1.37	3.25	14.7	7.35
EI55 × 28 × 21	6.64	1.87	3.54	5.5	3.75	2.78	1.85	1.72	2.1	2.78	12.0	6.0
EI42 × 27 × 9	1.38	1.28	1.08	4.3	2.95	2.1	1.48	1.22	0.9	2.1	0.7	3.35
EI42 × 21 × 9	1.37	1.28	1.07	4.3	2.95	2.1	1.48	1.22	0.9	2.1	10.8	5.4
EI42 × 21 × 20	3.02	1.28	2.36	4.3	2.95	2.1	1.48	1.22	2.0	2.1	9.8	4.9
EI42 × 21 × 15	2.32	1.28	1.82	4.3	2.95	2.1	1.48	1.22	1.52	2.1	9.7	4.85
EI36 × 18 × 11	0.9236	0.70	1.31	3.6	2.38	1.82	1.18	1.18	1.1	1.82	7.84	3.92
EI35 × 13 × 8	0.6401	0.82	0.77	3.52	2.5	1.46	0.9	1.03	1.2	1.46	6.73	3.4
EI33 × 29 × 13	0.8600	0.69	1.24	3.19	2.27	1.4	0.96	0.89	1.2	1.4	6.64	3.32
EI30 × 15 × 7	0.3561	0.59	0.59	3.0	1.95	1.95	0.66	0.71	1.07	1.95	5.07	2.53
EI28 × 20 × 11	0.2881	0.33	0.85	2.82	1.92	1.92	1.27	0.65	0.7	1.92	5.0	2.5
EI25 × 13 × 7	0.2464	0.44	0.55	2.5	1.75	1.26	0.87	0.75	0.75	1.26	5.74	2.87
EI25 × 9 × 6	0.1682	0.42	0.40	2.54	1.96	0.95	0.65	0.63	0.6	0.95	4.87	2.43
EI20 × 10 × 6	0.1035	0.30	0.33	2.0	1.41	1.01	0.7	0.59	0.59	1.01	4.49	2.24
EI20 × 10 × 5	0.0746	0.23	0.31	2.07	1.28	1.0	0.6	0.52	0.53	1.0	4.28	2.14
EI10 × 5 × 5	0.0118	0.12	0.09	1.0	0.7	0.55	0.42	0.24	0.5	0.55	2.54	1.27

Table 11.5 COSMO C-core Data

Core Number	A_p (cm^4)	W_a (cm^2)	A_c (cm^2)	*A* (cm)	*B* (cm)	*C* (cm)	*D* (cm)	*E* (cm)	*F* (cm)	*G* (cm)	*MPL* (cm)	*MLT* (cm)
UU 100	96.286	15.53	6.2	10.1	4.9	5.7	3.17	2.52	0	6.36	30.0	15
UU100A	49.85	15.53	3.21	10.1	4.9	5.7	3.17	1.27	0	6.34	30.8	15.4
UU9316	79.56	17.76	4.48	9.3	3.7	7.6	4.8	1.6	0	9.6	35.4	17.7
UU4628	17.36	4.375	3.97	4.6	1.75	3.95	2.5	2.8	0	5.6	18.2	9.1
UU2537	0.56	0.9765	0.57	2.4	0.99	1.84	1.08	0.75	0	1.5	8.69	4.34
UU2332	0.4148	0.68	0.61	2.3	0.8	1.57	0.85	0.75	0	1.5	7.4	3.7
UU2130A	0.3	0.55125	0.54	2.1	0.63	1.58	0.875	0.75	0	1.5	7.02	3.51
UU2130	0.28	0.5197	0.55	2.1	0.63	1.53	0.825	0.75	0	1.5	6.8	3.4
UU2036	0.345	0.96	0.36	2.0	0.80	1.8	1.2	0.6	0	1.2	8.28	4.14
UU1622	0.1225	0.49	0.25	1.6	0.70	1.1	0.7	0.6	0	1.2	5.52	2.76
UU1620	0.1134	0.42	0.27	1.6	0.70	1.06	0.6	0.6	0	1.2	5.2	2.6
UU1522	0.1015	0.3172	0.32	1.5	0.52	1.11	0.61	0.64	0	1.28	4.8	2.4
UU1116	0.3065	0.2809	0.13	1.05	0.53	0.79	0.53	0.5	0	1.0	4.0	2.0
UU1016	0.0178	0.2236	0.08	1.01	0.43	0.82	0.52	0.3	0	0.6	3.84	1.92

Index